T0260109

# From *c*-Numbers to *q*-Numbers

# From $c$-Numbers to $q$-Numbers

*The Classical Analogy in the*
*History of Quantum Theory*

Olivier Darrigol

UNIVERSITY OF CALIFORNIA PRESS
*Berkeley · Los Angeles · Oxford*

University of California Press
Berkeley and Los Angeles, California

University of California Press
Oxford, England

Copyright © 1992 by The Regents of the University of California

Library of Congress Cataloging-in-Publication Data

Darrigol, Olivier.
    From c-numbers to q-numbers : the classical analogy in the history
of quantum theory / Olivier Darrigol.
        p.   cm. — (California studies in the history of science)
    Includes bibliographical references and index.
    ISBN 0-520-07822-5
    1. Quantum theory—History.   I. Title.   II. Series.
QC173.98.D37   1992
530.1'2'09—dc20
                                                    91-42469
                                                    CIP

1 2 3 4 5 6 7 8 9

*A mes Parents*

# Contents

# Acknowledgments

I first conceived the project of this book three years ago, during a conversation with John Heilbron in the Berkeley hills. Since then, and before, John has encouraged me in many ways: commenting on my work, giving me professional advice, and inviting me for long stays at an ideal working place, the Office for the History of Science and Technology at Berkeley.

There I made friends with Edward Jurkowicz, without whom the project of writing a book in English would have been beyond my reach. Not only did Ed correct and clarify my English, but his highly detailed comments helped to structure and strengthen many of my arguments.

In France I received the highest intellectual stimulation from Catherine Chevalley, who discussed with me several issues of the history and philosophy of quantum theory, let me read her work on Bohr (Chevalley 1991a) prior to publication, and commented on large portions of my manuscript.

Since history of quantum theory is not a new field of research, I have sought advice from competent scholars. Norton Wise, a perceptive analyst of Bohr's early work and also an expert on analogical thinking, offered several suggestions for improving my manuscript. The convergence of our views on Bohr's role made our exchange particularly pleasant. The part of my manuscript on Dirac's quantum mechanics benefited from comments by the Danish historian Helge Kragh, who has recently published a major biography of Dirac, and by the philosopher Edward MacKinnon, who has long been interested in Dirac's methodology.

Some physicist friends have been kind enough to check the more technical content of my manuscript. Jean-Michel Raimond, an authority on atomic physics, read my considerations on the old spectroscopy of fine structure and anomalous Zeeman effects. Bruno Jech, a professor of physics and historian, went through the arcana of my presentation of Planck's radiation theory.

I have also tested the reaction of historians of science who have no specialized knowledge of the history of quantum theory. Mario Biagioli suggested improvements in the general introduction of this book; Mathias Dörries helped prune numerous obscurities in the nontechnical summaries.

All my research has been done under the auspices of the Centre National de la Recherche Scientifique, both in Paris and during my stays abroad. My research director, the philosopher-physicist Bernard d'Espagnat, has been, from the beginning, supportive of the type of work in which I was engaged. I have much benefited from his exceptionally deep understanding of the foundations of quantum theory. I owe special thanks to the authorities of CNRS who gave me the freedom to do my work in the best environments. In Paris I have profited from the intellectual ambience at REHSEIS (équipe du CNRS pour les Recherches en Epistémologie et Histoire des Sciences et des Institutions Scientifiques). In this group, I have especially appreciated the guidance of Michel Paty and Roshdi Rashed.

At the final stage of this project, the collaboration of my sponsoring editor, Elizabeth Knoll, of the University of California Press, has been pleasant and constructive. Some anonymous reviewers of her choosing made valuable suggestions for improving my manuscript.

Aage Bohr has kindly granted me permission to quote from his father's unpublished manuscripts and letters (which are deposited in the Bohr Archive in Copenhagen). For Heisenberg's unpublished letters (which belong to the Heisenberg Archive in Munich) I owe a similar favor to Helmut Rechenberg.

However small my project and modest the results, my debts appear to be extensive and numerous. I am very thankful to the colleagues and friends just mentioned, and to all those who, by their friendship or their research, helped me in an indirect, but invaluable, manner.

# Conventions and Notations

Vector notation is used throughout the book, even when it is anachronistic, for instance in Maxwell's and Planck's cases. Knowing that Maxwell used points in a geometric space (in the context of his dynamic theory of gases) and that Planck used Cartesian coordinates, my reader will easily imagine a more authentic form of their equations.

In conformity with the convention found in most works described in this book, Gaussian units are employed, which gives Maxwell's equations (in vacuo) the form

$$\mathbf{V} \cdot \mathbf{E} = 4\pi\rho \qquad \mathbf{V} \cdot \mathbf{B} = 0$$

$$\mathbf{V} \times \mathbf{E} = -\frac{1}{c}\frac{\partial \mathbf{B}}{\partial t} \qquad \mathbf{V} \times \mathbf{B} = \frac{4\pi}{c}\mathbf{j} + \frac{1}{c}\frac{\partial \mathbf{E}}{\partial t}$$

Then the Poynting vector is given by $(c/4\pi)\mathbf{E} \times \mathbf{B}$ and the energy density by $(1/8\pi)(E^2 + B^2)$.

For Hamiltonian systems collective coordinates $q$ and $p$ are introduced according to

$$q = (q_1, q_2, \ldots, q_s), \qquad p = (p_1, p_2, \ldots, p_s),$$

and the "dot product" of two such coordinates $q'$ and $q''$ is defined by

$$q' \cdot q'' = \sum_{i=1}^{s} q_i' q_i''.$$

Frequencies in the strict sense (cycles per unit time) are denoted by the letter $v$, whereas angular frequencies (radians per unit time) are denoted by the letter $\omega$ (therefore, $\omega = 2\pi v$).

Infinitesimal solid angles are denoted by $d\Omega$ and the corresponding direction by $\zeta$.

In order to distinguish them from radiation frequencies, orbital frequencies will be written with a bar: $\bar{v}$ instead of $v$.

As for physical constants, $c$ denotes the velocity of light, $k$ Boltzmann's constant, $h$ Planck's constant, $\hbar$ Planck's constant divided by $2\pi$, $e$ the arithmetic value of the electron charge ($e > 0$), $\mu$ the electron mass.

For quantum numbers I have mainly used Bohr's notation, $n$ for the principal quantum number, $k$ for the azimuthal one, $j$ for the inner one, and $m$ for the magnetic one. However, I have used other letters when the normalization was essentially different (as in Landé and Sommerfeld). The correspondence between Bohr's quantum numbers, those of his colleagues, and the modern ones will be given in footnotes.

Significant simplifications have been introduced in some original proofs, for instance in Boltzmann's and Planck's proofs of their $H$-theorems. Unless otherwise indicated, these simplifications are of a purely mathematical nature and do not alter the main logical arguments.

Citations of sources are in the author-date format and refer to works listed in one of the two bibliographies (primary or secondary literature). Square brackets ([ ]) enclosing a date indicate that the work in question is an unpublished manuscript and is listed in the bibliography of primary literature. Abbreviations used in citations and in the bibliographies are listed and explained on pp. 354–355 below.

Translations are generally mine, unless I am quoting from sources which are, or already contain, a translation (which is always the case for manuscripts included in Bohr's *Collected works.*)

# Introduction

In the radiation theory just as in the gas theory, one could determine a state of maximal probability.

*Boltzmann, 1897*

As we go from the kinetic theory of gases to the theory of thermal radiation . . . we come across relations which are very similar in a certain sense.

*Planck, 1901*

Using a metaphor, we may say that we are dealing with a translation of the electromagnetic theory into a language alien to the usual description of nature, a language in which continuities are replaced by discontinuities and gradual changes by immutability, except for sudden jumps, but a translation in which nevertheless every feature of the electromagnetic theory, however small, is duly recognized and receives its counterpart in the new conceptions.

*Bohr, 1924*

The quantum theory has now reached a form . . . in which it is as beautiful and in certain respects more beautiful than the classical theory. This has been brought about by the fact that the new quantum theory requires very few changes from the

classical theory, these changes being of a fundamental nature,
so that many of the features of the classical theory to which
it owes its attractiveness can be taken over unchanged into the
quantum theory.

*Dirac, 1927*

 

The genesis, maturation, and final formulation of quantum theory
owed much to analogies with classical theories.[1] Even modern quantum
mechanics is still an art of "quantization." Any application of it starts
with formally defining a classical system, and the quantum-theoretical
level is then reached by applying a precise mathematical procedure followed by interpretative rules. In the early history of quantum theory,
analogies with classical theory were not so sharply formulated, but they
already were fairly detailed and articulate. Not just a vague illustrative
resemblance, they concerned entire pieces of logical and mathematical
structures and were able to produce new laws and formalisms. The aim
of this book is to analyze the structure and development of such analogies
in three cases: Planck's radiation theory, Bohr's atomic theory, and Dirac's
quantum mechanics.

In 1926 Dirac introduced a parallel between "$c$-numbers" and "$q$-
numbers," capturing in symbols the correspondence between classical and
quantum (or "queer," as he humorously said) mechanics. The formal expression of ordinary dynamic laws was maintained in the new theory,
while the related quantities no longer behaved like ordinary numbers and
no longer received a space-time interpretation. The $c$ and $q$ in my title
alludes to this ultimate perception of the analogies under discussion.
Before quantum mechanics, analogies with classical theories were not
usually expressed in terms of the exact transference of mathematical formulae. However, one can still speak of the analogies as being *formal* in
Bohr's sense of the word: "We are . . . obliged to be modest in our de-

---

1. The references for the above quotations are: Boltzmann 1897c, *BWA* 3:621; Planck
1902, *PAV* 2:739; Bohr 1925b, *BCW* 5:[140]; Dirac [1927c], introduction. The dates given
under the quotations refer to the writing of the text, not to its publication.

mands and content ourselves with concepts that are formal in the sense that they do not provide a visual picture of the sort one is accustomed to in the explanations with which natural philosophy deals."[2] Bohr understood that the analogy involved in the description of the interaction between atoms and radiation could not be of a visual nature. Moreover, this analogy, originally combined with the orbital picture of atoms, ended up, at the dawn of quantum mechanics, being independent of any visual model of atomic motion. Accordingly, this book is not directly about classical models as visualization tools in the quantum theory, it is about *formal* classical analogies.

What is to be meant by classical theories in the historical episodes evoked in this book? The answer is immediate in the case of Bohr's quantum theory and Dirac's quantum mechanics, which are the main objects of parts B and C. A little before the beginning of Bohr's atomic theory, classical (or "ordinary" in Bohr's words) theory already had the meaning that is now familiar to us: it covered Newtonian mechanics, Lorentz's electrodynamics, and, if necessary (for instance in Dirac's case), Einstein's relativity. Obviously, these theories could not be called classical before a consensus had been reached, in the early 1910s, about the need for a radically new physics in the realm of atoms.[3]

At the turn of the century not only was such hindsight impossible but there was no uniform conception of mechanics, electrodynamics, and their relations to thermodynamics. At that time, opinions varied about the role to be played by microphysical entities in the organization of macroscopic physics. In this context the word "classical" is therefore misleading, unless it is used in a limited conventional way, referring to mechanical and electrodynamic laws commonly accepted at the macroscopic level. This remark has to be kept in mind in the first part of this book, which is dedicated to Planck's radiation theory.

The analogy that guided Planck's work around 1900 really was an analogy between Boltzmann's anterior gas theory and a new thermal radiation theory. In a sense Boltzmann's theory can be called classical, meaning that it subjected gas molecules to the well-established (in the macroscopic realm) laws of Newton's mechanics. But this should not hide the fact that the kinetic molecular theory was not universally accepted

---

2. Bohr 1923c, trans., 44.
3. For a criticism of the notion of "classical physics" see Needell 1988.

at the end of the nineteenth century. Planck himself converted to this theory only in 1896–1897, while developing his radiation theory.

Further classification of the analogies studied in this book is obtained by looking at the nature of the "target" theory, that is, the theory which the analogy helps to construct. In the case of Planck's celebrated work of 1900, the target theory was intended to consist of a simple extension or transposition of Boltzmann's methods—originally designed for gases—to a system made of electromagnetic radiation and sources. Until 1907–1908 Planck actually believed the sources of thermal radiation to comply with ordinary electrodynamic laws. In this approach the target theory was as classical as Boltzmann's gas theory could be, and we may call the relevant analogy "horizontal." Again, the present use of the word "classical" should not hide the fact that there was no universally accepted formulation of electrodynamics around 1900.[4]

The product of Planck's horizontal analogy was not meant to break with accepted theories. It was not a "quantum discontinuity."[5] This does not mean that "the father of the quantum theory" did not introduce anything substantial in 1900. He isolated the fundamental constant $h$, and he gave the formal skeleton of what could *later* be regarded as a quantum-theoretical proof of the blackbody law. This is just a first example of a recurrent characteristic of the history of quantum theory: the "correct" interpretation of new mathematical schemes generally came *after* their invention.

In the years 1905–1907 Einstein introduced an intrinsic discontinuity of the energy of microscopic entities, the so-called quantum discontinuity, and used it most successfully in a new theory of specific heats. In 1913 Bohr exploited the same discontinuity in his first atomic theory. By that time the rudiments of a radically new "quantum theory" were known, and the possibility was open for genuine "vertical" analogies, connecting the nascent theory to the now classical theory. However, such analogies could not flourish before the new "language of atoms" was sufficiently known, that is to say, after Sommerfeld generalized Bohr's original quantum condition in 1916.

Bohr then recognized that some laws of classical electrodynamics had a formal counterpart within the quantum theory. This analogy, which led

---

4. On the diversity of electromagnetic theories in the late nineteenth century, see Buchwald 1985.
5. See Kuhn 1978.

to what Bohr named the "correspondence principle" in 1920, was at least heuristically important, because the resulting quantum-theoretical laws could not be deduced from the general assumptions of the Bohr-Sommerfeld theory, which was clearly incomplete. In the absence of a rigorous deductive scheme, the qualitative or semiquantitative validity of these laws was essentially controlled by two conditions: one empirical, the compatibility with observed atomic spectra; and one intertheoretical, the asymptotic agreement with the corresponding classical laws in the case of relatively small quantum jumps.

An analogy between two theories, one of which is essentially incomplete and provisional, should not be expected to be unambiguous.[6] Accordingly, Bohr wished to formulate his correspondence principle in a not too sharp form. In part B we will observe the multiplicity of uses of this principle. A crucial ambiguity lay in the extent to which the analogy maintained the space-time description of the classical theory. In the spring of 1925, at the end of a crisis that started in 1922, Bohr and some of his disciples cut the by then dead branches of the correspondence principles, namely, its visual elements, and retained only the idea of a symbolic translation of classical laws. Within this stream of thought Heisenberg devised quantum mechanics.

What Heisenberg proposed in the summer of 1925 was a complete mathematical scheme interpreted in terms of the original postulates of Bohr's theory, that is, in terms of stationary states and atomic transitions. Formally, the analogy between this theory and classical mechanics could hardly be closer, since the formal expression of dynamic laws was integrally maintained (though transcribed in bold and gothic types). All the same, the distance between classical and quantum concepts was larger than ever, for the dynamic variables were now represented by infinite matrices instead of ordinary numbers.

At that stage, reference to the correspondence principle became unnecessary, because all properties of atomic spectra could be deduced from the new scheme. Yet an early follower of Heisenberg's ideas, Paul Dirac, found deeper connections between new and old mechanics, ones involving algebraic structures. He brilliantly exploited these structural connections in consolidating and developing Heisenberg's ideas. In part C the originality and power of Dirac's approach is shown to have also depended on a rather different kind of classical analogy. The latter was not so much

6. For a philosophical analysis of this type of ambiguity, see Meyer-Abich 1965.

between the formal contents of two theories as between characteristic strategies for theory-building. The model here was Einstein's general relativity, as perceived by Dirac's philosophy teacher, C. D. Broad, and England's foremost relativist, Arthur Eddington.

To summarize, four types of classical analogies will be described in this book: Planck's abusively conservative but formally suggestive "horizontal" analogy, Bohr's tentative "vertical" analogies between classical electrodynamics and an incomplete quantum theory, Heisenberg's and Dirac's analogies between the mathematical schemes of classical and quantum mechanics, and Dirac's reference to the relativistic strategy of theory-building.

The originators of each of these analogies all made general comments about the function they served. Planck declared he had reached, despite remaining obscurities in the proof of his blackbody law, a fundamental unification of gas theory and radiation theory; for he had managed to apply the same formula, $S = k \ln W$, in both theories, with the same fundamental constant $k$. Moreover, comparing the resulting expression of the blackbody law with empirical measurements provided the best available access to Avogadro's number, through the constant $k$. Who would not agree with Planck that horizontal analogies, if successful, bring unity in the architecture of physics?

When Bohr introduced the correspondence principle, he first emphasized its heuristic power: it was a means to compensate for the incompleteness of the quantum theory. Even the enemies of this principle came to agree with this. However, Bohr soon attributed more fundamental functions to his "principle": bring more structure into the quantum theory, and show the overall harmony of its various assumptions. Characteristically, he regarded the formal connections obtained by analogy as part of the quantum theory, even before these connections could be expressed in a precise quantitative way. In his opinion a reasonable degree of conceptual clarity and consistency could be achieved even before the advent of a more complete and definitive theory.

Sommerfeld and many other quantum theorists of lesser importance were out of sympathy with Bohr's strategy of rational guessing. Sommerfeld's concept of rationality demanded a sound complete mathematical framework or, as long as nothing better was available, a set of clear mathematical models. Yet, be it a historical contingency, a necessary outcome of a rational attitude, or something in between, the correspondence principle did play the most important part in the construction of the first version of quantum mechanics. Moreover, the main source of the early confidence in Heisenberg's strange kinematics of infinite matrices was its close formal analogy with classical mechanics. The precise expression of this analogy suggested the mathematical completeness and consistency of the new scheme even before a rigorous proof could be given; it automatically warranted the necessary asymptotic agreement with classical mechanics; and it integrated in a more quantitative form earlier-verified predictions of the correspondence principle. Last but not least, this analogy preserved what Dirac found to be the beauty of classical mechanics.

That the construction of quantum theory heavily relied on previous leading theories should not be a matter of surprise.[7] The modern theoretical physicist is almost more concerned with relations between different theories than with empirical data. Many theories are in fact now in use, some of which existed long ago; for mature science, by definition, saves most of its older theories.[8] Satellites' motion or trips to Jupiter are still calculated according to Newtonian mechanics; TV antennas and magnets in atom smashers are still designed according to ordinary electromagnetic theory. Admittedly, older theories have now come to be regarded as approximations of a more general superseding theory. In this process their conceptual foundation has been "filtered." Some concepts that originally seemed to be necessary to the formulation of these theories now appear to be irrelevant to their empirical content and to be incompatible with their integration in the superseding theory. Today's Newtonian mechanics does not rely on Newton's infinitesimal geometry, and modern electromagnetism does without ether. But, as Poincaré puts it, "ruins can still be useful":[9] Within their known fields of validity these theories are still in use, and not only for the sake of convenience.

More fundamentally, older theories are necessary to the empirical verifiability of the newer ones. If, as is now commonly accepted, no theory-bare facts exist, any fundamental theory must approximatively contain "observational" theories that are sufficient to describe the "circumstances" of all conceivable experiments within the field of the theory. The "circumstances" refer to all devices used to measure or define relevant parameters.

---

7. There is an abundant philosophical literature concerning this point, for instance Campbell 1920, Hesse 1966, Boyd 1979. The models or analogies analyzed by these authors are usually assumed to provide a visual (space-time) representation, which is not the case for the formal analogies treated in this book.
8. The definition is in Krajewski 1977, 7.
9. Poincaré 1902, 1968 reprint, 173.

These devices themselves should work within the range of validity of the observational theories, this range being determined both empirically and theoretically (by means of the superseding theory). No doubt, most physicists know in practice how to deal with this situation. Yet it brings about epistemological difficulties: In what sense can a theory approximate another one? How can a theory describe the circumstances of a possible experiment without applying to the phenomenon occurring within these circumstances?

These questions were on Bohr's mind very early (from 1916 on), as I will show in part B. His partial answers, before quantum mechanics, conditioned his use of the correspondence principle. On the one hand, he believed that the circumstances of atomic phenomena were defined through classical theory. On the other hand, he constantly emphasized that quantum discontinuity brought with it an irreducible conceptual gap between classical and quantum theory. Then, there could be no question of a mere inclusion of classical electrodynamics within the new theory. But the classical theory had to be an approximation of the new quantum theory in some (statistical) sense. In addition, there could be a formal analogy between them that preserved selected classical relations without excessive conceptual import.

The constructive virtues of formal analogies have not been the main object of interest in previous histories of quantum theory. Emphasis is usually put on model making—its commencement, crisis, and ultimate failure—as well as on the origins of the Copenhagen interpretation of quantum mechanics. Attention is also sometimes given to psychological, cultural, and institutional factors. Such approaches are certainly necessary and useful, and I have benefited from them, probably to a larger extent than is expressed in the footnotes. Yet I have not tried to give them a substantial weight in this book, because I believe the writing of "total" histories of quantum theory to be impossible. The subject is too complex to lend itself to such complete coverage. As the quantum physicist I have come to admire most would have said, there is here a need for complementary but not superposable perspectives.

The present approach originated in the conviction that formal constructive tools, particularly the correspondence principle, played an essential role in the history of quantum theory, and still do so in modern theoretical physics. Modern theorists live in a world of highly developed theories. In order to obtain new theories, they extend, combine, or transpose available pieces of theory. Phenomenology and the criticism of foundations do help in this process, but they are no longer self-sufficient (if they have ever been so).

In the theoretical construction game, mathematics plays a considerable part. It not only provides the form of available theories but allows the expression of systematic, detailed analogies. The formal expression of known equations can be saved while changing the meaning of the related symbols, or a structure (in the algebraic sense) of an older theory can be imported in the new theory. In this way new laws and formalisms are devised before the more difficult interpretative problems are solved.

This structuring role of mathematics cannot be further explained without entering technical arguments more than is usually done in historical studies. Here I must apologize for the effort required on the part of my reader. However, no more is needed than the Lagrangian and Hamiltonian formulations of mechanics, classical electrodynamics, and, perhaps, some quantum mechanics. Full mathematical demonstrations are given (in the physicists' style); I have avoided recourse to "technical appendices," but have indicated which sections can be skipped by hurried or learned readers. The notations are uniform throughout the book (although the last two parts can be read independently of the first one). Whenever such uniformity might erase a conceptual difference (for instance in the case of the normalization of quantum numbers in the Zeeman effects), I have mentioned it in the footnotes. Finally, I provide extensive nonmathematical summaries at the end of each chapter. Each summary can be read either before or after the corresponding chapter.

But the main difficulty experienced by my reader will perhaps be in the old conflict between normative reason and historical evidence. Under too-stiff rational prejudices, the creative function of formal analogies could be misjudged. Planck's radiation theory could be taken to innovate much more than it really did, Bohr's correspondence principle could be denied its fundamental meaning, and Dirac's spectacular success could be perceived as impenetrable magic. The elegance of the style of a top runner eludes a series of static pictures. It is best appreciated if one is willing to try running along with him.

# Planck's Radiation Theory

# Introduction

Most of Planck's early work was carried out with the principal goal of proving that the second law of thermodynamics was strictly valid and that the entropy of a closed system *always* increases. Accordingly, he first rejected kinetic gas theory, for it considered the second law to be only statistically valid—until he came to develop his radiation theory on the basis of an analogy with Boltzmann's gas theory. One may wonder how any reasonable theoretician could draw his inspiration from the theory which he is trying to disprove. A plausible answer could be that Planck was converted to Boltzmann's ideas. But he believed too much in the absolute validity of the entropy law to do so. The key that gave him access to the formal apparatus of Boltzmann's theory was in fact a reinterpretation of this theory in nonstatistical terms.[1]

According to Planck, the central concept of both radiation and kinetic gas theory had to be that of "elementary disorder." In his opinion the main difficulty encountered in these theories was that in the derivation of equations for the evolution of directly observable quantities from fundamental electrodynamic or mechanical processes, there were terms depending on uncontrollable details of the state of the system, for instance the position of individual molecules or the electromagnetic field at a precise point of space. According to Boltzmann, these terms really existed, but they could be neglected when considering the statistical behavior of a large

---

1. Allan Needell was the first to point out the central importance of the idea of absolute irreversibility in Planck's work before 1914 (Needell 1980, 1988).

number of exemplars of the system. The resulting evolution of the directly observable quantities (the one given by the Boltzmann equation) was irreversible, but only in a statistical manner. Instead, according to Planck's notion of elementary disorder the unknown structural details of the system, for instance the structure of the walls of the container or the internal structure of electric resonators, had to be adjusted in such a way that the unwanted terms completely disappeared. This warranted a strictly deterministic (and irreversible) evolution of "directly observable quantities."[2]

The relation between Planck's and Boltzmann's work in thermodynamics, then, is a subtle and intricate one, and an elucidation of it will be one of the principal goals of this chapter. Planck's appropriation of some of Boltzmann's computational methods has often misled his modern readers, who generally understand these problems from the point of view of statistical thermodynamics, which is essentially Boltzmann's. To a reader aware of the pitfall of incommensurability, Planck's approach will appear far more coherent and conservative than usually assumed.

The analogies used by Planck in his radiation theory were drawn from a *reinterpreted* version of Boltzmann's theory. Yet, in any analogy there is a risk of overestimating the similarities between the systems compared. Planck certainly did. In Boltzmann's irreversibility theorem, not only was irreversible behavior derived but the final equilibrium state of the system was shown to be unique. Planck initially believed that such uniqueness also held for the electrodynamic system which he considered in his radiation theory. More specifically, he thought he could show that Wien's law was the only possible distribution for thermal radiation. Under the pressure of new empirical data, however, he came to realize that *any* thermal radiation law was compatible with his irreversibility theorem.

At this stage, Planck thought of adapting the analogy between his and Boltzmann's theory to another method of determining the equilibrium state of a system, through Boltzmann's quantitative relation between entropy and "probability." Naturally, he did this within the context of his reinterpretation of Boltzmann's theory: he freed Boltzmann's "probability" from its original ties with the statistical conception of the entropy law

---

2. "Elementary disorder" was the generic expression used by Planck to characterize molecular chaos and natural radiation in his lectures on radiation theory: see Planck 1906, e.g., on 134: "The assertion that in nature every state and process involving a great number of uncontrollable elements is elementarily disordered provides the condition and also the strict guarantee for the unequivocal determination of measurable processes in both mechanics and electrodynamics and for the validity of the second principle of thermodynamics."

and interpreted it instead as a quantitative measure of the elementary disorder that warranted strict irreversible behavior.

Here the reinterpretation had considerable effects. Most important, we shall see that it permitted finite energy-elements to appear in the final expression of entropy (whereas they disappeared in Boltzmann's original method); at the same time it allowed maintaining the continuous equations for the evolution of the electrodynamic system, without apparent inconsistency. More generally, Planck's quantum hypothesis was meant to complete the existing electrodynamic theories, not to contradict them. Closely connected to elementary disorder, this hypothesis found its logical place in the uncontrollable details of electrodynamic systems and left untouched the laws ruling directly observable quantities.[3]

Altogether, tight connections existed between the central concepts of Planck's radiation theory, namely: absolute irreversibility, disorder, entropy, and energy quanta. By 1905, however, Einstein perceived an inconsistency in the corresponding reinterpretation of Boltzmann's theory. In his opinion the separation between directly observable quantities and internal structure necessary to Planck's idea of disorder could not be maintained. One then had to return to orthodox statistical thermodynamics, and this led, in the case of Planck's electrodynamic system, to absurd results. The observed properties of thermal radiation, Einstein concluded, could not be explained without a sharp break from ordinary electrodynamics. Nevertheless, the formal skeleton of Planck's derivation of his blackbody law remained valid. This should be seen as a virtue of the symbolic part of Planck's analogies, the resulting equations being "more clever than their inventor," as Born once put it.[4]

3. That Planck did not introduce a "quantum discontinuity" in 1900 was first asserted in Kuhn 1978.

4. Born to Bohr, undated (early 1926), AHQP: "For the time being our mathematical formulae tend to be more clever than we are. The formulae come to us quite naturally, but the interpretation is often difficult."

# CHAPTER I

# Concepts of Gas Theory

When Planck worked out his radiation theory, he relied on an analogy with Boltzmann's gas theory. The key conceptual issues in the latter theory are best understood in light of their historical source, Maxwell's kinetic theory of gases. The following is a critical discussion of some of Maxwell's and Boltzmann's results.

## MAXWELL'S COLLISION FORMULA

In the mid-nineteenth century James Clerk Maxwell was prominent in developing the kinetic theory of gases, a subject just then beginning to flourish. Like his precursor in the field, Rudolf Clausius, he conceived of a gas as a set of very small "molecules" animated with a continual motion. A molecule in a sufficiently dilute gas was supposed to travel along a straight line, except when it was redirected by short collisions with other molecules or with the walls of a container. Any quantitative theory of the observable effects of such collisions, for instance of pressure or of viscosity, required an evaluation of the number of collisions of a given kind.[5]

In 1866, through a seemingly obvious reasoning, Maxwell gave a precise mathematical expression for this number, later known to German-speaking theorists as the *Stosszahlansatz*. The corresponding formula turned out to provide the starting point for most subsequent kinetic

5. See, e.g., Brush 1976.

theories, and these confirmed its validity in many concrete cases. However, the precise formulation of the conditions of its applicability soon became an outstanding conceptual problem of physics. Not only the empirical predictions of the kinetic theory but also, as we shall see, the nature of thermodynamic irreversibility crucially depended on the solution of this problem.[6]

In his "dynamical theory of gases" Maxwell first examined the case of a chemically and spatially homogeneous gas and introduced the following hypothesis of dilution:

> We shall suppose that the time during which a molecule is beyond the action of other molecules is so great compared with the time during which it is deflected by that action, that we may neglect both the time and the distance described by the molecules during the encounter, as compared with the time and the distance described while the molecules are free from the disturbing force. We may also neglect cases in which three or more molecules are within each other's spheres of action at the same instant.[7]

In the lack of detailed information on intermolecular forces, Maxwell assimilated the molecules either to hard spheres or to centers of force. In the latter case a collision of two molecules, denoted "1" and "2," may be represented as a simple deflection of "2" in a reference frame fixed to "1" (fig. 1).[8] In the case of central forces the collision "kind" is characterized by the azimuth $\varphi$ of the plane of the trajectory of "2" in this reference frame, and by the angle $\theta$ between the initial and final relative velocities, which is a definite function of the parameter $b$ (and of the initial relative velocity $v_2 - v_1$).

Let it be agreed that a collision starts when "2" crosses a conventional plane $\Sigma$ perpendicular to the relative velocity $v_2 - v_1$. Then, for "2" to collide with "1" within the time $\delta t$, and with a kind $(\theta, \varphi)$, defined with the uncertainty $(d\theta, d\varphi)$, it must be located within the "efficient volume" shaded in figure 2. In order to obtain the number of such collisions occurring in a unit volume of a homogeneous gas of identical molecules, Maxwell simply multiplied the measure $|v_2 - v_1| \delta t \, d\varphi \, b \, db$ of this volume by the expression $f(v_1) \, d^3v_1 f(v_2) \, d^3v_2$, giving the number of pairs of molecules per square unit of volume with velocities $v_1$ and $v_2$, up to $d^3v_1$ and $d^3v_2$.

---

6. See ibid., 583–640.
7. Maxwell 1867, *Scientific Papers*, 37.
8. In his original reasoning Maxwell used the reference frame of the gravity center of the two molecules instead of the reference frame of the molecule "2." The notations used in the sequel are of course not Maxwell's.

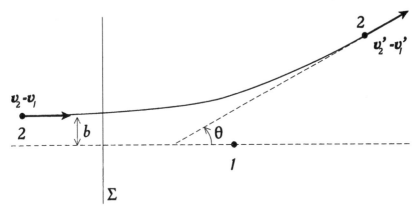

Figure 1. Deflection of a molecule "2" by a molecule "1," as seen from an observer fixed to "1."

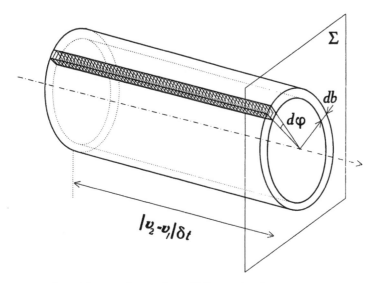

Figure 2. The "efficient volume" for colliding molecules.

This gives

$$dn = |\mathbf{v}_1 - \mathbf{v}_2| b \left| \frac{db}{d\theta} \right| d\theta \, d\varphi \, f(\mathbf{v}_1) f(\mathbf{v}_2) \, d^3 v_1 \, d^3 v_2 \, \delta t. \qquad (1)$$

In the case of more complex interactions Maxwell used a slightly more general form:

$$dn = |\mathbf{v}_1 - \mathbf{v}_2| \sigma(\zeta, |\mathbf{v}_1 - \mathbf{v}_2|) \, d\Omega \, f(\mathbf{v}_1) f(\mathbf{v}_2) \, d^3 v_1 \, d^3 v_2 \, \delta t, \qquad (2)$$

where $d\Omega$ is an element of solid angle in the direction $\zeta$ defined by $\theta$ and $\varphi$, and $\sigma$ has the dimension of a surface.[9]

The seeming naturalness of Maxwell's reasoning obscures an important difficulty. For a given target-molecule "1," the number of molecules in its "efficient volume" must almost always be zero, for it has been implicitly assumed that the time $\delta t$ is so small that no third molecule perturbs a molecule "2" traveling within this volume. Consequently, the relative spatial distribution of the molecules "2" cannot be considered to be uniform at the scale of the efficient volume. But this distribution depends on the choice of the molecule "1," and an average must be formed with respect to this choice (keeping however the velocity $v_1$ within $d^3v_1$). Maxwell's *Ansatz* implicitly assumes that the distribution resulting from this averaging is uniform, so that the number of pairs of molecules for which the second molecule belongs to the efficient volume of the first is proportional to the value of the efficient volume.

As Boltzmann and Maxwell's British successors would later find, the latter assumption is not always allowed. One can imagine microscopic configurations of the gas for which the number of collisions is not given by Maxwell's formula. For example, let us assume that at a given instant the velocities of nearest neighbor pairs of molecules point toward one another. Then, the number of collisions in a subsequent time interval will greatly exceed the value given by Maxwell, even if the spatial distribution of molecules is as uniform as possible.[10]

This example clearly shows the gap in Maxwell's reasoning: For every choice of the molecule "1" with a velocity within $d^3v_1$ and for a not too small value of $\delta t$, there is one molecule "2" in the total (integrated over $d\Omega$) efficient volume of "1." Therefore, the average spatial distribution of the molecules "2" (in the sense earlier defined) is far from being uniform; it is more concentrated in the efficient volume than elsewhere.

One would vainly seek for such critical considerations in Maxwell's writings. His *Ansatz* sounded obvious, and it had the essential advantage of making the number of collisions dependent only on a coarse description of the dynamic state of a gas, namely, a description by a continuous distribution, $f(v)$ or $f(r, v)$ in the configuration space of a molecule. Finer, physically inaccessible details of the molecular description were rendered irrelevant.

This state of affairs allowed Maxwell to draw important consequences from his collision formula. Every transport phenomenon (e.g., transport

9. Maxwell 1867, *Scientific Papers,* 44.
10. Brush 1976, 616–625; Boltzmann 1896b, trans., 40–41.

of heat, momentum, etc.) in a gas subjected to external constraints (temperature gradient, pressure gradient, etc.) could be calculated by simply multiplying the elementary transport produced by one collision of a given kind by the number of collisions of this kind, and summing over kind. Most fundamentally, the equilibrium distribution of molecular velocities could be derived from the collision formula. The resulting expression, the so-called Maxwell distribution, had already been obtained by Maxwell in 1860. The new proof of 1866 proceeded in the following way.

Consider two generic elements $d^3v_1$ and $d^3v_2$ in the abstract space of molecule velocities, respectively around the velocities $v_1$ and $v_2$. A sufficient condition of (kinetic) equilibrium is that the number of collisions $dn$ for which the *initial* velocities belong to these elements, of the kind $\zeta$, be equal to the number $dn'$ of collisions for which the *final* velocities belong to these elements, of the inverse kind $\zeta'$ ($\zeta'$ is obtained from $\zeta$ by changing the sign of the angle $\theta$ between initial and final relative velocities). For these numbers to be finite there must of course be a latitude $d\Omega$ in the definition of $\zeta$, but we will take it to be negligible in comparison with the latitude in the definition of the direction of $v_2 - v_1$ resulting from the finite extension of $d^3v_1$ and $d^3v_2$.[11]

In order to appreciate the consequences of Maxwell's equilibrium condition, one must first note that a precise choice of $v_1$, $v_2$, and $\zeta$ implies a definite value of the final velocities $v'_1$ and $v'_2$, if energy and momentum are conserved during the collision. Indeed, momentum conservation gives $v'_1 + v'_2 = v_1 + v_2$, the collision kind gives the orientation of $v'_2 - v'_1$ with respect to $v_2 - v_1$, and energy and momentum conservation give together $|v'_2 - v'_1| = |v_2 - v_1|$; the two last pieces of information give $v'_2 - v'_1$, which, combined with the first piece, gives $v'_1$ and $v'_2$. Consequently, direct collisions (contributing to $dn$) and reverse collisions (contributing to $dn'$) are simply related by permuting the roles of the initial and final velocities.

Let us denote by $d^3v'_1$ and $d^3v'_2$ the elements in the space of velocities respectively corresponding to the elements $d^3v_1$ and $d^3v_2$ for a sharply defined kind of collision $\zeta$. Then the number $dn'$ of inverse collisions is given by

$$dn' = |v'_1 - v'_2|\sigma(\zeta', |v'_1 - v'_2|)\, d\Omega\, f(v'_1)f(v'_2)\, d^3v'_1\, d^3v'_2\, \delta t, \qquad (3)$$

a result of Maxwell's *Ansatz* (2), when $v'_1$ and $v'_2$ are taken as initial velocities, and the selected kind of collision is $\zeta'$.

---

11. Maxwell 1867, *Scientific Papers*, 44–46.

As shown above, $|v_1 - v_2| = |v_1' - v_2'|$, which implies that during a collision the relative velocity $u$ merely rotates. Moreover, the velocity $V = (v_1 + v_2)/2$ of the center of gravity is conserved. Therefore the differential element $d^3u\,d^3V$ is conserved. Since $d^3u\,d^3V = d^3v_1\,d^3v_2$, the differential element $d^3v_1\,d^3v_2$ is also conserved. Finally, the sign of the angle $\theta$ is clearly irrelevant to the definition of $\sigma$, which implies $\sigma(\zeta') = \sigma(\zeta)$. According to these remarks, the number of inverse collisions can be rewritten as

$$dn' = |v_1 - v_2|\sigma(\zeta, |v_1 - v_2|)\,d\Omega\,f(v_1')f(v_2')\,d^3v_1\,d^3v_2\,\delta t. \qquad (4)$$

Consequently, the equality of $dn$ and $dn'$ occurs if and only if

$$f(v_1)f(v_2) = f(v_1')f(v_2'). \qquad (5)$$

Admitting with Maxwell that $f$ cannot depend on the direction of particle velocity, there must exist a function $\varphi$ of $v^2$ such that $f(v) = \varphi(v^2)$, and, for any positive numbers $x$ and $y$, $\varphi(x)\varphi(y) = \varphi(x')\varphi(y')$ if $x + y = x' + y'$. The latter property is characteristic of exponential functions; hence $f$ must have the form

$$f(v) = \alpha e^{-\beta v^2}, \qquad (6)$$

which is "Maxwell's distribution" of molecular velocities.[12]

Clearly, this distribution is not modified by collisions inside the gas, as long as these collisions occur at a pace ruled by Maxwell's *Ansatz*. But can there be other stationary distributions? Maxwell's answer to this question is so brief that it deserves full quotation:

> If there were any other [final distribution of velocities] the exchange of velocities represented by OA and OA' [v and v' in the above notation] would not be equal. Suppose that the number of molecules having velocity OA' increases at the expense of OA. Then since the total number of molecules having velocity OA' remains constant, OA' must communicate as many to OA'', and so on till they return to OA.
>
> Hence if OA, OA', OA'', &c. be a series of velocities, there will be a tendency of each molecule to assume the velocities OA, OA', OA'', &c. in order, returning to OA. Now it is impossible to assign a reason why the successive velocities of a molecule should be arranged in this cycle, rather than in the reverse order. If, therefore, the direct exchange between OA and OA' is not equal, the equality cannot be preserved by exchange in a cycle. Hence the direct exchange between OA and OA' is equal, and the distribution we have determined is the only one possible.[13]

12. The assumption of isotropy is in fact not necessary: to be a conserved additive function of molecular velocities, the function $\ln f$ must take the form $\ln f = -\beta v^2 + \lambda \cdot v$. This implies $f = \alpha e^{-\beta(v - v_0)^2}$, as expected from the Maxwell distribution of a gas in a container moving at the velocity $v_0$.

13. Maxwell 1867, *Scientific Papers*, 45.

Be it wrong, incomplete, or overly condensed, this argument is certainly one of the most impenetrable in all Maxwell's writings. In particular, it is difficult to understand why the balancing of the "direct exchange" between two values of the velocity should be equivalent to the condition $dn = dn'$, which involves four values of the velocity and pairs of molecules. Even though Maxwell's gas theory rested on sound intuition and skillful mathematics, it failed to provide a convincing proof of the uniqueness of the velocity distribution; and, as we first saw, it left another essential question open, the degree of validity of the collision formula.

## BOLTZMANN'S IRREVERSIBLE EQUATIONS

Boltzmann's first important works were dedicated to an extensive criticism of Maxwell's kinetic theory. In spite of some early doubts, Boltzmann soon adopted the *Stosszahlansatz* and quickly generalized it to more complex systems. In every application he could verify the validity of results obtained from this hypothesis by an independent method, one based on the general theory of Hamiltonian systems and what was later called the ergodic hypothesis.[14]

When in 1872 Boltzmann questioned Maxwell's laconic reasoning for the uniqueness of the velocity distribution in a dilute gas, he based his alternative proof on the collision formula. The basic idea was simple: knowing the number of collisions of each kind, one could calculate the evolution in time of an arbitrary distribution of velocities and could check whether or not this distribution tended toward Maxwell's. If it did, Maxwell's would then be the unique equilibrium distribution.[15]

### THE BOLTZMANN EQUATION

With the notation of the preceding section, the number $f(\mathbf{v}_1)\, d^3v_1$ of molecules in the element $d^3v_1$ of the space of velocities is decreased by all collisions for which one of the initial velocities belongs to $d^3v_1$; the other initial velocity and the kind of the collision may be arbitrary. Conversely, this number is increased by all collisions for which one of the final velocities belongs to $d^3v_1$; here the other final velocity and the kind of the collision may be arbitrary. In mathematical terms this gives

$$\frac{\partial}{\partial t} f(\mathbf{v}_1)\, d^3v_1 = \int_{\zeta, \mathbf{v}_2} (dn' - dn) \tag{7}$$

14. See Klein 1973; Brush 1976; Dugas 1959; Darrigol 1988a.
15. Boltzmann 1872.

or, substituting the expressions (2) and (4) for $dn$ and $dn'$:

$$\frac{\partial f}{\partial t}(\mathbf{v}_1) = \int_{\zeta, \mathbf{v}_2} |\mathbf{v}_1 - \mathbf{v}_2| \sigma \, d\Omega \, [f(\mathbf{v}_1')f(\mathbf{v}_2') - f(\mathbf{v}_1)f(\mathbf{v}_2)] \, d^3 v_2. \tag{8}$$

Here we need to remember that $\mathbf{v}_1'$, and $\mathbf{v}_2'$ are defined by $\mathbf{v}_1$, $\mathbf{v}_2$, and $\zeta$. This equation is the simplest case of a "Boltzmann equation." The practical importance of this type of equation was considerable, for it permitted Boltzmann to give a systematic derivation of the observable evolution of a thermodynamic system slightly out of equilibrium, including the transport properties already derived by Maxwell by means of a less general method.

THE $H$-THEOREM

However, the most fundamental consequence that Boltzmann drew from his equation was the following theorem on irreversibility: The function of time $H$ defined in terms of the distribution function, $f$, according to

$$H = \int f \ln f \, d^3 v \tag{9}$$

is a strictly decreasing function of time for any choice of $f$—except Maxwell's distribution, for which $H$ is stationary.[16] This theorem immediately accomplished Boltzmann's initial purpose, a proof of the uniqueness of the equilibrium distribution of velocities. The proof was as follows.

The time derivative of $H$ is given by

$$\frac{dH}{dt} = \int \frac{\partial f}{\partial t} \ln f \, d^3 v + \int \frac{\partial f}{\partial t} \, d^3 v. \tag{10}$$

The second term of this expression vanishes, since the derivative can be permuted with the integral sign. The substitution of (8) in the first term gives

$$\frac{dH}{dt} = \int d\mu \, (f_1' f_2' - f_1 f_2) \ln f_1, \tag{11}$$

where $d\mu$ is a positive measure in the $(\mathbf{v}_1, \mathbf{v}_2, \zeta)$-space, defined as

$$d\mu = d^3 v_1 \, d^3 v_2 |\mathbf{v}_1 - \mathbf{v}_2| \sigma(\zeta, |\mathbf{v}_1 - \mathbf{v}_2|) \, d\Omega, \tag{12}$$

---

16. Boltzmann's original notation for $H$ was "$E$." One might wonder why Boltzmann focused on this particular function. A plausible answer is that he already knew the expression $S = -k \int \rho \ln \rho \, d^N p \, d^N q$ for the entropy of a canonically distributed gas (in Gibbs's later terminology).

and where $f'_1 f'_2 - f_1 f_2$ stands for $f(v'_1)f(v'_2) - f(v_1)f(v_2)$. Boltzmann's original proof is considerably simplified by noticing that $d\mu$ is invariant under a permutation of 1 and 2, and by a permutation of primed and unprimed velocities.

The first of these symmetries gives

$$\frac{dH}{dt} = \int d\mu \, (f'_1 f'_2 - f_1 f_2) \ln f_2; \tag{13}$$

the second gives

$$\frac{dH}{dt} = -\int d\mu \, (f'_1 f'_2 - f_1 f_2) \ln f'_1; \tag{14}$$

and both together give

$$\frac{dH}{dt} = -\int d\mu \, (f'_1 f'_2 - f_1 f_2) \ln f'_2. \tag{15}$$

Adding (11), (13), (14), and (15) and dividing by four gives

$$\frac{dH}{dt} = -\frac{1}{4} \int d\mu \, (f'_1 f'_2 - f_1 f_2)[\ln (f'_1 f'_2) - \ln (f_1 f_2)]. \tag{16}$$

The integrand is always positive since the two factors in parentheses always have identical signs (the logarithm being an everywhere increasing function). Therefore $dH/dt$ is always negative, and $H$ is always decreasing. A stable condition occurs only if the integrand vanishes for all values of the variables, that is, $f(v_1)f(v_2) = f(v'_1)f(v'_2)$ for every possible collision. The latter condition is precisely the one leading to Maxwell's distribution. This ends the proof of Boltzmann's $H$-theorem.

Boltzmann did not fail to notice that $-H$, calculated for Maxwell's distribution, gave the entropy of a perfect gas (temperatures being measured in energy units). His theorem therefore reproduced the law of entropy increase, insofar as the function $-H$ represented an extension of entropy out of equilibrium. In this way the second principle of thermodynamics could be deduced from kinetic theory, at least for the case of a dilute gas.[17]

## THE NATURE OF IRREVERSIBILITY

However, Boltzmann soon had to answer an "extremely pertinent" objection raised by his friend and colleague Joseph Loschmidt. Assuming, as

17. Boltzmann 1872, *BWA* 1:345–346.

Maxwell and Boltzmann did, the reversibility in time of the dynamics ruling the microscopic evolution of a closed system, one could always imagine microscopic initial conditions leading to a violation of the entropy law. Indeed, to every evolution with increasing entropy there corresponded an evolution with decreasing entropy, obtained by inverting the velocities of all molecules in the final microscopic configuration. It therefore seemed hopeless to try to derive the entropy law from kinetic theory without an *ad hoc* selection of microscopic initial conditions.[18]

ENTROPY AND PROBABILITY

In the latter conclusion Boltzmann saw nothing but an "interesting sophism." To illustrate his point of view, he considered a gas of hard spheres uniformly spread within the volume of a container (which provides maximum entropy) at the initial time, and he went on to note:

> One cannot prove that for every initial choice of the positions and the velocities of the spheres the distribution will be always uniform after a very long time; one can only prove that the number of [microscopic] initial states leading to a homogeneous distribution after a given long time is infinitely greater than the number of initial states leading to a heterogeneous distribution; even in the latter case the distribution would return to homogeneity after a longer time.[19]

In this manner Boltzmann reconciled the second principle with the reversibility of molecular dynamics by arguing that extremely improbable initial conditions could be neglected. He underlined the role of probability considerations in this context: "Loschmidt's argument," he wrote, "shows how intimately the second principle is bound to probability calculus." He further suggested a possible extension: "One might even calculate the probability of various distributions [Zustandverteilung] on the basis of their relative numbers, which might perhaps give rise to an interesting method for the calculation of thermal equilibrium."[20]

In the same year 1877 Boltzmann gave a precise expression to this idea by introducing the quantitative relation between entropy and probability for which he is perhaps most famous.[21] The detailed account of this relation will be postponed to a later section. For the time being it is sufficient

---

18. Boltzmann 1877a, *BWA* 2: 117; Loschmidt 1876 and subsequent papers. See Brush 1976, 238–240.
19. Boltzmann 1877a, *BWA* 2:119.
20. Ibid., 119; on 121, one reads: "[Loschmidt's] argument shows how intimately the second principle is bound to probability calculus."
21. Boltzmann 1877b.

to mention a persistent ambiguity in Boltzmann's wording for the exclusion of antithermodynamic processes. In some places these are judged to be "infinitely" rare, and the entropy law is still formulated as being strict. In other places they are said to be "extremely improbable" or "impossible in practice."[22]

MOLECULAR CHAOS

In any case, during these years Boltzmann never explained the exact nature of the probability considerations necessarily entering the proof of this $H$-theorem. In 1894 he could hear the dissatisfaction of British kinetic theorists concerning this lack of clarity. The discussion led to the concept of "molecular chaos," which had only a transitory importance in Boltzmann's work but a central one in the evolution of Planck's ideas.[23]

For a given partition of the space of molecular velocities into cells, the distribution $f(\mathbf{v})$ can be given a more definite meaning as the list of the numbers $N_i$ representing the number of molecules in the cell $i$. The corresponding value of $H$ is then defined as the sum

$$H = \sum_i N_i \ln N_i. \qquad (17)$$

In principle, the molecular dynamics allows one to calculate the evolution in time of these numbers $N_i$, and therefore the exact value of $H$ at every instant, independent of the Boltzmann equation. The corresponding curve $H(t)$ is extremely chaotic, with a very rapid alternation of increasing and decreasing phases. However, Boltzmann (and, later, more rigorously, Ehrenfest) could show that in a certain sense, which does not need to be specified here, the $H$-function was more probably decreasing than increasing.[24]

Boltzmann's equation gives the secular evolution of $H$, that is to say, the average evolution, smearing out local irregularities; it does so thanks to the introduction of a special hypothesis hidden in Maxwell's *Ansatz*. According to this assumption, called "molecular chaos" by Burbury, certain microscopic "ordered" configurations of the gas must be excluded—for instance, the one already described in the previous section, for which the velocities of closest neighboring molecules point toward each other. Beyond this intuitive but vague characterization, neither Boltzmann nor

22. Boltzmann 1877a, *BWA* 2:120; ibid., 121.
23. Burbury 1894 and other British contributions discussed by Brush; Boltzmann 1896b, trans., 40–41. See Brush 1976, 616–625.
24. On the $H$-curve, see Klein 1970c, 114–119.

his British friends could provide a general a priori definition of molecular chaos. In the end, they had to content themselves with defining disordered configurations *ad hoc,* as those for which Maxwell's *Ansatz* is valid.[25]


RECURRENCE

In 1896 Planck's assistant Ernst Zermelo published a new objection to the *H*-theorem, which left Boltzmann no time to dwell on molecular chaos. Poincaré had proved a few years earlier that any mechanical system confined to a finite region of space would, after a sufficiently long time, return to a configuration arbitrarily near the original one. According to Zermelo, this contradicted not only the *H*-theorem but also any kinetic theory of heat, since recurrence was never observed in thermodynamics. Boltzmann was particularly irritated by this objection: his previous description of the *H*-curve did not exclude recurrence; it just made it very improbable. In his opinion, all the objections to the *H*-theorem could be turned into harmless comments on the statistical validity of the entropy law.[26]

As for Zermelo's contention that recurrence in kinetic theory would contradict the entropy law of thermodynamics, Boltzmann remarked that a typical recurrence time as estimated from kinetic theory was so enormous that "in this length of time, according to the laws of probability, there [would] have been many years in which every inhabitant of a large country committed suicide."[27]

Answering Zermelo's more detailed objections to the shape of the *H*-curve, Boltzmann improved the description of it by distinguishing eras of entropy increase from eras of entropy decrease in terms of the secular trend of *H*. For a large number of particles the eras were so long that human observers were confined to a single one, that is, to one direction of time (this direction being defined as that of entropy change).[28]

Coming back to a more realistic time scale, at the end of the century these episodes left no doubt about Boltzmann's understanding of thermodynamic irreversibility: irreversible behavior could be deduced from the kinetic molecular model, but only as a statistical property of the system under consideration. In this way the second principle of thermodynamics lost its absolute character. This had been foreseen by Maxwell in the famous "demon argument" of 1867, and assented to by Gibbs in 1875 in

25. Boltzmann 1896b, trans., 40–41.
26. Zermelo 1895, 1896; Boltzmann 1896a, 1897a. See Brush 1976, 627–640.
27. Boltzmann 1898a, trans., 444.
28. Boltzmann 1897a. See Klein 1970c.

the following terms: "The impossibility of an uncompensated decrease of entropy seems to be reduced to an improbability." In 1898 Boltzmann quoted this thought at the start of the second part of his *Gastheorie*.[29]

In the foreword of the same book Boltzmann lamented on the general resistance his ideas met: "I am conscious of being an individual struggling weakly against the stream of the times. But it still remains in my power to contribute in such a way that, when the theory of gases is again revived, not too much will have to be rediscovered."[30] Rather than converting physicists to the kinetic theory, the notion of statistical irreversibility gave new weapons to those, energetists and positivists, who believed that the concepts of energy and entropy were irreducible.

## SUMMARY

Maxwell based his kinetic theory of 1866 on a seemingly obvious derivation of a central quantity: the number of encounters between the molecules of a homogeneous dilute gas. An important property of the resulting expression was that it did not depend on the detailed configuration of the molecules but only on the main quantity of physical interest, the distribution of molecular velocities. In thus eliminating the uncontrollable features of the microscopic description, Maxwell reaped a rich crop: he derived the equilibrium distribution of molecular velocities ("Maxwell's distribution law") and calculated how, through collisions, momentum, kinetic energy, and so on are transferred between contiguous layers of a gas (which accounts for viscosity, thermal conductivity, and other transport phenomena). However, the derivation of the collision formula entailed a hidden assumption of disordered motion, which Maxwell's followers later tried to explain.

Impressed by Maxwell's considerations, Boltzmann greatly extended the generality of the collision formula and the breadth of its applications. Whereas Maxwell had contented himself with the derivation of the final equilibrium distribution (and had given no satisfactory proof of the uniqueness of this distribution), in 1872 Boltzmann managed to derive the time evolution of the distribution of molecular velocities. The corresponding differential equation, which results from Maxwell's collision formula and Conservation laws, involved only the distribution of velocities

29. Gibbs 1878, *Collected works* 1:167; Boltzmann 1896b, trans. 215. For Maxwell's demon, see pp. 22–23 below.
30. Boltzmann 1896b, trans., 216.

and its time derivative: every uncontrollable feature of the microscopic model was, again, eliminated.

The Boltzmann equation not only simplified access to transport phenomena but also—more fundamentally—implied that the velocity distribution evolves irreversibly toward Maxwell's equilibrium form. Specifically, from the velocity distribution Boltzmann built a certain quantity (later named $H$ by Burbury) which, as a result of the Boltzmann equation, steadily decreases in time until it reaches a minimum value, one that corresponds to Maxwell's distribution. Moreover, the negative of this quantity provided a natural extension of the concept of entropy for a system out of equilibrium; for it increased during irreversible evolution, and, in the equilibrium state, it was identical with the entropy of a perfect gas.

The above result, which constitutes the $H$-theorem, was repeatedly criticized for conflicting with the general principles of mechanics. In 1876 Loschmidt enunciated his famous paradox: The Boltzmann equation produced irreversible changes, while the equations controlling the underlying molecular dynamics were reversible (symmetric with respect to time reversal). To resolve the conflict, Boltzmann explicitly limited the validity of his equation. He admitted the existence of special molecular configurations for which his equation did not hold; but, he added, such configurations were highly improbable, for they represented only an extremely small fraction of those compatible with given initial macroscopic conditions. In 1877, as a confirmation of this probabilistic view of irreversibility, Boltzmann proposed a direct quantitative relation between entropy and probability, to which I will return later.

In 1894, stimulated by his British colleagues' questions, Boltzmann discussed more precisely the nature of the statistical assumption necessary for the derivation of the Boltzmann equation. This led him to the concept of "molecular chaos" (so named by Burbury). For the Boltzmann equation to hold, excessively "ordered" configurations of the molecules had to be excluded (for instance, those in which every molecule flies toward its nearest neighbor). Aside from such intuitive remarks, Boltzmann contended himself with defining disordered states as states to which Maxwell's collision formula applies. In such a nominalistic guise the concept of molecular chaos could play only a minor role in Boltzmann's writings.

A late attack on the $H$-theorem came from Planck's assistant Zermelo in 1896. As follows from Poincaré's "recurrence theorem," any finite molecular system has to return to its original macroscopic configuration after a sufficiently long time; therefore, Zermelo concluded, the $H$-theorem could not be true, even if special "ordered" configurations of the molecules were initially excluded. This objection irritated Boltzmann but did not embarrass him: his idea of disorder was meant to be statistical, and average recurrence times were far beyond human observation. But to those who believed in a complete generality of the principles of thermodynamics, Zermelo's argument showed that kinetic theory had to be abandoned, for it was incompatible with strict irreversible behavior.

# CHAPTER II

# Planck's Absolute Irreversibility

## AGAINST ATOMS

Among the outspoken adversaries of Boltzmann's notion of irreversibility stood one of the most respected German thermodynamicists: Max Planck. In his dissertation work of 1879 this son of a law professor had raised the entropy law to the status of an absolute principle. Two years later he explicitly rejected the molecular hypothesis, precisely because it contradicted his conception of the second principle of thermodynamics:

> When correctly used, the second principle of the mechanical theory of heat is incompatible with the hypothesis of finite atoms [here a footnote refers to Maxwell's demon argument]. One should therefore expect that in the course of the further development of the theory, there will be a fight between these two hypotheses that will cost the life of one of them. It would be premature to predict the outcome of this fight now; but for the moment it seems to me that, in spite of the great successes of the atomic theory in the past, we will finally have to give it up and to decide in favor of the assumption of continuous matter.[31]

Planck's argument derived from Maxwell's earlier demon argument. The superhuman demon described in Maxwell's letter to Tait of December 1867 was able to discriminate between slow and fast molecules and, without expense of energy, to direct these two sorts of molecules toward different containers. The temperature gradient created in this way of course

---

31. Planck 1882, 475; for Planck's background see, e.g., Heilbron 1986; for his anti-atomistic positions see Kuhn 1978.

contradicted the second principle of thermodynamics. The conflict between the molecular hypothesis and the entropy law shown in this thought experiment convinced Maxwell that entropy was a subjective notion, namely, one depending on the physicist's incapability of controlling the motion of individual molecules. The same conflict led Planck to exclude the molecular hypothesis.[32]

During the years following this first public condemnation of kinetic molecular theory, Planck successfully applied macroscopic thermodynamics to various systems (chemical equilibrium, solutions, thermocouples, etc.), which strengthened his opinion that atoms were superfluous. In 1891 he found in the meeting of German scientists at Halle a public occasion to denigrate Boltzmann's and Maxwell's kinetic theory. The results of this theory, he declared, were not in suitable proportion to the mathematical effort expended.[33]

Boltzmann did not answer Planck's attack directly, but in 1894 he found a proper opportunity to confound his adversary. Planck had accepted the editorship of the posthumous fourth volume of Kirchhoff's lectures on physics. The last part was dedicated to a detailed exposition of Maxwell's kinetic theory, and it contained, among other things, a presentation of the proof of Maxwell's distribution based on collision numbers. In a paper published in *Annalen der Physik*, Boltzmann criticized the negligence of Kirchhoff's editor. Whereas Maxwell had equated the number $dn$ of collisions with initial velocities belonging to two given elementary domains of velocity space to the number $dn'$ of collisions with final velocities in these domains, the proof by Kirchhoff-Planck equated two different expressions of the same $dn$: the one given by Maxwell, and another, $dn^*$, obtained by multiplying "the probability of collision of two molecules" $\sigma \, d\Omega \, |v_1 - v_2|$ by the population of the final state $f(v_1') \, f(v_2') \, d^3v_1' \, d^3v_2'$. Boltzmann protested that the probability of collision of the two molecules equaled $\sigma \, d\Omega \, |v_1 - v_2|$ *only* if the past history of the molecules was unknown; a quite different expression had to be used when the molecules were known to have interacted in a given way.[34]

Planck acknowledged this point but attributed responsibility for the mistake to Kirchhoff. He further claimed that the objection equally invalidated Maxwell's original reasoning (which reveals his ignorance of it). As a token of his good will, however, he offered an alternative deduction

32. Maxwell to Tait, 11 Dec. 1867, Cambridge University Archive; Maxwell 1871, 308–309. See Klein 1970b.
33. Planck 1891.
34. Kirchhoff 1894, 134; Boltzmann 1894.

based on time reversal. When changing the direction of time, a stationary distribution of velocities is left unchanged and stationary; therefore, the number $dn$ of collisions of a given kind must be equal to the number of collisions of the reverse kind, which is exactly Maxwell's $dn'$.[35]

Boltzmann appreciated this argument, for it made the equality of $dn$ and $dn'$ necessary and thus gave a new proof of the uniqueness of Maxwell's distribution without recourse to the $H$-theorem. In his published comments, Boltzmann emphasized the compatibility of the proof with the notion of molecular chaos. Indeed this concept, a fresh outcome of his discussions with British kinetic physicists, was precisely the one supposed to control the applicability of Maxwell's collision probabilities. On this occasion, in spite of his persisting dislike of atomism, Planck could gather that in kinetic theory irreversibility was intimately connected with some assumption of molecular chaos, whatever it meant. Besides, this incident may have convinced him that a new explanation of irreversibility was urgently needed to secure the foundations of thermodynamics.[36]

## BLACKBODY RADIATION

Planck's belief in an absolute entropy principle excluded atomism but not Boltzmann's (and Clausius's) endeavor to give a mechanical explanation of entropy. With respect to mechanical reduction he believed entropy to be comparable to energy. Both quantities had to be determined not only by the thermodynamic state of the system but also by the underlying mechanical state. The problem was to find the proper mechanical basis. While in his opinion molecules would not do, the possibility of a continuum with well-chosen properties, the prime example being the electromagnetic ether, merited further exploration.

There was another reason to examine the relation between electrodynamics and thermodynamics: In the preceding years, several thermodynamic arguments had been successfully applied to the light emitted by heated bodies, and, since Maxwell's "dynamical theory of the electromagnetic field" (1864), the nature of light was often believed to be electromagnetic. Perhaps most important for someone in search of universal properties of nature, Kirchhoff had proved in 1859 an important theorem, a consequence of which was the universality of the spectrum of thermal radiation emitted by what he called a (perfect) blackbody.[37]

35. Planck 1895.
36. Boltzmann 1895.
37. Kirchhoff 1859. See Jungnickel and McCormmach 1986, 1:299–301; Jammer 1966, 2–6; and Kangro 1970.

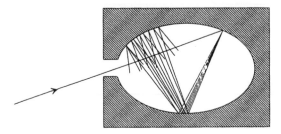

Figure 3.   Absorption of a radiation beam entering a pierced cavity.

By definition a blackbody absorbs any radiation falling upon it, which implies that it looks black in the common sense of the word as long as it is cold enough not to emit much thermal radiation. An excellent concrete realization of a blackbody is obtained by piercing a hole in a container whose walls are at least partially absorbing at all frequencies. Indeed, a light ray penetrating through the hole is both reflected and attenuated a great number of times by the inside walls, until it practically vanishes (see fig. 3). Let us now maintain the walls of this cavity at a constant high temperature. The radiation emitted by a portion of the walls interacts a great number of times with other portions of the walls, which leads to thermal equilibrium for the energy per unit volume $u_v \, dv$ corresponding to a frequency interval $dv$.[38] A simple proof of the universality of the spectral density $u_v$ (not Kirchhoff's) goes as follows.

Consider, *ab absurdo,* two cavities at the same temperature but with different spectral densities $u_v^{(1)}$ and $u_v^{(2)}$. Connect them through a small tube in which is placed a filter F at the frequency $v$ (fig. 4). If $u_v^{(1)}$ is greater than $u_v^{(2)}$, there must be a flow of radiation from 1 to 2; and this flow must be permanent since the excess of radiation energy in 2 must be re-absorbed in order to maintain the thermal equilibrium in 2. But a permanent energy flow between sources at equal temperature is incompatible with the second principle of thermodynamics. Therefore, it must be that $u_v^{(1)} = u_v^{(2)}$.

In 1884 Boltzmann assumed the electromagnetic nature of thermal radiation to prove Stefan's empirical law (1879): The total energy density $u$ (given by $\int_0^{+\infty} u_v \, dv$) is proportional to the fourth power of the absolute

38. The first definition of a blackbody is in Kirchhoff 1860. As Jammer explains, Kirchhoff's considerations originated in studies of solar radiation (Jammer 1966, 2–6). The use of a pierced cavity for the empirical study of blackbody radiation was first suggested by Christiansen 1884 and first realized by Lummer and Wien 1895.

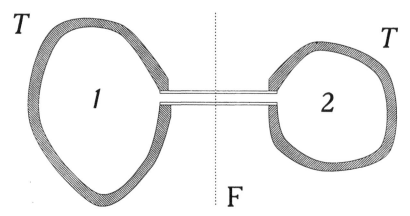

Figure 4.    Thought experiment leading to the universality of the blackbody spectrum. *T* is the temperature of a thermal bath, and F a monochromatic filter.

temperature *T*. A slightly modernized version (in Planck's manner) follows.[39]

According to Maxwell, an electromagnetic plane wave falling normally upon a perfectly reflecting surface exerts a mechanical pressure, the intensity of which is given by twice the energy density of the wave. Consider now the case of an isotropic radiation falling upon a reflecting plane $\Pi$. The pressure exerted by the fraction of this radiation oriented in the direction $\zeta$, within the solid angle $d\Omega$, is given by $p = 2u \cos\theta\, d\Omega/4\pi$, where $u$ is the total energy density, and $\theta$ is the angle between the direction $\zeta$ and the normal to $\Pi$. Since this pressure is directed along $\zeta$, its component normal to $\Pi$ is $p\cos\theta$, or $2u\cos^2\theta\, d\Omega/4\pi$. The average pressure, $P$, exerted on $\Pi$ is obtained by integrating the latter expression over all solid angles pointing toward $\Pi$:

$$P = (2u/4\pi) \int_0^{\pi/2} 2\pi \sin\theta \cos^2\theta\, d\theta = u/3. \tag{18}$$

It should be further noted that in the case of blackbody radiation this pressure is independent of the reflecting quality of $\Pi$. Indeed, if it were dependent, one could build a perpetual motion of the second kind by inserting in a blackbody cavity a plate silvered on one side and blackened on the other. One may therefore legitimately speak of a "radiation gas" with a definite pressure $P$, as Boltzmann did in his paper.

In this circumstance, the entropy variation

$$dS = [d(uV) + P\, dV]/T, \tag{19}$$

39. Boltzmann 1884a, 1884b. For the context of this work see Carazza and Kragh 1989.

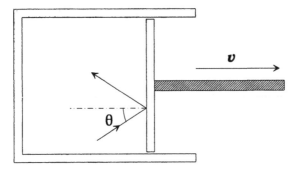

Figure 5.   Wien's perfectly reflecting piston, as used in the proof of the displacement law.

where $V$ is the volume of the cavity, must be a differential. This implies (after subtraction of $d(uV/T)$ from $dS$)

$$\frac{1}{T^2}\frac{\partial(uV)}{\partial V}\bigg|_T = \frac{\partial(P/T)}{\partial T}\bigg|_V \qquad (20)$$

or, using (18) and Kirchhoff's law (which makes $u$ a function of $T$ only),

$$\frac{u}{T^2} = \frac{1}{3}\frac{d(u/T)}{dT}. \qquad (21)$$

Integrating the latter equation gives

$$u = \sigma T^4, \qquad (22)$$

which is Stefan's law.

Boltzmann's original reasoning involved, for historical reasons not worth mentioning here, an adiabatic displacement of a reflecting piston inserted in a cylindrical blackbody cavity. During this displacement the spectrum of the reflected radiation differs slightly from that of the incident radiation, as a result of the Doppler effect. In 1894 Wilhelm Wien cleverly exploited this thought experiment to restrict the form of the function $u_\nu(T)$.[40]

Wien's argument, or more exactly Planck's version of it,[41] starts with the consideration of a cavity with perfectly reflecting walls, one of which is part of a mobile piston (fig. 5). At the initial time this cavity contains isotropic electromagnetic radiation with the spectral density $\rho_\nu$. Under a slow displacement of the piston at a constant speed $v$, the light of frequency

40. Boltzmann 1884b; Wien 1894.
41. Planck 1906, 68–82.

$v$ in an incident beam is Doppler-shifted by

$$\delta v = -2v(v/c) \cos \theta, \tag{23}$$

where $\theta$ is the angle of incidence, and $c$ the velocity of light. In a time interval $\delta t$ and for the same direction of incidence, the energy flux impinging upon the surface area $S$ of the moving mirror is given by

$$\phi = c\rho_v \, dv \, S(d\Omega/4\pi) \cos \theta \, \delta t. \tag{24}$$

As a result of the Doppler shift of the reflected radiation, there is a variation $\delta(\rho_v V)$ of the spectral energy at frequency $v$; its value may be obtained by subtracting the flux at the frequency $v$, whose frequency will be increased above $v$, and adding the flux at the frequency $v - \delta v$, whose frequency will be increased just to $v$ by the reflection:

$$\delta(\rho_v V) = c(\rho_{v-\delta v} - \rho_v)(d\Omega/4\pi)S \cos \theta \, \delta t, \tag{25}$$

or, using (23) for $\delta v$,

$$\delta(\rho_v V) = 2v \frac{\partial \rho_v}{\partial v} \cos^2 \theta \ (d\Omega/4\pi)Sv \, \delta t. \tag{26}$$

Noting that $Sv \, \delta t$ is equal to the variation in volume $\delta V$ of the cavity, and integrating over $\theta$ yields

$$\delta(\rho_v V) = v \frac{\partial \rho_v}{\partial v} \delta V \int_0^{\pi/2} \sin \theta \cos^2 \theta \, d\theta = \frac{1}{3} v \frac{\partial \rho_v}{\partial v} \delta V, \tag{27}$$

and

$$\delta \rho_v = \left( \frac{1}{3} v \frac{\partial \rho_v}{\partial v} - \rho_v \right) \frac{\delta V}{V} \tag{28}$$

for the variation $\delta \rho_v$ of the spectral density.

If the radiation initially contained in the cavity is blackbody radiation at a given temperature $T$ $(\rho_v = u_v(T))$, then the radiation obtained after the adiabatic displacement of the piston must also be in thermal equilibrium (but at a different temperature). Indeed, if the opposite were true, one would have a state out of equilibrium connected to a state in equilibrium through an adiabatic reversible transformation, which is absurd.[42] In this

42. Planck (1906, 69–70) demonstrates the absurdity by considering the following cycle. The piston, initially containing blackbody radiation, is moved adiabatically to a different position (step 1); a little piece of a "black" substance is introduced inside the piston and removed (step 2); the piston is brought back adiabatically to its initial position (step 3); a little piece of "black" substance is introduced inside the piston (step 4). This is a cycle because both the total heat and the total work exchanged are zero; energy densities and temperatures are therefore equal in the initial and final states. If the radiation was not at equilibrium after step 1 or after step 3, entropy would be created during step 2 and step 4, without compensation during the adiabatic processes (1) and (3).

case the variation of the spectral density can be obtained in a second independent way, as the variation of the universal function $u_\nu(T)$ for the variation $\delta T$ of temperature corresponding to the adiabatic transformation:

$$\delta u_\nu = \frac{\partial u_\nu}{\partial T}\, \delta T. \tag{29}$$

$\delta T$ itself is readily obtained by equating the variation of the total radiation energy as given by the Stefan-Boltzmann law (equation 22) to the work performed by the piston against the radiation pressure (given by equation 18):

$$\delta\, (\sigma T^4 V) = -P\, \delta V = -\tfrac{1}{3}\sigma T^4\, \delta V, \tag{30}$$

which implies

$$\frac{\delta T}{T} = -\frac{1}{3}\frac{\delta V}{V}. \tag{31}$$

Equating the two expressions (28) and (29) for $\delta u_\nu$ now gives

$$\left(T\frac{\partial}{\partial T} + \nu\frac{\partial}{\partial \nu}\right)(u_\nu/\nu^3) = 0, \tag{32}$$

with the general integral:

$$u_\nu = \nu^3 f(\nu/T), \tag{33}$$

where $f$ is an arbitrary function. This is the so-called "displacement law" which allows the derivation of the blackbody spectrum at any temperature, once it is known at a given temperature.


## PLANCK'S RESONATORS

The empirical validity of Kirchhoff's law, Stefan's law, and Wien's displacement law left no doubt about the legitimacy of combining electrodynamic and thermodynamic laws; and Planck was well aware of these developments. In 1895 he decided to examine the electrodynamic mechanisms responsible for the thermalization of radiation, hoping to find in them the ultimate source of irreversibility and to perhaps determine the arbitrary function in Wien's displacement law. This grand program focused on a very simple system, a Hertz resonator: that is, a small, nonresistive, oscillating electric circuit interacting with electromagnetic waves, the characteristic wavelength of the oscillator being much larger than the oscillator. The simplest choice was considered the most adequate by Planck,

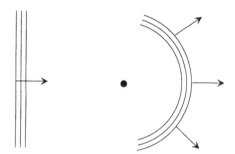

Figure 6. Planck's intuition of the source of thermodynamic irreversibility: the diffusion of a plane electromagnetic wave by a resonator (at the frequency of the wave).

for in light of Kirchhoff's theorem, the properties of thermal radiation could not depend on the specific properties of the thermalizing system.[43]

What first attracted Planck's attention was the apparent irreversibility of the interaction between radiation and resonator: a plane monochromatic wave falling upon a resonator forces vibrations of the resonator when the condition of resonance is approximately met; in turn these vibrations emit secondary waves over a wide angle (fig. 6). Also, an excited resonator left to itself emits radiation at its characteristic frequency and thereby gradually loses its energy. Such processes, resonant scattering or radiation damping, looked essentially irreversible, even though the total energy (that of the resonator plus that of radiation) was strictly conserved (in the absence of the Joule effect in the circuit). Planck concluded:

> The study of conservative damping seems to me highly important, because it opens a new perspective on the possibility of a general explanation of irreversible processes through conservative interactions, a more and more pressing problem in contemporary theoretical physics.[44]

These results of classical electrodynamics are now very well known; they are usually obtained through a specific model of the resonator, for instance an elastically bound electron. In 1895, however, the existence of the electron had not been proved; and Lorentz's formulation of electrodynamics, with its detailed analysis of microscopic sources, was not yet currently known (the famous *Versuch* was published in the same year).

43. For earlier accounts of Planck's program, see Brush 1976, 628–640; Klein 1962, 1963a; Kuhn 1978; Needell 1980. A general survey of Planck's thermodynamic views is in Hiebert 1968.
44. Planck 1896, 1897a, *PAV* 1:470.

Planck was therefore confined to a different method, first used by Hertz in 1889 for a calculation of the energy radiated by an oscillating dipole. In this method *the detailed structure of the resonator was irrelevant.*[45]

As we shall observe in the following, Planck maintained this generality throughout his program and even gave it an essential role. It is therefore useful at this point to explain how the equation of a resonator in an electromagnetic field can be established with this method. Another reason for analyzing Planck's reasoning is that it is typical of a style of theoretical physics, namely, concentrating on the features of physical systems which can be determined on the basis of general principles only, without recourse to detailed microscopic assumptions. Nevertheless, hurried readers may be content with these general comments and may jump to the equation (62), p. 36, for the evolution of the dipolar moment $f$ of a resonator.

### THE RESONATOR EQUATION

A variable distribution of charge and current is located around the origin O of coordinates, and its extension does not exceed the length $d$. According to Hertz, the simplest possible form of the electromagnetic field at a large distance $r$ from O $(r \gg d)$ derives from a vector potential directed along the $z$-axis

$$A_z = \frac{1}{cr} f'\left(t - \frac{r}{c}\right), \tag{34}$$

(here $f'$ denotes the derivative of the function $f$) and from a scalar potential

$$\varphi = -\frac{\partial}{\partial z}\left[\frac{1}{r} f\left(t - \frac{r}{c}\right)\right]. \tag{35}$$

Indeed, $A_z$ is isotropic, and the fields

$$\mathbf{E} = -\nabla\varphi - \frac{1}{c}\frac{\partial \mathbf{A}}{\partial t}, \qquad \mathbf{B} = \nabla \times \mathbf{A} \tag{36}$$

satisfy Maxwell's equations in an empty portion of space, as results from[46]

$$\Box A_z = 0, \qquad \Box\varphi = 0, \qquad \text{and} \qquad \nabla \cdot \mathbf{A} + \frac{1}{c}\frac{\partial\varphi}{\partial t} = 0. \tag{37}$$

---

45. Hertz 1889; Planck 1897a.
46. Planck used Gaussian units (see Conventions and Notations at the front of this book). Hertz introduced the function $f$ (which he denoted $\Pi$) but did without the potentials $\varphi$ and $\mathbf{A}$.

A physical meaning is given to this solution by considering the particular case of a monochromatic source, for which $f = f_0 e^{i\omega(t - r/c)}$. If the wavelength $\lambda$ exceeds the dimensions of the source $d$ by a large amount, the scalar potential in the region defined by $d \ll r \ll \lambda$ is given by

$$\varphi \sim \frac{f}{r^2}\frac{z}{r}. \tag{38}$$

This is precisely the form of the electric potential that would be created by a static dipole with an amplitude $f$ (in Gaussian units). By a natural extrapolation, Hertz's solution should be regarded as the radiation created by an electric dipole with the varying amplitude $f$.

We now return to the general case, but with a limitation on the variations of the dipole $f$: these can be neither too fast nor too slow. More specifically, it will be necessary to assume that there exist two characteristic lengths $\underline{\lambda}$ and $\bar{\lambda}$ such that $\bar{\lambda} > \underline{\lambda} \gg d$ and, at any time,

$$\frac{c}{\bar{\lambda}} < \left|\frac{f'}{f}\right| < \frac{c}{\underline{\lambda}}, \qquad \frac{c}{\bar{\lambda}} < \left|\frac{f''}{f'}\right| < \frac{c}{\underline{\lambda}}, \qquad \frac{c}{\bar{\lambda}} < \left|\frac{f'''}{f''}\right| < \frac{c}{\underline{\lambda}}. \tag{39}$$

The energy $dY/dt$ emitted in a time unit by the variable dipole can be derived by calculating the flux of the Poynting vector $(c/4\pi)\, \mathbf{E} \times \mathbf{B}$ across a sphere centered in O with a radius $r \gg \bar{\lambda}$. On this sphere the potentials (34) and (35) in spherical coordinates $r$, $\theta$, $\alpha$ (see fig. 7) are approximatively given by:

$$A_r = \frac{f'}{cr}\cos\theta, \qquad A_\theta = -\frac{f'}{cr}\sin\theta, \qquad A_\alpha = 0, \qquad \varphi = \frac{f'}{cr}\cos\theta. \tag{40}$$

The resulting field strengths are (using (39))

$$\left.\begin{array}{llll} B_r = 0, & B_\theta = 0, & B_\alpha = (f''/rc^2)\sin\theta \\ E_r = 0, & E_\theta = (f''/rc^2)\sin\theta, & E_\alpha = 0 \end{array}\right\} \tag{41}$$

so that the Poynting vector is a radial vector with the amplitude $(c/4\pi)$ $(f''/rc^2)^2 \sin^2\theta$. Its flux through the sphere is

$$\phi(r) = \frac{c}{4\pi}\left(\frac{f''}{rc^2}\right)^2 2\pi r^2 \int_0^\pi \sin^2\theta \sin\theta\, d\theta, \tag{42}$$

or

$$\phi(r) = \frac{2}{3c^3}f''^2\left(t - \frac{r}{c}\right). \tag{43}$$

In a second step of his considerations, Planck superposed on Hertz's solution an external field $\mathbf{E}_e$, $\mathbf{B}_e$. This field must be a solution of Maxwell's

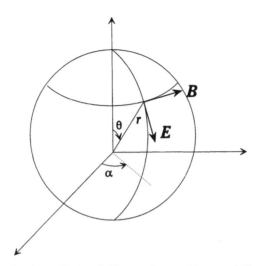

Figure 7.   The dipolar radiation field according to Hertz and Planck.

equations in empty space; and at any time the characteristic distance $\lambda_e$ over which it varies is assumed to greatly exceed the size $d$ of the charge distribution. Then the energetic coupling between the variable charge distribution around O and the external field can be determined *without knowing the internal structure of this distribution*. To this end Planck cleverly considered the energy flux, $\Phi$, through a sphere $\Sigma$ centered on O, much larger than the charge distribution but small enough to make the variations of the enclosed fields negligible ($r \gg d$, $r \ll \lambda_e$, $r \ll \underline{\lambda}$).

Assuming that the charge distribution exchanges energy only with the electromagnetic field (there is no Joule effect and no internal electromotive force), this flux must be equal to the diminution rate $-dY/dt$ of the energy of the charge distribution plus the diminution rate $-dW_e/dt$ of the energy of the external field enclosed by $\Sigma$ in the absence of a charge distribution:

$$\Phi = -\frac{dY}{dt} - \frac{dW_e}{dt}. \tag{44}$$

Another expression of this flux results from Poynting's formula:

$$\Phi = \frac{c}{4\pi} \int_{\Sigma} [(E + E_e) \times (B + B_e)] \cdot dS. \tag{45}$$

This integral may be split into four terms according to the development of the vector product. The one corresponding to $E \times B$ is the flux $\phi_0$ across $\Sigma$ resulting from the field created by the charge distribution, and it is simply obtained by noting that the flux $\phi(r)$ calculated above (formula 43) must be equal to its retarded value (the propagation time being $r/c$):

$$\phi_0 = \frac{2}{3c^3} f''^2(t). \tag{46}$$

The term corresponding to $\mathbf{E}_e \times \mathbf{B}_e$ exactly compensates $-d\mathbf{W}_e/dt$. The coupling terms remain:

$$\Phi_1 = \frac{c}{4\pi} \int_\Sigma (\mathbf{E}_e \times \mathbf{B}) \cdot d\mathbf{S}, \qquad \Phi_2 = \frac{c}{4\pi} \int_\Sigma (\mathbf{E} \times \mathbf{B}_e) \cdot d\mathbf{S}. \tag{47}$$

Since the radius $r$ of $\Sigma$ is much smaller than the length $\underset{\sim}{\lambda}$, according to (34), (35), (36), and (39) the fields $\mathbf{E}$ and $\mathbf{B}$ on $\Sigma$ are approximatively given by

$$\left.\begin{array}{lll} E_r = (2f/r^3) \cos\theta, & E_\theta = (f/r^3) \sin\theta, & E_\alpha = 0 \\ B_r = 0, & B_\theta = 0, & B_\alpha = (f'/cr^2) \sin\theta \end{array}\right\} \tag{48}$$

This leads to

$$\Phi_1 = \frac{c}{4\pi} \int E_z^e f' \frac{\sin\theta}{cr^2} (\sin\theta) 2\pi r^2 \sin\theta \, d\theta = -\frac{2}{3} E_z^e(O) f', \tag{49}$$

since the value of $\mathbf{E}_e$ at any point of the sphere may be replaced by its value in O.

One might easily believe that $\Phi_2$ vanishes, because for a uniform $\mathbf{B}_e$ the corresponding integral vanishes: one has

$$\int [\mathbf{E} \times \mathbf{B}_e(O)] \cdot d\mathbf{S} = \mathbf{B}_e(O) \cdot \int d\mathbf{S} \times \mathbf{E}, \tag{50}$$

and the latter integral is a zero vector, since $d\mathbf{S} \times \mathbf{E}$ is always tangent to the parallels of $\Sigma$ and has a constant modulus on a given parallel. Yet, $\Phi_2$ is not zero, because the *gradient* of $\mathbf{B}_e$ contributes a term of the same order (in $r/\underset{\sim}{\lambda}$) as $\Phi_1$. This term is

$$\Phi_2 \sim \frac{c}{4\pi} \int_\Sigma [\mathbf{E} \times (\mathbf{r} \cdot \mathbf{V})\mathbf{B}_e] \cdot d\mathbf{S}. \tag{51}$$

Because of the rotational invariance of $d\mathbf{S} \times \mathbf{E}$, only the part $\frac{1}{2}\mathbf{r} \times (\mathbf{V} \times \mathbf{B}_e)$ of $(\mathbf{r} \cdot \mathbf{V})\mathbf{B}_e$ contributes to this integral. Taking into account Maxwell's equation $\mathbf{V} \times \mathbf{B}_e = (1/c)(\partial \mathbf{E}_e/\partial t)$, an elementary calculation then yields

$$\Phi_2 = \tfrac{1}{3}\dot{E}_z^e(O) f. \tag{52}$$

Equating the two expressions (44) and (45) of $\Phi$ now gives

$$-\frac{dY}{dt} = \phi_0 + \Phi_1 + \Phi_2, \tag{53}$$

or

$$\frac{dY}{dt} + \frac{2}{3c^3}f''^2 - \frac{1}{3}(2E_z^e f' - \dot{E}_z^e f) = 0. \tag{54}$$

In the latter equation one can consider the time to be the only variable since, according to (39), $f(t - r/c)$ can be replaced with $f(t)$ whenever $r \ll \lambda$. This gives

$$\frac{d}{dt}\left(Y + \frac{1}{3}E_z^e f + \frac{2}{3c^3}f\ddot{f}\right) = \frac{2}{3c^3}\dot{f}\ddot{f} + \dot{f}E_z^e. \tag{55}$$

A further simplification of this equation is obtained by noticing that the terms accompanying $Y$ in the parentheses are negligible, as results from the following remarks.

From the expressions (48) for $\mathbf{E}$ and $\mathbf{B}$ on $\Sigma$ result the orders of magnitude $f^2/r^6$ for the electric and $f'^2/c^2r^4$ for the magnetic energy density on $\Sigma$. Since $r$ is much larger than the dimensions of the charge distribution, these densities must be very small compared with the contribution $Y/(4/3)\pi r^3$ of the charge distribution to the average energy density inside $\Sigma$. This gives the inequalities

$$f^2/r^3 \ll Y, \qquad f'^2/c^2r \ll Y. \tag{56}$$

Besides, $E_z^e$ must have the same order of magnitude as $f'''/c^3$ in order that, in equation (55), the variations of $f$ may be coupled to those of $E_z^e$. Finally, the inequalities (39) give, for $r \ll \lambda$,

$$|f^{(n)}|/c^n \ll |f|/r^n \qquad (n = 1, 2, 3). \tag{57}$$

Taken together, these remarks allow us to write

$$|E_z^e f| \sim |f'''f|/c^3 \ll f^2/r^3 \ll Y, \tag{58}$$

$$|\dot{f}\ddot{f}|/c^3 \ll f'^2/c^2r \ll Y, \tag{58'}$$

as previously asserted. Thanks to this property, one may, without any loss of information, redefine the energy of the charge distribution as the expression in the parentheses in equation (55), which gives the simpler equation

$$\frac{dY}{dt} = \frac{2}{3c^3}\dot{f}\ddot{f} + \dot{f}E_z^e. \tag{59}$$

At this point Planck specified the form of $Y$ in order to represent the case of a resonator, for which the variable distribution of charge and current is comparable to an oscillating circuit. From an analogy with the theory of such circuits he set

$$Y = \tfrac{1}{2}(Kf^2 + L\dot{f}^2), \tag{60}$$

wherein the two terms correspond to electric and magnetic energies that can be periodically transformed into each other. This hypothesis implies an equation for the evolution of $f$:

$$Kf + L\ddot{f} - \frac{2}{3c^3}\dddot{f} = E_z^e(O). \tag{61}$$

In the absence of an external excitation the term proportional to $\dddot{f}$ produces a spontaneous (conservative) damping of the resonator oscillations as a result of the emission of radiation. In the presence of an external field, the electric field $E_z^e(O)$ at the resonator along the direction of its axis acts on its electric moment $f$ if the Fourier spectrum of this field contains frequencies that are sufficiently close to the eigenfrequency $\omega_0 = \sqrt{K/L}$ of the resonator. If so, then the resulting field, which is the sum of the external field and the Hertz field produced by $f$, spreads out in every direction of space even if the external field propagates in a definite direction. This is the feature perceived as irreversible by Planck.

Since radiation damping is always very small (it takes a great number of periods), the previous equation is approximately equivalent to a simpler one:[47]

$$\ddot{f} + \rho\omega_0\dot{f} + \omega_0^2 f = E/L, \tag{62}$$

where $E$ is an abbreviation for the exciting field $E_z^e(O)$, and

$$\omega_0^2 = K/L, \qquad \rho\omega_0 = \frac{2}{3c^3} K/L^2, \tag{63}$$

or

$$K = \frac{2}{3c^3} \omega_0^3/\rho, \qquad L = \frac{2}{3c^3} \omega_0/\rho. \tag{64}$$

Planck's subsequent studies of radiation processes were based on these equations and on the energy formula (60).

## SUMMARY

Planck believed that the *two* principles of thermodynamics were absolutely valid. Just as certainly as the energy of a closed system remained constant,

---

47. In the Fourier space, the Green functions of equations (61) and (62) are respectively $\omega_0^2 - \omega^2 + i\rho\omega^3/\omega_0$ and $\omega_0^2 - \omega^2 + i\rho\omega\omega_0$. They differ appreciably from zero only for $\omega \sim \omega_0$, and then take approximately equal values.

so too for Planck its entropy could *only* increase in time. Every new theory had to be compatible with both principles—or it had to be rejected. As Planck repeatedly asserted in the years 1880–1895, kinetic molecular theories did not pass the test, for they implied the possibility of entropy-decreasing processes in closed systems. Such violations of the second principle had been shown to occur by Maxwell in his demon argument (1867), and again by Zermelo (1896) with the help of Poincaré's recurrence theorem.

Planck nevertheless studied Maxwell's kinetic theory, if only as a part of his duties as editor of Kirchhoff's lectures on thermodynamics, which included this topic. The resulting book, published in 1894, contained an unfortunate mistake in the proof of Maxwell's distribution law, which stirred a short but instructive polemic with Boltzmann. Planck proposed a better proof and in the process became acquainted with, though not convinced by, Boltzmann's idea of molecular chaos. On this occasion he might also have felt the need of an alternative microscopic foundation of thermodynamics. The following year he engaged upon a new program, the principal aim of which was to provide an *electromagnetic* explanation of the principles of thermodynamics.

Relations between electrodynamics and thermodynamics had already been found in the study of the so-called blackbody radiation, that is, radiation in thermal equilibrium with the walls of a cavity. That thermodynamics applied to this radiation was clear from the experimental verification of one of this theory's consequences: the universality of the blackbody spectrum (proved by Kirchhoff in 1859). That electromagnetic theory also applied was suggested by Boltzmann's proof (1884) of Stefan's law (1879), which rested upon the use of Maxwell's expression for radiation pressure. Finally, in 1894 Wien had shown that Boltzmann's latter argument could be extended to derive the so-called displacement law, which expresses the blackbody spectrum at an arbitrary temperature in terms of that for any given temperature. By then Hertzian waves had been discovered (1888), and most physicists believed in the electromagnetic nature of blackbody radiation.

Well aware of these developments, in 1895 Planck proposed that the electromagnetic interaction between matter and radiation could explain both thermodynamic irreversibility and the observed value of the black-body spectrum. As the archetype of an energy-conserving, irreversible process, he imagined the scattering of a plane electromagnetic wave by a miniature version of a perfect Hertz resonator (small oscillating circuit with neither dissipation nor internal electromotive force). Planck then proceeded to a quantitative evaluation of this effect. Unlike Lorentz, who had calculated electromagnetic scattering with the help of a specific model of elastically bound ions, Planck favored Hertz's phenomenological approach to electrodynamics and left the internal structure of the resonators undetermined. Just by balancing the energy flux through cleverly chosen surfaces enclosing the resonator, Planck deduced the mathematical form of the interaction between the electric moment of a resonator and the surrounding radiation. The resulting differential equation involved a damping term, which Planck interpreted as the sought-after source of irreversibility.

# CHAPTER III

# On Irreversible Radiation Processes

## A POLEMIC WITH BOLTZMANN

In early 1897 Planck published the first of a series of five memoirs entitled "On irreversible radiation processes." His aim was to exploit the preceding equations in a systematic investigation of the uniformizing action of resonators on the radiation enclosed in a cavity with ideal (non-thermalizing) reflecting walls.[48]

In the first memoir he contented himself with some general considerations announcing the three main points to be proved:

1. the asymmetry of the system under time reversal

2. the absence of Poincaré recurrences

3. the evolution of the system toward a unique stationary final state, for which radiation would be homogeneous and isotropic and would have a definite spectrum.[49]

Boltzmann, as a specialist in both thermodynamics and electrodynamics, protested immediately and vigorously. Planck's objectives, he argued, could never be reached. Maxwell's equations—even in the presence of a resonator (without the Joule effect)—were just as reversible as

48. Planck 1897b, 1897c, 1897d, 1898, 1899; also in PAV 1:493–600.
49. The last point (the definite final spectrum) was never established by Planck. In reality, ideal resonators cannot change at all the spectrum of cavity radiation, as proved in Ehrenfest 1906 and remarked by Einstein in a letter to Marić of 10 Apr. 1901 (Einstein 1987). See also Planck 1906, 220–221.

were the equations of mechanics. More generally, the analogy between electrodynamics and mechanics was sufficient to exclude Planck's objectives. For instance, Boltzmann explained, to the scattering of a wave by a resonator corresponded, in mechanics, the scattering of a parallel beam of particles by a fixed target. Nobody would have questioned the reversibility of the latter process; therefore the former also had to be reversible. In both cases the seeming irreversibility came from an arbitrary selection of a certain class from among the possible initial conditions, one excluding beams or waves converging toward the target.[50]

In his second memoir Planck replied to Boltzmann's criticism. From the beginning of his considerations, he stated, the external electromagnetic field was assumed to vary by a negligible amount over the dimensions of the resonator. This excluded singular fields converging toward the resonator. More generally, Planck denied physical character to solutions of this type, for they could not be realized except by extreme contortions.[51]

This answer failed to satisfy Boltzmann. The "anti-Planck" solutions (obtained from Planck's solutions by time reversal) were singular only in the physically inaccessible limit of an infinitely small resonator, so that the exclusion from nature of certain types of singularities could not affect the question of reversibility. In order to understand the thermodynamic irreversibility of radiation processes, Boltzmann suggested, one had to imitate the example of kinetic theory and introduce adequate concepts of probability and disorder:

> "Just as in gas theory, in radiation theory one could define a state of maximal probability, more exactly a general formula that would include all states for which the waves are not ordered, but cross each other in the most diverse way."[52]

Planck could not easily adopt suggestions that reduced to naught his original project of deducing from radiation theory a strict, nonprobabilistic entropy law. In his third memoir, communicated in December 1897, he analyzed in some detail the case of a resonator placed at the center of a spherical (perfectly reflecting) cavity. On the basis of the exclusion of singular external fields he believed that he had proved an asymmetry of this system under time reversal. In order to reach his second objective, the exclusion of Poincaré's recurrences, he made a slight concession to

50. Boltzmann 1897b.
51. Planck 1897c.
52. Boltzmann 1897c, BWA 3:621.

Boltzmann's viewpoint by introducing a notion of disordered radiation. But this notion was not statistical; it just meant the exclusion of some states of radiation "synchronized" with the resonator, on account of their being nonphysical.[53]

Boltzmann protested for a last time: Planck's expression for the time-reversed version of an equation used in his proof of irreversibility was simply wrong, for it "reversed" the external field without reversing the field of the resonator![54]

In his fourth memoir (1898) Planck humbly acknowledged his mistake and turned to a more systematic exploitation of the analogy between gas theory and radiation theory, as Boltzmann had earlier advised. Within the context of this program the central concept became that of "natural radiation"—the counterpart of molecular chaos; and the main theorem became a theorem of irreversibility, which was the counterpart of the $H$-theorem. In the fifth memoir Planck generalized his results to an arbitrarily shaped cavity, in the manner that will now be described in detail.[55]

## NATURAL RADIATION

The exciting field $E(t)$ and the electric moment $f(t)$ of the resonator are conveniently described by their Fourier transforms $\tilde{E}(\omega)$ and $\tilde{f}(\omega)$ such that[56]

$$E(t) = \int_{-\infty}^{+\infty} \tilde{E}(\omega)e^{i\omega t}\,d\omega, \qquad f(t) = \int_{-\infty}^{+\infty} \tilde{f}(\omega)e^{i\omega t}\,d\omega. \tag{65}$$

In these terms the resonator equation (62) becomes

$$Z(\omega)\tilde{f}(\omega) = \tilde{E}(\omega), \tag{66}$$

where

$$Z(\omega) = L(\omega_0^2 - \omega^2 + i\rho\omega_0\omega). \tag{67}$$

As a preparatory step, Planck characterized what he called the "directly measurable" quantities connected with the resonator and with its exciting field. There was first the secular average $U(t)$ of the energy $Y(t)$ of the resonator, which can be obtained by retaining only the slow frequencies

53. Planck 1897d.
54. Boltzmann 1898b.
55. Planck 1898, 1900a.
56. Until 1899, instead of the more convenient Fourier integrals Planck used Fourier series over a time interval largely exceeding all characteristic times of the system. He never used the complex notation, although it greatly simplifies the calculations.

in the Fourier spectrum of the function $Y(t)$. More specifically, the Fourier transform $\tilde{U}(\omega)$ will be defined as the part of the Fourier transform $\tilde{Y}(\omega)$ for which the frequency $\omega$ is inferior to a cutoff $\bar{\omega}$, with $\bar{\omega} \ll \omega_0$.[57] Then Planck defined the spectral (electric) intensity $J_{\omega_0}(t)$ of the exciting field as a quantity proportional to the secular average of the energy of a test-resonator with central frequency $\omega_0$, and a frequency width $\rho\omega_0$ larger than $\bar{\omega}$ but much smaller than $\omega_0$. Physically, this means that the test-resonator must be damped in such a way that it can "follow" the secular variations of the field without, however, losing the quality of resonating at the frequency $\omega_0$. This can be achieved only by introducing a resistance $\rho$ much larger than the one implied by pure radiation damping.

The secular energy $U(t)$ of the test-resonator is simply given by the secular average of $Kf^2$, for the two terms in the energy formula (60) have equal secular averages. Then, the form (66) of the resonator equation gives, for a frequency $\mu$ less than the cutoff $\bar{\omega}$:

$$\tilde{U}(\mu) = K \int_{-\infty}^{+\infty} \tilde{E}(\omega + \mu)\tilde{E}^*(\omega)Z^{-1}(\omega + \mu)Z^{-1*}(\omega)\, d\omega. \tag{68}$$

From the condition $\bar{\omega} \ll \rho\omega_0$ it can be inferred that $Z^{-1}(\omega + \mu) \sim Z^{-1}(\omega)$ (which means that the response of the resonator is not affected by a frequency shift of the excitation by an amount small compared with the width of the resonator). Consequently, the Fourier transform of the spectral intensity has the form

$$\tilde{J}_{\omega_0}(\mu) = \int_{-\infty}^{+\infty} \tilde{E}(\omega + \mu)\tilde{E}^*(\omega)\chi_{\omega_0}(\omega)\, d\omega, \tag{69}$$

where $\chi_{\omega_0}(\omega)$ is proportional to $|Z^{-1}(\omega)|^2$. The normalization of $J_{\omega_0}(\mu)$ is obtained by requiring the integrated spectral intensity to be identical with the (secular) intensity $\widetilde{E^2}(\mu)$:

$$\int_{-\infty}^{+\infty} \tilde{J}_{\omega_0}(\mu)\, d\omega_0 = \int_{-\infty}^{+\infty} \tilde{E}(\omega + \mu)\tilde{E}^*(\omega)\, d\omega, \tag{70}$$

with the result

$$\int_{-\infty}^{+\infty} \chi_{\omega_0}(\omega)\, d\omega_0 = 1. \tag{71}$$

Consequently, equation (69) just means that $\tilde{J}_{\omega_0}(\mu)$ is an average of $\tilde{E}(\omega + \mu)\tilde{E}^*(\omega)$ for frequencies $\omega$ close to $\omega_0$, with the weight function $\chi_{\omega_0}$.[58]

57. This definition of a secular average is the one best suited to the following calculations. However, Planck preferred a more direct definition, as a time average over an interval $[t, t + \tau]$.
58. Planck 1898, 1899.

## THE FUNDAMENTAL EQUATION

We now return to a genuine Planck resonator, that is, a resonator with *purely* radiative damping, and immerse it in cavity radiation, the walls of the cavity being perfectly reflecting (so that they have no thermalizing effect). Planck supposed the secular energy $U(t)$ of this resonator and the spectral electric intensity $J_{\omega_0}(t)$ of the radiation to be the only directly measurable quantities at the place of the resonator. Accordingly, he looked for an equation directly relating these quantities, one that did not involve a more detailed mathematical description based on the functions $f(t)$ and $E(t)$. In this respect Planck's approach had an antecedent in Boltzmann's work: the Boltzmann equation directly rules the evolution of the velocity distribution of a gas, without requiring a detailed description of the configuration of individual molecules.[59]

The exact value of $U(t)$ as a function of $E(t)$ results from the low-frequency part ($\mu \ll \bar{\omega}$) of equation (68):

$$\tilde{U}(\mu) = K \int_{-\infty}^{+\infty} \tilde{E}(\omega + \mu)\tilde{E}^*(\omega)Z^{-1}(\omega + \mu)Z^{-1*}(\omega)\,d\omega.$$

Unfortunately, this expression depends on the detailed form of the function $\tilde{E}(\omega + \mu)\tilde{E}^*(\omega)$ of the variable $\omega$. In order to counter this inconvenience, Planck introduced the hypothesis of "natural radiation," according to which the $\tilde{E}(\omega + \mu)\tilde{E}^*(\omega)$ "perceived" by the resonator can be replaced by its measurable average $\tilde{J}_{\omega_0}(\mu)$.

We remember that, in the absence of a better definition, Boltzmann had characterized molecular chaos as what legitimates Maxwell's collision formula; in a similar manner, Planck *defined* natural radiation as what legitimates the above mathematical prescription:

> In the sequel I shall assume the validity of the following hypothesis, which is most natural and presumably unavoidable: In the calculation of $U$ from equation [68], the quickly varying factor $[\tilde{E}(\omega + \mu)\tilde{E}^*(\omega)]$ in the integral can be replaced, without appreciable error, with its slowly varying average $[\tilde{J}_{\omega_0}(\mu)]$. The problem of calculating $U$ from $[\tilde{J}_{\omega_0}]$ thus receives a perfectly determinate solution, to be verified by measurement.

On the qualitative side, Planck described natural radiation in a manner reminiscent of Boltzmann's reference to the irregularities of molecular motion: "We may grasp the concept of natural radiation in a less direct ... but more intuitive manner: the deviations of the nonmeasurable, quickly

59. Planck 1898, PAV 1:551–552; Planck 1899, PAV 1:571–573.

varying quantity $[\tilde{E}(\omega + \mu)\tilde{E}^*(\omega)]$ from its slowly varying average $[\tilde{J}_{\omega_0}(\mu)]$ are small and irregular."[60]

The corresponding formal substitution transforms equation (68) into

$$\tilde{U}(\mu) = K\tilde{J}_{\omega_0}(\mu) \int_{-\infty}^{+\infty} Z^{-1}(\omega + \mu)Z^{-1*}(\omega)\, d\omega. \tag{72}$$

The resonance being very narrow, one can use an approximative expression for $Z(\omega)$:

$$Z(\omega) \sim 2L\omega_0(\omega_0 - \omega + i\rho\omega_0/2). \tag{73}$$

The integral in (72) becomes[61]

$$\frac{1}{4L^2\omega_0^2} \int_{-\infty}^{+\infty} \left(\omega + \mu - \omega_0 - i\frac{\rho\omega_0}{2}\right)^{-1} \left(\omega - \omega_0 + i\frac{\rho\omega_0}{2}\right)^{-1} d\omega$$

$$= \frac{1}{4L^2\omega_0^2} \frac{2\pi i}{-\mu + i\rho\omega_0}. \tag{74}$$

Using the relations (64), the following equation results:

$$(i\mu + \rho\omega_0)\tilde{U}(\mu) = \frac{3\pi c^3\rho}{4\omega_0} \tilde{J}_{\omega_0}(\mu). \tag{75}$$

Planck's "fundamental equation" was just the Fourier transform of this equation:[62]

$$\frac{dU}{dt} + \rho\omega_0 U = \frac{3\pi c^3\rho}{4\omega_0} J_{\omega_0}(t). \tag{76}$$

Accordingly, the energy of a resonator is increased at a rate given by the spectral electric intensity of the surrounding field at the frequency of the resonator, but also damped with a time constant $(\rho\omega_0)^{-1}$ identical with that controlling the damping of the dipolar moment of the resonator.

THE ELECTROMAGNETIC $H$-THEOREM

From this equation Planck could derive the measurable effects of a resonator on cavity radiation.[63] For this purpose he borrowed from optics the notion of a radiation beam, which he considered to provide the finest information about the radiation field that was accessible to measurement. An elementary conical beam is characterized by its direction $\zeta$, by the solid angle $d\Omega$ in which it is confined, by its frequency, by its spectral

60. Planck 1899, PAV 1:573.
61. The integration can be easily performed by the method of residues.
62. Planck 1898, PAV 1:552; Planck 1899, PAV 1:575.
63. Planck 1898, PAV 1:543–556; Planck 1899, PAV 1:575–592.

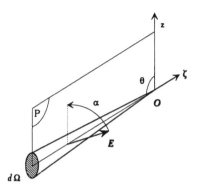

Figure 8.   Geometric parameters for a beam converging toward a resonator (the electric field E must be orthogonal to the beam).

intensity (of energy flow) $I_\nu$, and by its state of polarization. The latter property is given by the two principal directions (along which the intensity across a polarizer is extremal) and the corresponding principal intensities $I'_\nu$ and $I''_\nu$.

In this representation the radiation falling on a resonator at a time $t$ from the direction $\zeta$ is characterized by two functions $I'_\nu(\zeta, t)$ and $I''_\nu(\zeta, t)$ giving the principal intensities at the frequency $\nu$ (with $\nu = \omega/2\pi$), and by the angle $\alpha(t)$ between one of the principal directions and the plane P defined by $\zeta$ and the electric moment of the resonator (see fig. 8). Since the part of the radiation with a frequency close to $\nu_0(= \omega_0/2\pi)$ is the only one affected by the resonator, we need only consider the intensities $I'_{\nu_0}$ and $I''_{\nu_0}$, and we can simplify the notation by dropping the index $\nu_0$ in the sequel. Then the intensity in the plane P is

$$I_{\|} = I' \cos^2 \alpha + I'' \sin^2 \alpha, \tag{77}$$

and the intensity in the direction normal to P is

$$I_{\perp} = I' \sin^2 \alpha + I'' \cos^2 \alpha \tag{78}$$

(one must have $I_{\|} + I_{\perp} = I' + I'' = I$).

The contribution of this beam to the electric intensity $J_{\nu_0}$ (taking $J_{\nu_0} = 2\pi J_{\omega_0}$) is to the flux $I_{\|} \, d\Omega$ what the squared component $E_z^2$ of the electric field of a plane wave traveling in the direction $\zeta$ and polarized in P is to the modulus of the Poynting vector, $(c/4\pi)|E \times B|$. Therefore,

$$J_{\nu_0} = \frac{4\pi}{c} I_{\|} \sin^2 \theta \, d\Omega, \tag{79}$$

where $\theta$ denotes the angle between $\zeta$ and the electric moment of the resonator. The corresponding energy absorbed by the resonator in a time unit,

$\mathscr{E}_A \, d\Omega$, is given by the right-hand side of the "fundamental equation" (76):

$$\mathscr{E}_A \, d\Omega = \frac{3c^3\rho}{16\pi v_0} J_{v_0} = \frac{3c^2\rho}{4v_0} I_{\parallel} \sin^2 \theta \, d\Omega. \tag{80}$$

The total energy radiated by the resonator in a time unit is given by the damping term $\rho\omega_0 U$ in the fundamental equation. As results from the expression (41) of dipolar radiation, the radiation emitted in the direction $\zeta$ is completely polarized in the plane defined by this direction and the electric moment, and its intensity varies with the azimuthal angle $\theta$ like $\sin^2 \theta$. Consequently, the energy radiated in the solid angle $d\Omega$ around $\zeta$ amounts to

$$\mathscr{E}_R \, d\Omega = \frac{3}{8\pi} \rho\omega_0 U \sin^2 \theta \, d\Omega, \tag{81}$$

where the factor $3/8\pi$ gives the proper normalization, as results from

$$\int_0^\pi \sin^2 \theta \, d\Omega = 2\pi \int_0^\pi \sin^3 \theta \, d\theta = 8\pi/3. \tag{82}$$

For the *intensity* of this emitted radiation to be defined, the finite breadth (*Breite*) of the source, that is, its efficient cross section times its spectral width, must be taken into account. Planck identifies this breadth, $\sigma$, with the ratio of the absorbed energy to the active part (the one contributing to $J_{v_0}$) of the incoming flux:[64]

$$\sigma = \mathscr{E}_A \, d\Omega / I_{\parallel} \sin^2 \theta \, d\Omega = 3c^2\rho/4v_0. \tag{83}$$

The intensity emitted in the direction $\zeta$ is therefore, using (81),

$$I_R(\zeta) = \mathscr{E}_R/\sigma = (v_0^2 U/c^2) \sin^2 \theta. \tag{84}$$

However, this radiation is only one part of the outgoing radiation (at a frequency $v \sim v_0$) in the direction $\zeta$. Another contribution comes from the part of the incoming radiation in the same direction $\zeta$ which is not absorbed by the resonator. The latter radiation is a mixture of a radiation polarized in the $\perp$ direction with the intensity $I_\perp$, and of a radiation polarized in the $\parallel$ direction with the intensity $I_{\parallel} \cos^2 \theta$. Consequently, the principal directions of the total emerging radiation are these two directions, and its two principal intensities are

$$\bar{I}' = I_\perp, \qquad \bar{I}'' = I_{\parallel} \cos^2 \theta + I_R. \tag{85}$$

Calling $\Delta I = \bar{I}' + \bar{I}'' - I' - I''$ the intensity balance corresponding to a given direction $\zeta$ of absorption or emission, and using the identities (79),

---

64. This identification might seem somewhat arbitrary, but it is the only one leading to the natural form of energy conservation expressed in equation (86).

(84), and (85), the fundamental equation (76) can now be rewritten as

$$\frac{dU}{dt} + \int \sigma \, d\Omega \, \Delta I(\zeta, t) = 0. \tag{86}$$

This equation means that the increase of the (secular) energy of the resonator is equal to the balance of the ingoing and outgoing fluxes of radiation energy.

Having expressed energy conservation in this form, Planck looked for a similar form for the total entropy variation of the system:

$$\frac{dS_T}{dt} = \frac{dS}{dt} + \int \sigma \, d\Omega \, \Delta L, \tag{87}$$

where $S$ would represent the entropy of the resonator, and $L$ an "intensity of entropy" corresponding to the intensity $I$ of energy. Presumably guided by a formal analogy with Boltzmann's $H$-function, Planck posited (taking $\nu = \nu_0$, to simplify the notation)[65]

$$S = -\frac{U}{a\nu} \ln \frac{U}{b\nu}, \qquad L = -\frac{I}{a\nu} \ln \frac{I(c^2/\nu^2)}{b\nu}. \tag{88, 88'}$$

That $S$ depended only on $U/\nu$ was a consequence of Wien's displacement law, as will be seen later; and the factor $c^2/\nu^2$ next to $I$ was suggested by the relation $I_R(c^2/\nu^2) = U \sin^2 \theta$. For this choice Planck's "electromagnetic $H$-theorem" holds: *the total entropy variation $dS_T/dt$ is always positive, and it vanishes if and only if the distribution of intensities is isotropic and unpolarized.* Planck's derivation of this result will now be given.

Using equation (86), one has

$$\frac{dS}{dt} = \frac{dS}{dU} \frac{dU}{dt} = \frac{1}{a\nu} \left( \ln \frac{U}{b\nu} + 1 \right) \int \sigma \, d\Omega \, \Delta I, \tag{89}$$

and, inserting this and (88') into (87),

$$\frac{dS_T}{dt} = -\int \frac{\sigma}{a\nu} \, d\Omega \, \Delta \left[ I \left( \ln \frac{Ic^2}{\nu^2 U} - 1 \right) \right]. \tag{90}$$

We are now left with the study of the sign of an expression of the form $\Delta y = \bar{y}' + \bar{y}'' - y' - y''$, where $y = x(\ln x - 1)$, and $x = Ic^2/\nu^2 U$. Writing the relations (77), (78), (84), and (85) in terms of the $x$ variables gives

$$x_{||} = x' \cos^2 \alpha + x'' \sin^2 \alpha, \qquad \begin{cases} \bar{x}' = x' \sin^2 \alpha + x'' \cos^2 \alpha \\ \bar{x}'' = x_{||} \cos^2 \theta + \sin^2 \theta. \end{cases} \tag{91}$$

65. Planck 1899, PAV 1:585.

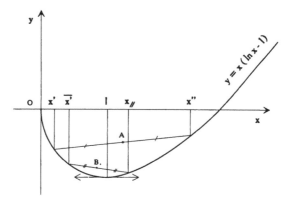

Figure 9.   Diagram for the proof of the electromagnetic $H$-theorem. The point A has the ordinate $(y' + y'')/2$, and the point B, $(\bar{y}' + y_{||})/2$. The concavity of the curve implies that A is always "higher" than B, and therefore that $y' + y'' - \bar{y}' - y_{||} \geq 0$.

As a first consequence, $\bar{x}''$ lies between $x_{||}$ and 1. Since the function $y = x(\ln x - 1)$ is a concave function reaching its minimum for $x = 1$, this implies that $y_{||}$ is larger than $\bar{y}''$ and that

$$-\Delta y \geq y' + y'' - \bar{y}' - y_{||}. \tag{92}$$

The positivity of the right-hand side results (fig. 9) from the concavity of the function $x(\ln x - 1)$ and from the fact that $\bar{x}'$ and $x_{||}$ always lie between $x'$ and $x''$, as can be seen from (91). Therefore, the total entropy variation is always positive.[66]

The total entropy variation vanishes only if $\Delta y = 0$ for any $\zeta$, since $\Delta y$ is always negative. In the above chain of reasoning, the inequalities can be replaced with equalities only if $x_{||} = \bar{x}''$ and also $x' = \bar{x}'$, $x'' = x_{||}$, up to a permutation of $x'$ and $x''$, which play symmetric roles. Once having eliminated the singular cases $\alpha = \pi/2$, $\theta = 0$, these three equalities can be satisfied only if $x' = x'' = \bar{x}' = \bar{x}'' = 1$, which implies that incoming and outgoing radiations are isotropic and unpolarized. This concludes the proof of what might be regarded as Planck's greatest theorem.

Once uniformity has been reached, the exciting intensity appearing in the fundamental equation (76) can be expressed in terms of the spectral density $\rho_\nu$ of the radiation field. Following Planck, $\rho_\nu$ is defined by applying to the energy density $(E^2 + B^2)/8\pi$ the same operations as those that

66. Planck's original proof was heavier here, since it did not exploit the concavity of $x(\ln x - 1)$.

lead from $E_z^2$ to $J_v$. Writing symbolically $J_v = E_z^2|_v$, the density may be written

$$\rho_v = (E^2|_v + B^2|_v)/8\pi. \tag{93}$$

Isotropy implies

$$E_x^2|_v = E_y^2|_v = E_z^2|_v = B_x^2|_v = B_y^2|_v = B_z^2|_v, \tag{94}$$

and

$$\rho_v = \frac{3}{4\pi} J_v. \tag{95}$$

The fundamental equation (76) in the stationary case therefore reads:[67]

$$U = \frac{c^3}{8\pi v^2} \rho_v. \tag{96}$$

## THE BLACKBODY LAW

None of the equations reached by Planck was yet able to determine the equilibrium spectrum of radiation; they just gave a relation between the average energy of a resonator and the density of the surrounding radiation. As noted in Planck's third memoir, a resonator acts only on radiation with frequencies close to its own resonance frequency and therefore cannot change the spectrum of cavity radiation, as long as its more detailed structure is not taken into account. Planck therefore gave up his original hope (see p. 39, point (3)) to describe with the help of his resonators the evolution of the radiation spectrum toward its equilibrium value. Nevertheless, he found another access to the equilibrium spectrum, by identifying the entropy function used in the electromagnetic $H$-theorem to the real thermodynamic entropy. This would be a legitimate procedure only if the form

$$S = -\frac{U}{f(v)} \ln \frac{U}{g(v)} \tag{97}$$

of the resonator entropy was the sole form compatible with the irreversibility theorem. In his fifth memoir (1899) Planck asserted without proof this uniqueness of form and derived the equilibrium spectrum in the following manner.[68]

---

67. Planck 1899, *PAV* 1:581.
68. Ibid., 596: "I have made repeated efforts to modify the expression for the electromagnetic entropy of a resonator in such a manner ... that it remains compatible with all well-founded electromagnetic and thermodynamic theoretical laws; but I have not succeeded."

He first evoked Wien's displacement law to limit the choice of the arbitrary functions $f$ and $g$ in the above entropy formula. Thanks to the fundamental relation (96), this law can be expressed directly in terms of the resonator energy as

$$U = v\varphi(v/T). \tag{98}$$

This implies $dS = (v/T)\varphi'(v/T)d(v/T)$, which means that $S$ must be a function of $v/T$, or another function of $U/v$, since $U/v = \varphi(v/T)$. Consequently, the resonator entropy must have the form

$$S = -\frac{U}{av} \ln \frac{U}{bv}. \tag{99}$$

The absolute temperature of a resonator in thermodynamic equilibrium is then given by the relation $dS/dU = 1/T$, with the result

$$\frac{1}{T} = -\frac{1}{av}\left( \ln \frac{U}{bv} + 1 \right). \tag{100}$$

Inverting this equation provides[69]

$$U = b've^{-a(v/T)}. \tag{101}$$

Finally, the "fundamental equation" (96) gives for the spectral density $u_v$ of the blackbody spectrum

$$u_v = \frac{8\pi b'v^3}{c^3} e^{-av/T}. \tag{102}$$

This law was not new. It had been proposed two years earlier by Wien on the basis of a fragile analogy with the exponential form of Maxwell's law. At the date of Planck's fifth memoir (early 1899) it was well confirmed by experimental measurements. At the price of incompletely justified assumptions about natural radiation and entropy, Planck therefore believed that he had reached the main objectives of his program: a deduction of irreversibility from electrodynamic processes, and a derivation of the universal law of blackbody radiation. As a clear manifestation of his trust in the fundamental character of his theory, he suggested that the constants $a$ and $b$ appearing in the resonator entropy be considered new fundamental constants and recommended natural units of length, time, mass, and temperature built from $a$, $b$, $c$, and the universal gravitation constant.[70]

69. Planck 1899, PAV 1:591.
70. Wien 1896; Planck 1899, PAV 1:599–600.

## PLANCK VERSUS BOLTZMANN

From a systematic comparison of Planck's reasonings with those which led Boltzmann to the $H$-theorem we may surmise that Planck benefited at various steps from suggestive analogies. Table 1 summarizes the correspondence between Planck's and Boltzmann's arguments. The first horizontal section refers to the most detailed, microscopic level of description, while the second refers to the "directly observable quantities." The third and fifth sections respectively give the equations ruling the microscopic evolution of the system and those ruling the directly observable quantities. The transition between these two types of equations is expressed in the fourth section, with the idea of disorder, and the corresponding simplification of the interaction within the system. The fifth section gives the functions of the directly observable quantities which always increase or decrease. The last section describes the final state of the system in terms of the directly observable quantities.

As appears from this table, the relevant analogies concerned general concepts or categories rather than specific mathematical expressions, except for that connecting the formulae for $H$ and $S$. This should not

TABLE ONE      ANALOGY BETWEEN PLANCK'S AND
BOLTZMANN'S IRREVERSIBILITY THEOREMS

| Boltzmann | Planck |
|---|---|
| Molecular velocities: $\mathbf{v}_1, \mathbf{v}_2, \ldots, \mathbf{v}_N$ | External fields: $\mathbf{E}_e(\mathbf{r}, t)$, $\mathbf{B}_e(\mathbf{r}, t)$ Electric dipole: $f(t)$ |
| Velocity distribution: $f(\mathbf{v})$ | Secular resonator energy: $U(t)$ Beam intensities: $I(\zeta, t)$ Electric intensity: $J_\omega(t)$ |
| Equations of molecular dynamics | Resonator equation (62) Maxwell's equations |
| Maxwell's *Ansatz* (molecular chaos) | Planck's fundamental equation (natural radiation) |
| Boltzmann's equation for $f$ | Energy balancing (86) for $U$ and beam intensities |
| $H = \int f \ln f d^3 v$ | $S = -\dfrac{U}{av} \ln \dfrac{U}{bv}$ (for the resonator) |
| Maxwell's distribution | Isotropic unpolarized radiation |

surprise us, because the original dynamic systems, the molecular gas on the one hand, the resonator in cavity radiation on the other, exhibited strong qualitative differences. As vague as it was, the idea of a selection of disordered states for which the evolution of all measurable quantities would not depend on finer uncontrollable details was all Planck needed to establish his "fundamental equation." The rest, as we saw, proceeded from the autonomous development of Planck's program, with the exception of the entropy formula.

Did such procedural parallelism imply that Planck would also accept Boltzmann's conception of irreversibility? Somewhat surprisingly, the answer is no. For about fifteen years Planck maintained his idea of an absolute entropy law. As we have seen, Boltzmann's notion of molecular chaos was intrinsically probabilistic: only with a certain probability did it grant a deterministic evolution of the velocity distribution. Moreover, an initially disordered state had to evolve toward ordered states after a sufficiently long time, as implied by Poincaré's recurrence theorem (since perpetual disorder would exclude recurrence). In contrast, Planck's natural radiation had to remain natural forever, and ordered states were absolutely (nonstatistically) excluded from his theory as being nonphysical. In his fourth memoir Planck made this point completely explicit:

> If one wishes to apply to nature the preceding theory of irreversible radiation processes, one must admit that these processes, especially those which provide temperature equilibrium, have the properties of natural radiation in any circumstance and for an unlimited amount of time.[71]

Planck nevertheless observed that his differential equation of the dipole $f(t)$ combined with Maxwell's equations led to the Poincaré recurrence (at least in the case of a resonator placed at the center of a spherical mirror-cavity). This difficulty did not stop him: at the time scale at which Poincaré recurrences would occur, the deviations in the equation resulting from the detailed structure of the resonator had to play a role, and the dipolar differential equation would no longer be adequate for describing the evolution of the system. Planck commented:

> This indetermination lies in the nature of things. Indeed, the physical problem has no definite solution as long as nothing is known about the resonator but its eigenfrequency $\omega_0$ and its damping constant $\rho$; that our theory is able to determine the approximate course of phenomena while only these two constants

---

71. Planck 1898, *PAV* 1:556. Needell 1980 first exhibited Planck's persistent belief that the entropy law was strictly valid and the central role of this belief in his early conception of quantum theory. I have independently reached the same conclusion in Darrigol 1988a.

are given, must be regarded as a great advantage. For the same reason this theory cannot tell us more about the resonator than the determination of $\omega_0$ and $\rho$. Precisely in this gap [*Lücke*] does the hypothesis of natural radiation find its place; otherwise, this hypothesis would be either superfluous or impossible, for the processes would be completely determined without it.[72]

Thus Planck believed that the indetermination of the internal structure of his resonators made room for an everlasting disorder, which would make the evolution of the accessible properties of the system strictly irreversible. Not only did he not admit a probabilistic interpretation of natural radiation, but he reinterpreted Boltzmann's concept of molecular chaos in a way that would make kinetic theory acceptable to him. On this occasion he emphasized the analogy between gas theory and radiation theory:

> Our electrodynamic interpretation of the second principle of thermodynamics suggests a brief comparison with the *mechanical* interpretation of the same principle, that is, with the corresponding questions in the kinetic gas theory. As is well known, here also we encounter the same, often-noted conflict between the fundamental equations of mechanics, which are perfectly reversible, and the second principle [of thermodynamics], which demands irreversibility for all real processes. But here also the conflict can be resolved in a very similar way by the introduction of a special hypothesis, which, as long as it remains valid, implies all consequences of the second principle. Boltzmann calls this "molecular disorder." This hypothesis is a necessary and sufficient condition for the existence of a definite function of the instantaneous state that perpetually increases in time and therefore shares the essential properties of entropy. However, the hypothesis of molecular disorder, once applied not only to the initial state but also to any subsequent time, has been shown to be incompatible with the assumption of a finite number of simple atoms confined within rigid walls; this circumstance impedes the introduction of the second principle, as a general principle, in [kinetic] gas theory. Some perceive here an objection to the legitimacy of [kinetic] gas theory, while others question the general validity of the second principle of thermodynamics. In reality one is not at all confined to these two alternatives. For whoever is willing to give up only one of the above assumptions, the existence of rigid walls—which, strictly taken, seems to be very improbable—there seems to be no obstacle whatsoever to a general application of the hypothesis of molecular disorder, and one is free to extend the second principle to arbitrarily long times even from the viewpoint of kinetic gas theory.[73]

As was the indefinite structure of resonators, the complexity of the walls of a container was supposed to maintain a perpetual chaos warranting

72. Planck 1898, *PAV* 1:557.
73. Planck 1900a, *PAV* 1:619–620.

an absolute irreversibility of thermodynamic phenomena. In this way the assumption of disordered states, originally meant by Boltzmann as a marginal commentary on a certain collision formula, became central to Planck's conception of thermodynamics, which sought to maintain strict irreversibility.

Pursuing the comparison between natural radiation and molecular chaos, Planck gained some intuition about the nature of the disorder intervening in his radiation theory. He emphasized that his equations led to irreversible behavior for a *single* resonator, while Boltzmann's considerations required a very large number of molecules, and explained: "The principle of disorder *on which every notion of irreversibility seems to rest* steps in at very different moments [of the reasonings] in gas theory and thermal radiation theory" (emphasis added). Disorder in a gas consisted of the irregular multitude of molecular velocities and positions, while in the resonator case it had to do with the irregular multitude of Fourier components of the electric moment. There was spatial disorder in one case, temporal disorder in the other.[74]

## SUMMARY

In 1897 Planck defined the aim of his grand program in the first of a series of five memoirs "on irreversible radiation processes": to show that a system of ideal resonators acted irreversibly on radiation enclosed in a cavity with perfectly reflecting walls, leading eventually to a uniform, isotropic distribution of radiant energy. Moreover, he hoped that the spectrum of this radiation would evolve toward a well-defined final state, which, according to Kirchhoff's law, could only be the universal blackbody spectrum.

Planck's announcement of this project triggered a public polemic with Boltzmann. The latter argued that the laws of electrodynamics were just as reversible as the laws of molecular dynamics, so they could not be used to derive irreversible thermodynamic behavior. The only possible escape from this conclusion, Boltzmann suggested, would be—as he had done in gas theory—to shift to a statistical conception of irreversibility and to introduce a notion of "disordered" radiation that would be the counterpart of molecular chaos.

Eventually, after a few misconceived attempts to counter Boltzmann's objection, Planck did adopt a notion of disordered radiation, which he

74. Planck 1900b, *PAV* 1:673–674.

called "natural radiation." As a first step toward defining this notion, he introduced the relevant "directly measurable quantities," that is, certain quadratic time-averages, one for the electric moment of the resonator and one for the electric field at the place of the resonator. These were indeed accessible to measurement, for instance, in the case of the field, through the resonance of a damped test-resonator. But the differential equation relating the resonator moment and the electric field did not imply a definite relation between these quadratic expressions, unless certain "cross-terms" were set to zero. The vanishing of cross-terms is precisely what defines Planck's natural radiation; it leads to an equation involving only the directly observable quantities, which Planck called the "fundamental equation." The analogy with molecular chaos was transparent enough. In both cases there were two levels of description: the detailed micro-level, which includes uncontrollable features of the model (dynamics of molecular collisions/electrodynamic interaction between resonator and radiation), and the "physical" level of description, which involves only physically meaningful quantities (Maxwell's collision formula/Planck's fundamental equation). In order to deduce the second level of description from the first a special assumption must be made, molecular chaos in one case, natural radiation in the other.

Planck then proceeded to derive a counterpart to Boltzmann's *H*-theorem. To this end he discussed the effect of a resonator on radiation beams and introduced expressions for the entropy of resonators and of beams in terms of the corresponding "measurable quantities." These expressions mirrored very closely Boltzmann's *H*-function, and, as in Boltzmann's case, they led to a perpetual increase of the total entropy. By this means Planck could prove that resonators made the surrounding radiation increasingly (spatially) uniform and unpolarized. However, the analogy with Boltzmann's *H*-theorem was imperfect in at least one respect: while the Boltzmann equation led to a definite final distribution of velocities, Planck's equations said nothing about the time evolution of the spectrum of radiation. In fact, contrary to his original hope, Planck gradually realized that his resonators were unable to redistribute radiation from one frequency to another.

Nevertheless, in his concluding memoir of 1899 (the fifth!) Planck managed to derive the blackbody spectrum. He just had to assume that his expression for the resonator entropy was the *only* one compatible with a global entropy increase and that it was identical with the thermodynamic entropy. Then standard thermodynamic reasoning, Wien's displacement law, and the "fundamental equation" led to a definite spectrum. To

Planck's satisfaction the resulting law was already well known (previously having been proposed by Wien on a frail theoretical basis) and fitted available observations excellently. Not doubting the fundamental character of this derivation, Planck extracted universal constants from his entropy formula (including $h$, under a different letter) and even combined them with the universal constant of gravitation to produce absolute units of length, time, and mass.

Apparently Planck had achieved his original aims: the demonstration that resonators could produce the irreversible change required by the second law, and a derivation of the blackbody spectrum. His introduction of "natural radiation" and the overall analogy of his irreversibility theorem with Boltzmann's $H$-theorem might suggest that he had in the process become a convert to the statistical conception of irreversibility and thus given up the motivating force behind his program, a nonstatistical foundation of the entropy law. This, however, was not the case. Planck maintained his absolute conception of irreversibility for fifteen more years. In his mind the "naturalness" of radiation was *not* a statistical property; instead, it applied to *individual* states of radiation. The internal structure of resonators, which Planck deliberately left undetermined in all his reasonings, was in charge of perpetually maintaining this naturalness. Similarly, in gas theory Planck imagined an indeterminate structure of the walls of containers that would maintain molecular chaos forever—and make kinetic theory acceptable to him. Planck's "elementary disorder" (a generic name for molecular chaos and natural radiation) was essential for strict irreversibility and thus became the central concept of his thermodynamics.

# CHAPTER IV

# The Infrared Challenge

The pretense of Planck's theory of radiation to determine the blackbody law was soon contradicted by Berlin's best spectroscopists. In 1899 Paschen, Lummer, and Pringsheim observed violations of Wien's blackbody law in the infrared part of the spectrum. Even though he did not immediately take this result as irreproachable, Planck came to recognize that the proof of the electromagnetic $H$-theorem was compatible with an infinite number of choices for $S$, the resonator entropy.[75] As will presently be seen, the only necessary restriction on this choice was that the second derivative of $S$ be negative: $d^2S/dU^2 < 0$. Subsequently, Planck found a physical meaning for this derivative and used it to justify the choice of $S(U)$ that led to Wien's law.[76]

## THE SECOND DERIVATIVE OF THE RESONATOR ENTROPY

The meaning of $d^2S/dU^2$ derives from Planck's following consideration. Having already quickly increased (or decreased) the energy of a resonator initially in equilibrium with thermal radiation by an amount $\delta U$, one allows this energy to relax by $dU$ (with $dU \ll \delta U$) toward its equilibrium

75. This has to do with the fact that ideal resonators cannot change at all the spectrum of cavity radiation (see n. 49). An infinite number of radiation spectra are compatible with Planck's system, as implied by the infinite number of possible choices for the resonator entropy.
76. Planck 1900b. On infrared blackbody measurements see Jammer 1966, 16–17; Mehra and Rechenberg 1982a, 39–43.

value. The derivative $d^2S/dU^2$ is then proportional to the total entropy change, $dS_T$, occurring during the relaxation process. More exactly, Planck proved that

$$dS_T = \frac{3}{5} \frac{d^2S}{dU^2} dU \, \delta U \tag{103}$$

in the following manner.[77]

During the increment of $U$ by $\delta U$ the beam intensity balance $\Delta I$ (as defined on p. 46) goes from zero (equilibrium) to

$$\Delta I = \delta \bar{I}'' = \frac{v^2}{c^2} \sin^2 \theta \, \delta U, \tag{104}$$

since, according to (85), $\bar{I}''$ is the only intensity that depends on the energy of the resonator. Accordingly, at the second order of approximation the entropy-intensity balance goes from zero to

$$\Delta L = \delta \bar{L}'' = \frac{dL}{dI}\bigg|_0 \Delta I + \frac{1}{2} \frac{d^2L}{dI^2}\bigg|_0 (\Delta I)^2. \tag{105}$$

In this development the index zero refers to the original state of equilibrium, for which all $I$'s are equal to $v^2U/c^2$. Now let a time $dt$ elapse after the excitation of the resonator. The corresponding relaxation of the resonator energy, $dU$, is obtained by substituting the above $\Delta I$ into the equation (86) of energy conservation:

$$dU = -\sigma \, dt \int d\Omega \, \Delta I = -2\pi\sigma \frac{v^2}{c^2} dt \, \delta U \int_0^\pi \sin^2 \theta \, d\theta = -\frac{8\pi}{3} \sigma \frac{v^2}{c^2} \delta U \, dt. \tag{106}$$

In the same time interval the total entropy variation (from (90)) is

$$dS_T = dS + \sigma \, dt \int d\Omega \, \Delta L. \tag{107}$$

Inserting the development (105) in the latter equation and using again the equation (86) of energy conservation yields

$$dS_T = dS - \frac{dL}{dI}\bigg|_0 dU + \frac{\sigma \, dt}{2} \frac{d^2L}{dI^2}\bigg|_0 \int (\Delta I)^2 \, d\Omega. \tag{108}$$

The first term can be rewritten as

$$dS = \frac{dS}{dU}\bigg|_{U+\delta U} dU = \frac{dS}{dU}\bigg|_0 dU + \frac{d^2S}{dU^2}\bigg|_0 dU \, \delta U, \tag{109}$$

77. Planck 1900b, PAV 1:679.

and the integral in the last term can be evaluated using (104), (106), and

$$\int \sin^4 \theta \, d\Omega = 2\pi \int_0^\pi \sin^5 \theta \, d\theta = 32\pi/15. \tag{110}$$

The resulting expression for $dS_T$ reads

$$dS_T = \left(\left.\frac{dS}{dU}\right|_0 - \left.\frac{dL}{dI}\right|_0\right) dU + \left(\left.\frac{d^2S}{dU^2}\right|_0 - \frac{2}{5}\frac{v^2}{c^2}\left.\frac{d^2L}{dI^2}\right|_0\right) dU \, \delta U. \tag{111}$$

So that the system may evolve toward equilibrium, $dU \, \delta U$ must be negative; but the individual signs of $dU$ and $\delta U$ are not fixed. Consequently, the sign of the above expression of $dS_T$ is definite if and only if

$$\left.\frac{dS}{dU}\right|_0 = \left.\frac{dL}{dI}\right|_0. \tag{112}$$

This identity must hold for all values of $I$ and $U$ related through $I = Uv^2/c^2$, which implies

$$L(I) = \frac{v^2}{c^2} S\left(\frac{c^2}{v^2} I\right) + \text{const.} \tag{113}$$

and

$$\left.\frac{d^2L}{dI^2}\right|_0 = \frac{c^2}{v^2} \left.\frac{d^2S}{dU^2}\right|_0. \tag{114}$$

Taking into account the latter remarks, the expression (111) collapses into

$$dS_T = \frac{3}{5} \left.\frac{d^2S}{dU^2}\right|_0 dU \, \delta U,$$

which was to be proved.

This calculation gave two necessary conditions for entropy increase: the form (113) for the function $L(I)$ corresponding to the function $S(U)$, and the convexity condition for the latter function, $d^2S/dU^2 < 0$, which provides $dS_T > 0$ (since $dU \, \delta U < 0$). As Planck now noticed, these conditions were also *sufficient* for entropy increase. Indeed, the proof of the electro-magnetic $H$-theorem earlier given may be adapted in the following way.

Starting from the equations (86) and (87) for energy and entropy balance, one can use the first condition (113) to derive

$$\frac{dS_T}{dt} = \sigma \int d\Omega \, \Delta \left[ L(I) - I \frac{dS}{dU} \right]. \tag{115}$$

Then, the two conditions together imply that $L(I) - I \, dS/dU$ (at a constant $U$) is a convex function of $I$, with an absolute maximum for $I = v^2U/c^2$. These properties are the only ones necessary to the rest of the proof of entropy increase. That Planck originally believed the form $S = -(U/f) \ln (U/g)$ to be the only one possible may be explained by the intricacy of his own proof of the electromagnetic $H$-theorem, which did not exploit the convexity of $S(U)$.

Once aware of the physical meaning of $d^2S/dU^2$, Planck considered a set of $n$ resonators (separated from one another by large distances) immersed in the same thermal radiation and submitted to the same perturbation $\delta U$. The total entropy variation during a common relaxation $dU$ of the energy of these resonators is $n\,dS_T$. Planck then equated this variation with the one obtained in the case of a single resonator with the initial energy $nU$ submitted to the perturbation $n\,\delta U$ and relaxing by $n\,dU$, which gives

$$n \frac{d^2S}{dU^2}\bigg|_U dU\,\delta U = \frac{d^2S}{dU^2}\bigg|_{nU} nUn\,\delta U \qquad (116)$$

or

$$\frac{d^2S}{dU^2}\bigg|_{nU} = \frac{1}{n} \frac{d^2S}{dU^2}\bigg|_U . \qquad (117)$$

This property implies the form $d^2S/dU^2 = -\alpha/U$, which integrates into Planck's original formula $S = -(U/f) \ln (U/g)$ and therefore implies Wien's law.[78]

A NEW RADIATION LAW

Planck's colleagues quickly perceived the mistake in the above reasoning: a single resonator with the energy $nU$ is not in equilibrium with the same thermal radiation as the $n$ resonators with the energy $U$, so that there is no reason to equate the relaxation rates in the two cases. Moreover, by the summer of 1900 the violations of Wien's law in the infrared range had become incontestable. On 19 October, Planck acknowledged his mistake at the Berlin Academy. His methods were really unable to provide a precise form of the resonator entropy and were in fact compatible with an infinity of different equilibrium distributions. He nevertheless "guessed" a new blackbody law, starting from an expression for the second derivative of the resonator entropy $S(U)$.[79]

78. Of course Planck's presentation of this argument does not necessarily reflect the chronological order of his considerations. Perhaps he first noticed the extreme simplicity of the second derivative of the entropy corresponding to Wien's law and then sought a physical interpretation of this derivative that would justify the property of homogeneity.

79. Planck 1900c. See Kuhn 1978, 96–97.

This expression, he argued, had to be negative in order not to contradict the entropy law; it had to give back the form $-1/\alpha U$ leading to Wien's law for small values of $U$ (that is to say for large frequencies); and, for the sake of simplicity, it had to be easily integrable. Planck therefore conjectured the form

$$d^2S/dU^2 = -\alpha\left(\frac{1}{U} - \frac{1}{U+\beta}\right), \tag{118}$$

whose integration gives

$$S = \alpha \ln\left[(\beta + U)^{\beta + U}/U^U\right]. \tag{119}$$

This formula, once combined with $dS/dU = 1/T$, Wien's displacement law, and the fundamental equation $u_\nu = (8\pi\nu^2/c^3)U$, leads to "Planck's law":

$$u_\nu = \frac{A\nu^3}{e^{\beta\nu/T} - 1}. \tag{120}$$

Within a few days blackbody spectroscopists could verify how excellently this law fitted their experimental data.

In the same communication Planck noted that the form of $S(U)$ was logarithmic, "as suggested by probability calculus." Presumably he was already hoping for a deeper justification of this law based on Boltzmann's relation between entropy and combinatorial probability. He was here confronted with a situation different from that occurring in Boltzmann's gas theory. In the gas case the Boltzmann equation provided by itself the equilibrium distribution, and the method of complexions gave hardly more than "an illustration of the mathematical meaning of the quantity $H$" of the $H$-theorem, as Boltzmann wrote in his Gastheorie. Instead, in Planck's radiation theory the electromagnetic $H$-theorem had nothing to say about the equilibrium spectrum, while the method of complexions seemed to offer a new hope of a derivation of this spectrum.[80]

This is why Planck decided to turn to the combinatorial method. However, he consistently rejected the probabilistic context of Boltzmann's original considerations. One could well call a certain mathematical function of the state of a system a "probability" without having to consider the increase of this function in time as a matter of probability. Such was Planck's

80. Planck 1900c, PAV 1:689; Boltzmann 1896, trans., 55. In Boltzmann's $H$-theorem the temperature is unambiguously defined through the average kinetic energy corresponding to the distribution $f(v)$. In the electromagnetic $H$-theorem the only available definition of temperature is through the relation $dS/dU = 1/T$.

opinion, as already expressed in a letter to Graetz of 1897:

> Probability calculus can serve, if nothing is known in advance, to determine
> a state [of equilibrium] as the most probable one. But it cannot serve, if an
> improbable state is given, to compute the following state. That is determined
> not by probability but by mechanics.[81]

The relation between entropy and probability had been introduced by
Boltzmann in 1877 in the context earlier described (see pp. 16–17). It is
now time to specify the mathematical content of these considerations, on
which Planck's success very much depended.[82]

## BOLTZMANN'S COMBINATORICS

The basic object of Boltzmann's combinatorics of 1877 is a perfect gas
of point molecules, a microstate of which is characterized by the set of
molecule velocities and positions. In a first simplifying step, Boltzmann
considers only the kinetic energies of the molecules and tries to define the
probability of an energy partition (*Energieverteilung*), that is, of a distribution of a given total energy $E$ over the molecules. Since energy is a
continuous variable, there is no obvious definition of such a probability.

As in his combinatorial considerations of 1868, Boltzmann starts
with a "fiction" wherein molecules can take only discrete energy values
$0, \varepsilon, 2\varepsilon, \ldots, i\varepsilon, \ldots$. Then, if molecules are labeled by the index $\alpha$
($\alpha = 1, 2, \ldots, N$), a microstate of the system, or "complexion," is de-
fined by attributing to each molecule a given energy:

$$\alpha \to i_\alpha \varepsilon, \tag{121}$$

where $i_\alpha \varepsilon$ is an integral multiple of $\varepsilon$. An "energy partition" is given by
a sequence of integers $N_0, N_1, \ldots, N_i, \ldots$, where $N_i$ is the number of
molecules carrying the energy $i\varepsilon$. To a given partition correspond $\mathscr{P}$ different *Komplexionen*. Boltzmann calls $\mathscr{P}$ the "permutability," since it is
equal to the number of permutations of the $N$ molecules that transfer at
least one molecule from one discrete energy value to another:

$$\mathscr{P} = N!/N_0!N_1! \ldots N_i! \ldots . \tag{122}$$

The probability $W$ of a partition $(N_0, N_1, \ldots, N_i, \ldots)$ is obtained
through division of $\mathscr{P}$ by the normalization factor; Boltzmann gave an
explicit formula for this divisor:

$$\sum \mathscr{P} = (N + P - 1)!/(N - 1)!P!, \tag{123}$$

---

81. Planck to Graetz, 23 Mar. 1897 (Deutsches Museum), quoted in Kuhn 1978, 265–266.
82. Boltzmann 1877b.

where the sum is taken over all distributions $(N_0, N_1, \ldots, N_i, \ldots)$ such that

$$\sum_i N_i = N, \qquad \sum_i N_i i\varepsilon = E, \tag{124}$$

and $P$ is obtained by dividing the total energy $E$ by $\varepsilon$. It might be worth mentioning that this combinatorial formula was the one later used by Planck.

If the $N_i$'s are large enough to allow the Stirling approximation,

$$\ln N_i! \sim N_i(\ln N_i - 1), \tag{125}$$

$W$ reaches its maximum value with the constraints (124) on the total number of particles and energy if for any $i$, and for Lagrange multipliers $\alpha$ and $\beta$.

$$\frac{\partial}{\partial N_i} \sum_j [N_j(\ln N_j - 1) - \alpha N_j - \beta N_j j\varepsilon] = 0. \tag{126}$$

This equation implies that $N_i$ must be proportional to $e^{-i\beta\varepsilon}$. Maxwell's distribution, or a discrete imitation of it, appears to be the "most probable" one, in the sense that it has the greatest number of complexions.

So much for the fiction. Boltzmann then turned to the more realistic continuous case. This was readily achieved by supposing that the energy unit $\varepsilon$ was small enough to consider that molecules whose energies lie between $i\varepsilon$ and $(i + 1)\varepsilon$ have the same energy. The numbers $N_i$ now count the molecules in the various energy intervals. Provided that the sums over $i$ can be approximated by integrals, the most probable number of molecules whose energy $K$ lies within an infinitesimal energy interval $dK$ is proportional to $e^{-\beta K} dK$.

This is not yet Maxwell's law. To get it Boltzmann had to cut up the velocity space, instead of the energy axis, into uniform cells. Then Maxwell's expression

$$f(\mathbf{v}) = Ce^{-\beta mv^2/2}d^3v = C'e^{-\beta K}\sqrt{K}\, dK, \tag{127}$$

was found to represent the most probable distribution of molecular velocities. Boltzmann also considered the positions $\mathbf{r}$ together with the velocities of the molecules. In this case the $(\mathbf{r}, \mathbf{v})$-space has to be cut up into uniform cells, and $N_i$ gives the number of molecules in the cell $i$. In the continuous limit the most probable distribution $f(\mathbf{r}, \mathbf{v})$ is uniform in the r-space and results in Maxwell's distribution in the velocity space. Furthermore, the logarithm of permutability may be calculated in this case to give

$$\ln \mathscr{P} = \ln N! - \sum_i \ln N_i! \sim N \ln N - \sum_i N_i \ln N_i \tag{128}$$

or, in the continuous approximation,

$$\ln \mathscr{P} = -\int f \ln f \, d^3r \, d^3v + \text{const.} \qquad (129)$$

In the case of maximal probability this expression of $\ln \mathscr{P}$, Boltzmann noted, is identical with the function $-H$ and therefore gives the entropy of a perfect gas up to an additive constant (function of $N$).

These calculations were simple enough, but their point of departure, the expression of the probability of a state distribution, needed further justification, as Boltzmann himself recognized: "I do not think that one is allowed to set this forth [that the equilibrium state is the most probable one] as something obvious, at least not without having first defined very precisely what is meant by the most probable state-distribution." He tried to justify the two main assumptions leading up to his expression for permutability, namely, the possibility of cutting up, and the uniformity of, (r, v)-space.[83]

This uniformity, he said, resulted from the invariance of the differential element $d^3r \, d^3v$ during a Hamiltonian evolution of a molecule. But the accompanying proof was either incomplete or wrong. In any case it could not fill the conceptual gap later emphasized by Einstein: a proper connection between the evolution in time of a system and the probability of its state was needed to justify not only uniformity in (r, v)-space but also, more generally, the relevance of combinatorial probabilities to thermodynamics.[84]

Boltzmann's combinatorics, if not fully justifiable, had, at least, to be consistent. In this respect the recourse to finite cells could seem problematic. The number of molecules in a given cell had to be large (more precisely, there had to be a negligible number of cells for which $N_i$ is neither zero nor very large) so that the Stirling approximation could be applied. At the same time the size of the cells had to be small enough that the sums over $i$ could be approximated by integrals. Instead of directly investigating the consistency of these assumptions, Boltzmann preferred an analogy with familiar problems of kinetic theory:

Nevertheless, after closer inspection, this assumption must be regarded as obvious. Indeed, any application of differential calculus to gas theory rests on the same assumption. If for instance one wishes to calculate diffusion, viscosity, conductivity, etc., one has to admit in the same way that in every infinitesimal element of volume $dx\,dy\,dz$ there is still an infinite number of gas molecules

83. Ibid., 193.
84. See Kuhn 1978, 55.

with velocity components lying between the limits $u$ and $u + du$, $v$ and $v + dv$, $w$ and $w + dw$. This assumption means only that one can choose the limits for $u$, $v$, $w$, so that they include a very large number of molecules and that one may nonetheless regard all these molecules as having the same velocity components.[85]

In the content of the 1877 memoir, one can easily check the legitimacy of this hypothesis for the most probable distribution of molecules. For instance, in the simplest case in which the energy axis is cut up into intervals of equal size $\varepsilon$, the numbers $N_i$ are given by

$$N_i = N(1 - e^{-\beta\varepsilon})e^{-i\beta\varepsilon}. \tag{130}$$

It can easily be seen that the condition for the Stirling approximation to be valid is a large value of the number $N_0$ of molecules in the zero-energy interval:

$$N(1 - e^{-\beta\varepsilon}) \sim N\beta\varepsilon \gg 1. \tag{131}$$

The other condition, that the sums can be replaced with integrals, reads:

$$(N_{i+1} - N_i)/N_i \sim \beta\varepsilon \ll 1. \tag{132}$$

The two conditions are both met if

$$1/N\beta \ll \varepsilon \ll 1/\beta. \tag{133}$$

Consequently, for any value of the available energy $(E = N/\beta)$ the size of the cells can be chosen consistently, and it then disappears from the final result for the most probable distribution and the corresponding entropy.

Since in Planck's later combinatorics $\varepsilon$ is not always a negligible fraction of $1/\beta (= kT)$ and appears in the final entropy formula, it is important to understand what makes it disappear in Boltzmann's case. The reason is not that $\varepsilon$ is infinitesimal in the mathematical sense; indeed, it must be larger than $1/N\beta$. At a purely formal level, the elimination occurs when sums like the one giving the entropy are replaced with integrals. Boltzmann had to take this formal step because the main physical quantity of interest, the distribution of molecules over cells $(N_0, N_1, \ldots N_i, \ldots)$, was expected to be well approximated by a continuous distribution, the most probable of which is Maxwell's law. We will find that neither this circumstance nor its formal corollary occurs in Planck's combinatorics.

To summarize, Boltzmann's memoir of 1877 on entropy and probability was not very explicit about the physical meaning of its main procedural elements, the method of dividing up the space of configurations and the uniformity of this space. It was clear to him, however, that combinatorial methods were relevant insofar as they were able to reproduce the "continuous" entropy formula

$$S = -\int f \ln f \, d^3r \, d^3v, \qquad (134)$$

which had already been founded on what he considered to be more fundamental bases, that is, on the ergodic hypothesis or on the methods of the $H$-theorem. On the contrary, Planck, still unfamiliar with the foundations of Boltzmann's theory, would venture to confer physical meaning on the artifact of energy elements.

## QUANTIFIED CHAOS

When in late 1900 Planck tried to apply Boltzmann's combinatorial method to his resonator, he naturally drew his inspiration from the marked analogy between the $H$-theorem for gases and that for radiation. The pivot of this analogy had to be the principle of disorder, since it was at the center of his conception of irreversibility. In this regard a very typical statement of Planck reads, "One can speak of a disorder, therefore of an entropy of a resonator" (March 1900).[86] As we saw, Planck identified the kind of disorder affecting a resonator as the unobservable temporal fluctuation of the energy $Y(t)$ of this resonator around its secular average $U$. More generally, disorder was what made the microscopic details of the description of a system irrelevant to the evolution of the really accessible parameters of the system, namely spatial (in the gas case) or secular (in the radiation case) averages.

In harmony with his idea of the centrality of disorder, Planck perceived Boltzmann's permutability as a quantitative measure of molecular chaos, whereas Boltzmann never tried to make such a direct connection between entropy and disorder. For a gas disorder is what makes the detailed microscopic configuration irrelevant to the evolution of the distribution $f(v)$ or $f(r, v)$ entering the Boltzmann equation. Permutability, being a measure of the number of microscopic configurations compatible with the distribution $f$, appeared to Planck as a natural quantitative measure of this disorder. In the resonator case, the counterpart of the permutability had to be something like the number of functions $Y(t)$ compatible with a given average energy $U$, since the disorder lay in the uncontrollable irregularity of the instantaneous energy.

86. Planck 1900b, *PAV* 1:674.

There is no doubt that this relation between entropy and disorder was the key point of Planck's published reasoning. His famous communication of 14 December 1900 to the Berlin Academy introduced the new determination of the entropy of a resonator with the words "Entropie bedingt Unordnung," that is, "Entropy presumes disorder." It continued with a description of the nature of the disorder of a resonator drawn from his previous theory that justified his subsequent computation of the number of complexions, or "probability," leading to the entropy of a resonator. Planck's use of the word "probability" in this context should not confuse the reader: he just meant it in the mathematical sense of an abstract lottery game leading to combinatorial expressions.[87]

Admittedly, Planck's presentation did not necessarily reflect the way he really reached his derivation of the blackbody law. It would seem plausible that he worked backward from the radiation law and guessed the proper combinatorics from the form of the resonator entropy (as in Rosenfeld's reconstruction, for instance). According to a letter from Planck to Lummer of 26 October 1900, the truth seems to have lain somewhere in between; that is to say, Planck simultaneously used inductive (from the blackbody law to entropy) and deductive (from entropy to the blackbody law) considerations:

> If the prospect should exist at all of a theoretical derivation of the radiation law, which I naturally assume, then, in my opinion, this can be the case only if it is possible to derive the expression for the probability of a radiation state, and this, you see, is given by the entropy. Probability presumes disorder, and in the theory I have developed, this disorder occurs in the irregularity with which the phase of the oscillation changes even in the most homogeneous light. A resonator, which corresponds to a monochromatic radiation, in resonant oscillation will likewise show irregular changes of its phase [and also of its instantaneous energy, which was more important to Planck's subsequent derivation], and on this the concept and the magnitude of its entropy are based. According to my formula [the blackbody law communicated on 19 October to the German Academy], the entropy of the resonator should come to:
>
> $$S = \alpha \ln \left[ (\beta + U)^{\beta + U} / U^{U} \right]$$
>
> [formula (119)], and this form very much recalls expressions occurring in the probability calculus. After all, in the thermodynamics of gases, too, the entropy $S$ is the log of a probability magnitude, and Boltzmann has already stressed the close relationship of the function $\chi^{\chi}$, which enters the theory of combinatorics, with the thermodynamic entropy. I believe, therefore, that the prospect would certainly exist of arriving at my formula by a theoretical route, which

87. Planck 1900d.

would then also give us the physical significance of the constants $C$ and $c$ [of Planck's law].[88]

In any case, the final justification of Planck's calculation in terms of quantified chaos certainly determined his opinion about the status of the finite energy elements.

Having described the nature of the disorder to be found in a resonator, Planck continued his communication as follows. In order to give a definite meaning to the number $W$ of evolutions $Y(t)$ of the energy compatible with a given temporal average $U$, he replaced $Y(t)$ with its value at $N$ different instants of time, or, more exactly, with the energy values of $N$ independent (far removed from one another) identical resonators at one given instant. Then $W$ is represented by the number of distributions of a total energy $E = NU$ over these $N$ resonators.[89]

For the rest, Planck proceeded in exact analogy with Boltzmann's "fiction." That is to say, he divided the energy $E$ into finite elements $\varepsilon$:

> If $E$ is taken to be an indefinitely divisible quantity, the distribution is possible in an infinite number of ways. But I regard $E$—and this is the essential point of the whole calculation—as made up of a completely determinate number of finite equal parts, and for this purpose I use the constant of nature $h = 6.55 \times 10^{-27}$ (erg·sec). This constant, once multiplied by the common frequency of the resonators, gives the energy element $\varepsilon$ in ergs, and by division of $E$ by $\varepsilon$ we get the number $P$ of energy elements to be distributed over the $N$ resonators. When this quotient is not an integer, $P$ is taken to be a neighboring integer.[90]

According to this hypothesis, $W$ becomes the total number of complexions compatible with the total energy $E$, wherein the word "complexion" is defined strictly in Boltzmann's sense, by attributing to each resonator a given discrete energy (as specified in (121)). One could calculate this number by adding the permutabilities of all the distributions $(N_0, N_1, \ldots, N_i, \ldots)$ (in Boltzmann's notation) such that [91]

$$\sum_i N_i = N, \qquad \sum_i N_i i\varepsilon = E. \tag{135}$$

For the sake of simplicity, however, Planck preferred to compute $W$ directly as the "number of distributions of $P$ energy elements over $N$ resonators," it being understood that only the *number* (not the identity) of the energy

88. Planck to Lummer, 26 Oct. 1900, quoted in Jungnickel and McCormmach 1986, 2:261–262.

89. Planck's $W$ stands for *Wahrscheinlichkeit* (probability).

90. Planck 1900d, *PAV* 1:700–701.

91. $N$ can always be chosen so great that the constraint does not play any role. In Planck 1906, 151, the characterization of $W$ as a sum of permutabilities comes first; then Planck switches to the "faster and easier" method of the energy elements.

elements attributed to a given resonator is considered. The latter stipulation surprised some of Planck's readers (Ehrenfest and Natanson), but it was in fact implied by the analogy with Boltzmann's fiction.[92]

For the W formula Planck referred his reader to the calculus of combinations. Here follows an elegant proof, due to Ehrenfest and Kamerlingh Onnes (1914). A complexion may be represented as a symbol

$$\varepsilon/\varepsilon\varepsilon\varepsilon/\varepsilon\varepsilon/ \ldots /\varepsilon\varepsilon\varepsilon$$

containing $P$ times the symbol $\varepsilon$ and $N - 1$ times the symbol $/$. The number of complexions is therefore equal to the number $(N + P - 1)!$ of all these symbols regarded as different, divided by the number, $P!$, of permutations of the $\varepsilon$ symbols and by the number, $(N - 1)!$, of permutations of the $/$ symbols:[93]

$$W = (N + P - 1)!/(N - 1)!P!. \tag{136}$$

Adapting Boltzmann's relation between entropy and probability to this problem, Planck wrote

$$S = \frac{k}{N} \ln W \tag{137}$$

for the entropy $S$ of a single resonator. The constant $k$, Planck emphasized, had to be the same in gas theory and in radiation theory (Boltzmann did not need such a constant, since he measured temperatures in energy units). But contrary to Boltzmann's case, no procedure of extremum was here necessary: as a consequence of the different type of disorder, the average energy $U$, not the more detailed distribution $(N_0, N_1, \ldots N_i, \ldots)$, characterizes the "macroscopic" state of the resonators.[94]

As $N$, the number of exemplars of the resonator, can be taken as great as one wishes, the Stirling approximation applies:

$$S = \frac{k}{N} [(N + P) \ln (N + P) - N \ln N - P \ln P]$$

$$= k \left[ \left(1 + \frac{U}{\varepsilon}\right) \ln \left(1 + \frac{U}{\varepsilon}\right) - \frac{U}{\varepsilon} \ln \frac{U}{\varepsilon} \right]. \tag{138}$$

92. On Ehrenfest's and Natanson's reactions and Planck's reply, see Darrigol 1988b, 52.
93. Ehrenfest and Kamerlingh Onnes 1915.
94. Here I have followed Planck's subsequent *Annalen* paper: Planck 1901. In the original communication Planck distributes energy over resonators of different frequencies and forms the product $\Pi\, W_\nu$, where $W_\nu$ is calculated in the manner explained in this paragraph. This procedure parallels Boltzmann's distribution of energy over molecules. The analogy between gas and resonators is thereby improved, but in an unessential way: the maximizing of $\Pi\, W_\nu$ just implies that the resonators at various frequencies have the same temperature.

From the relation between entropy and temperature, $1/T = dS/dU$, results

$$U = \frac{\varepsilon}{e^{\varepsilon/kT} - 1}. \tag{139}$$

The fundamental relation (96) and $\varepsilon = h\nu$ finally give:

$$u_\nu = \frac{8\pi h\nu^3}{c^3} \frac{1}{e^{h\nu/kT} - 1}, \tag{140}$$

which is the canonical form of Planck's law.

In a subsequent publication Planck explained that the proportionality of the energy element $\varepsilon$ to the frequency was implied by Wien's displacement law (expressed in the form (98)). Later, in 1906, he showed that this property and also the uniformity of the cutting up of the energy axis resulted from Boltzmann's general assumption of uniformity in configuration space (here the $(f, L\dot{f})$-plane) and from the quadratic form of the energy of a resonator. These remarks made the analogy with Boltzmann's combinatorics even closer.[95]

## QUANTUM CONTINUITY

Table 2 summarizes the formal correspondence between Boltzmann's "fiction" and Planck's combinatorics of $N$ exemplars of a resonator.

Despite the exact transposition of the definition of a complexion, this correspondence is certainly not the most direct that one could imagine. Had he not been guided by his interpretation of $W$ as a measure of disorder, Planck would no doubt have characterized the macrostate by a distribution $(N_0, N_1, \ldots N_i, \ldots)$, in the resonator case as in the gas case. This would have led to

$$W = N!/N_0!N_1! \ldots N_i! \ldots \tag{141}$$

and

$$S = \frac{k}{N} \text{Max} [\ln W] \sim -\frac{k}{N} \sum_i \bar{N}_i \ln \frac{\bar{N}_i}{N}, \tag{142}$$

---

95. Planck 1901. The uniformity of quantum cells in the $(f, L\dot{f})$-plane ($L\dot{f}$ being the momentum conjugated with $f$) is equivalent to the uniformity of energy intervals because of the identity

$$\int_{Y=nh\nu}^{Y=(n+1)h\nu} df \, d(L\dot{f}) = h$$

for $Y = \frac{1}{2}(Kf^2 + L\dot{f}^2)$ and $2\pi\nu = \sqrt{K/L}$.

TABLE TWO ANALOGY BETWEEN PLANCK'S AND<br>BOLTZMANN'S COMBINATORICS

| | Gas | Resonators |
|---|---|---|
| Microstate | Complexion:<br>$\alpha \to i_\alpha \varepsilon$ | Complexion:<br>$\alpha \to i_\alpha \varepsilon$ |
| Macrostate | Energy partition:<br>$(N_0, N_1, \ldots, N_i, \ldots)$ | Total energy:<br>$E = NU$ |
| Number of complexions | Permutability:<br>$\mathscr{P} = N!/N_0!N_1! \ldots N_i! \ldots$ | "Probability"<br>$W = (N + P - 1)!/(N - 1)!P!$ |
| "Boltzmann's principle" | $S = k \ln \mathscr{P}$ | $S = k \ln W$ |
| Uniformity | in $(r, v)$-space | in $(f, L\dot{f})$-plane |

where the distribution $(\bar{N}_0, \bar{N}_1, \ldots, \bar{N}_i, \ldots)$ is the one for which $W$ reaches its maximum, with the constraints (135)

$$\sum_i N_i = N, \sum_i N_i i\varepsilon = E.$$

An elementary calculation by the method of Lagrange multipliers (see equation 126) gives

$$S = k \left[ \left(1 + \frac{U}{\varepsilon}\right) \ln \left(1 + \frac{U}{\varepsilon}\right) - \frac{U}{\varepsilon} \ln \frac{U}{\varepsilon} \right],$$

which is identical with Planck's expression (138) (this identity results from the fact that most complexions belong to distributions that are very close to the most probable one).[96]

However, if the distributions $(\bar{N}_0, \bar{N}_1, \ldots, \bar{N}_i, \ldots)$ really played similar roles in the gas case and in the resonator case, they would have to be replaced in both cases by their continuous limits, formally obtained by setting $\varepsilon \to 0$ in the above expressions. The expression (139) for the resonator energy would therefore become $U = kT$, and instead of Planck's law one would have

$$u_v = \frac{8\pi v^2}{c^3} kT, \tag{143}$$

96. The above calculation provided the formal basis for the derivation of Planck's law first given in Lorentz 1910 (see Darrigol 1988b, 62–63) and interpreted in terms of an intrinsic discontinuity of the resonator energy.

a result incompatible with experiments, and even absurd since it leads to an infinite energy for the entire spectrum.[97]

Because his conception of disorder implied a combinatorics different from Boltzmann's, Planck avoided this catastrophic conclusion. However, a deeper understanding of the foundations of Boltzmann's theory would have left him no choice. As Einstein correctly pointed out in 1905, Planck's resonators could be brought to interact with the molecules of a gas, without their interaction with radiation being substantially modified. In such a system, the energy distribution of the resonators and the velocity distribution of the molecules play parallel roles. Consequently, the reasoning just given should apply, and the formula $U = kT$ should hold, as a particular case of energy equipartition.[98]

In conformity with the original object of this program, Planck reasoned purely in terms of radiation theory and did not consider such an admixture of molecular and electrodynamic systems. Furthermore, like most of his colleagues, he did not believe in the generality of the equipartition law, from which the relation $U = kT$ trivially resulted. This law generally led to much too high values of specific heats of materials, and the best specialists, including Lord Kelvin and Boltzmann (with the exceptions of Gibbs and Maxwell, who died prematurely), attributed this failure to some unknown intricacies of molecular dynamics.[99]

Planck's only guide was his characterization of disorder in radiation theory, which deprived the energy distribution $(N_0, N_1, \ldots N_i, \ldots)$ of a direct physical meaning and made the secular energy $U$ the only observable property of resonators (except for their frequency, of course). The energy elements had therefore no reason to disappear from the end results. Being the gauge of elementary disorder, they fitted harmoniously, like the hypothesis of natural radiation, in the logical "gap" open because of the indetermination of the detailed structure of resonators; they started to play a role where ordinary electrodynamics ceased to provide definite information. In other words, electrodynamic laws and Planck energy elements did not contradict each other, they complemented each other.

In this context Planck could not possibly have understood the introduction of energy elements as a discrete selection of the admissible energy

97. This result is called the "Rayleigh-Jeans" law. It was first obtained by Rayleigh in 1900, up to a numerical mistake corrected by Jeans in 1905. The original derivation rested on an application of the equipartition theorem to the stationary modes of a cavity. However, Rayleigh and Jeans did not believe in the validity of this theorem for large frequencies. See Kuhn 1978.

98. Einstein 1905.

99. See Brush 1976, 356–363, and Kuhn 1978.

values of a resonator. Such a discontinuity would have contradicted the rest of his theory, especially the proof of the fundamental equation (96), which entered the derivation of the blackbody law. Moreover, Planck's own wording of the "essential point" of his communication of 14 December leaves no room for doubt. Immediately after introducing the energy elements, he wrote: "When this quotient [E/ε] is not an integer, P is taken to be a neighboring integer." This by itself shows that the energy of N independent resonators, and *a fortiori* the energy of a single resonator, was not thought to be restricted to multiples of ε.[100]

To be the counterpart of Boltzmann's "fiction," Planck's discrete complexions also had to be fictitious. This point was made entirely explicit in the lectures on the theory of thermal radiation of 1905–6: a complexion, Planck said, signified the attribution to each resonator of an "energy domain" delimited by the energy values iε and (i + 1)ε, not of a discrete energy value iε. In conformity with this viewpoint, the fundamental notion of Planck's subsequent theory of quantization was that of "elementary domains of probability," a generalization of the energy domains in configuration space. This conception stood against the notion of discrete quantum state introduced by Einstein, raised to a fundamental postulate by Niels Bohr and adopted by most early quantum theorists.[101]

## SUMMARY AND CONCLUSIONS

The Berlin spectroscopists did not let Planck rejoice for long about his fundamental derivation of Wien's law. In the very year Planck completed his program, 1899, they began to observe systematic deviations from Wien's law in the infrared part of the blackbody spectrum. This helped Planck realize that, contrary to his earlier conviction, there were an infinite number of expressions for the resonator entropy compatible with his electromagnetic H-theorem, and thus an infinity of corresponding blackbody laws. In fact, in order for the total entropy to increase, the only constraint on the expression for the resonator entropy was that its second derivative (with respect to energy) should be negative. Then, on the basis of a new independent argument, Planck imposed an additional constraint on this derivative and recovered Wien's law, to the experimenters' great disbelief.

100. See n. 90. The absence of a quantum discontinuity in Planck's early derivations of his blackbody law was first observed by Kuhn 1978 against a long historical tradition that asserted the contrary.

101. Planck 1906, 135: ". . . resonators that carry a given amount of energy (better: that fall into a given 'energy domain') . . . " The notion of *Elementargebiete der Wahrscheinlichkeit* was systematically developed in the second edition (1913) of Planck 1906.

The new argument was wrong, and Planck publicly withdrew it in October 1900, after the experimental violations of Wien's law had become more obvious. In the same communication he proposed an alternative blackbody law, a happy guess based on a simple modification of the expression for the second derivative of the resonator entropy corresponding to Wien's law. The new blackbody law immediately proved to fit empirical data quite well, and Planck started to think about a more fundamental derivation. This led him to consider the relation between entropy and probability which Boltzmann had introduced in 1877.

According to the relevant memoir of Boltzmann, in a dilute gas the equilibrium distribution of velocities—that is, Maxwell's distribution— was also the most "probable"; and the entropy (or the function -$H$) was given by the logarithm of the (unnormalized) "probability." Calling (according to modern terminology) the exact microscopic configuration of the molecular model a microstate, and the distribution of velocities a macrostate, Boltzmann's (unnormalized) "probability" was defined as the number of microstates compatible with a given macrostate. Of course, this definition has problems since there is a continuous infinity of microstates corresponding to every macrostate. To solve this difficulty, Boltzmann divided up the configuration space of a molecule into cells and regarded all configurations belonging to a given cell as one single configuration. For instance, in a simple model for which the configuration of a molecule is completely determined by its energy, the energy axis is cut up into equal intervals or energy elements, and a microstate is obtained by assigning to each molecule one of these intervals. Boltzmann's subsequent calculations required the energy elements to be finite (so that the number of molecules in an energy interval could be very large) but small enough not to blur the definition of macrostates, to which the quantities of physical importance pertained. On this condition the energy elements disappeared from the end results; and Maxwell's distribution and the corresponding entropy were recovered. In other words, Boltzmann employed the energy elements as a mathematical artifice, for the purpose of giving a definite meaning to the "probability" of a macrostate. They did not belong to the microscopic model, nor could they enter macroscopic laws; for these could be reached independently of the relation between entropy and "probability," through the $H$-theorem or the ergodic hypothesis.

The relation between entropy and probability played only a minor role in Boltzmann's subsequent work. For instance, in his *Gastheorie* it appeared only as a "mathematical illustration" of the expression for the $H$- function. Boltzmann (rightly) believed that derivations of thermodynamic quantities and laws through the *H*-theorem or through the ensemble technique were more fundamental. In 1900 Planck faced a different situation: his electromagnetic *H*-theorem had proved useless in determining the entropy of a resonator, so that the relation between entropy and probability, far from being superfluous, seemed to be the only available access to the blackbody law. Planck accepted the relation but not its original context, which was a probabilistic interpretation of the irreversibility theorem. Instead he reinterpreted Boltzmann's "probability" as a quantitative measure of elementary disorder, a notion that was at the core of his (Planck's) non-probabilistic conception of irreversibility. Such reinterpretation also had a practical advantage: it provided some guidance about how to extend the analogy between gas theory and radiation theory.

Planck first discussed the type of disorder to be found in a resonator, knowledge gleaned from the requirements of derivation of the electromagnetic *H*-theorem. In this way he determined what played the roles of microstates and macrostates, as the states of the system respectively in the detailed and the physical levels of description. Next, following Boltzmann, he introduced finite energy elements in order to obtain a definite value for the "probability," that is, the number of microstates in a given macrostate. The logarithm of this "probability" gave him the entropy of a resonator, which leads to Planck's new blackbody law—if only the energy elements can be taken to be proportional to the frequency of the resonator.

Contrary to Boltzmann's case the energy elements now appeared in the final thermodynamic expressions. Planck attributed this peculiarity to a difference in the type of disorder. Indeed, his understanding of the disorder in a resonator led to a notion of macrostate (characterized by the total energy of an ensemble of resonators) that was insensitive to the introduction of energy elements; therefore, Boltzmann's condition that the energy elements should be small enough not to blur the definition of macrostates had no counterpart in Planck's case, and nothing seemed to forbid the appearance of the energy element in the final entropy formula.

In this situation Planck had no reason to question the continuity of the resonator energy. Moreover, such a step would have contradicted, among other things, his derivation of the "fundamental equation," which was necessary for his proof of the blackbody law. In his mind the energy elements were something like the gauge of elementary disorder; they therefore pertained to the indeterminate internal structure of resonators, and they did not contradict his electrodynamic reasonings, which were independent of this structure. In short, Planck relaxed Boltzmann's connection between microworld and macroworld by leaving part of the micromodel indeterminate. This allowed him to maintain strict irreversibility in the macroworld, by adjusting the indeterminate part of the micromodel (introduction of elementary disorder). In turn, this adjustment permitted his derivation of the blackbody law, without contradicting the determinate part of the micromodel.

As is well known, a few years ago Thomas Kuhn published an in-depth study of blackbody theory at the turn of the century. I will briefly indicate how my account may differ from his. Kuhn concludes, as I do, that Planck did not restrict the energy of his resonators to discontinuous values. His reasoning may be summarized as follows: Boltzmann introduced finite energy elements with no intention of jettisoning the continuity of molecular dynamics; Planck reached his expression for the resonator entropy working in close analogy with Boltzmann's method; therefore, despite some delusive formal manipulations, he did not quantize the energy of the resonators. As convincing as it might be, this argument does not say why Planck did not feel compelled, within the framework of his own thermodynamics, to imitate Boltzmann's procedure even more closely, which would have led to an absurd blackbody law (the so-called Rayleigh-Jeans law). My explanation for this rests on the idiosyncratic nature of Planck's conception of the microscopic foundations of thermodynamics. Kuhn describes Planck's conversion to Boltzmann's views and methods as quasi-complete (as starting with the introduction of "natural radiation"). In fact, as Allan Needell first demonstrated, Planck did not renounce his nonstatistical conception of irreversibility until much later (around 1914). This in turn explains the role elementary disorder played in orienting Planck's use of analogies in his derivation of the blackbody law in 1900. It also explains why Planck's early readers (and a good number of later ones) found his derivation either obscure or implicitly based on an intrinsic quantization of resonators: they were wearing Boltzmann's spectacles.[102]

During the first ten years of this century, Planck's theory of radiation, and more generally the problem of thermal radiation, became the object of critical investigations by unusually penetrating minds, among whom were two young physicists, Ehrenfest and Einstein, and the venerated H. A. Lorentz. Some of Planck's results survived: the electromagnetic $H$-theorem (so named by Ehrenfest) proving the spatial uniformizing effect of resonators, and the blackbody law with its characteristic energy elements and the new fundamental constant $h$.[103]

102. Kuhn 1978; Needell 1980, 1988.
103. See Kuhn 1978; Klein 1963b, 1967; Darrigol 1988b.

However, the central concept of Planck's theory, namely his notion of elementary chaos, appeared to be untenable. According to Einstein, no coherent conception of microscopic dynamics was able to provide a strict and indefinite increase of entropy. On the contrary, microscopic disorder implied observable effects like the perpetual agitation of Brownian particles and mirrors. Within Boltzmannian orthodoxy, Planck's assumption of finite energy elements proved to be incompatible with the foundation of electrodynamics. No interpretation of the blackbody law could be given without emancipating the resonators from their classical (even secular) behavior.

In 1906 Einstein reinterpreted the formal skeleton of Planck's derivation of the blackbody law on the basis of a discrete quantization of resonators. In other words, he turned Boltzmann's "fiction" into a reality, interpreting the energy unit as the minimal amount of energy that resonators could exchange with radiation. This idea of a radical quantum discontinuity was certainly paradoxical, for no one (not even Einstein) could imagine a satisfactory mechanism of the quantum jumps. Nevertheless, it quickly led Einstein to a successful theory of specific heat. By the Solvay congress of 1911 an increasing number of specialists (but not Planck) were convinced that the energy of atomic entities could only take discrete values. The ground was ready for even sharper departures from classical theory, which Bohr soon brought with his atomic theory.[104]

To conclude, the retrospective successes and defects of Planck's program can be largely understood as deriving from certain powerful analogies with Boltzmann's theory, these analogies being constrained by a belief in the absolute validity of the entropy principle (which was *not* Boltzmann's). One of these successes, the electromagnetic *H*-theorem, depended upon a formal analogy between the notions of natural radiation and of molecular chaos. Further, the conception of disorder bound to this analogy guided Planck in his exploitation of another analogy, that between Boltzmann's combinatorics and resonator combinatorics. The resulting derivation of the blackbody law happened to be *formally* meaningful, even though its conservative interpretation would not survive the quantum revolution initiated by Einstein.

104. Einstein 1906. See Kuhn 1978, and Klein 1965.

# The Correspondence Principle

# Introduction

Analogical thinking usually works with a touch of blindness: formal relations of a given theory are tentatively applied to new objects, and if the operation is empirically successful, the concepts originally underlying these relations are assumed to extend to these objects. The eventual need for a reinterpretation of the extended theory in terms of new concepts appears only at a later stage. In the previous chapter, we saw a good example of this typical process in Planck's formal adaptation of Boltzmann's discretization of mechanical states. Planck's procedure preserved the continuity of energy exchanges between resonators and radiation, as Boltzmann's original procedure presupposed the continuity of the dynamics of gas molecules; the necessity of quantum discontinuity appeared only a few years later.

With the correspondence principle Niels Bohr has given us a most remarkable counterexample: a principle of analogy which never concealed the contrast between the old and the new theory. In this instance, the old theory was "ordinary" electrodynamics, while the new one was an atomic theory that from the start flatly contradicted some basic principles of electrodynamics. The analogy was explicitly formal and was certainly never intended to include the old theory in the new one.

However, Bohr's description of atomic phenomena retained classical concepts like electromagnetic field, electrons' position, momentum, and energy. This could give the impression that his quantum theory was self-contradictory, drawing its success from clever empirical considerations amidst a cloud of illusory depths.

In this chapter I will document the opposite thesis. Bohr was never a narrow empiricist (and never became a positivist either). His quantum theory, far from being contradictory, provided at any stage an analysis of its relation to classical theory that conciliated the persisting recourse to classical concepts with quantum discontinuity. Most important, Bohr realized that certain fundamental concepts could still be used in the quantum theory because they could be defined through an application of classical theory, *within its accepted range of validity*. For instance, the frequency of the emitted radiation could be defined through a legitimate application of wave optics to spectrometers, and the energy of stationary states could be defined through an application of ordinary mechanics (according to the adiabatic principle) to slow deformations of atomic systems.

Quantum-theoretical relations like "$\Delta E = h\nu$" were allowed to relate classically defined *concepts*; but they could not be explained in terms of a mere extension of classical *laws*, which would have brought contradiction. For instance, the mechanism of quantum transition had to be left undetermined, at least until proper quantum concepts could be built. This is why Bohr insisted on the incompleteness of his theory. Any further recourse to classical concepts or laws in the atomic realm had to be of a "formal" nature. So was the recourse to electronic orbits in stationary states, since these orbits did not interact with radiation according to ordinary electrodynamics. In other words (not Bohr's), there was no valid optical theory providing a means of observation of electrons at the atomic scale.

There was, moreover, no warranty that the formal use of classical concepts or laws would remain a lasting feature of quantum theory. For instance, the application of classical mechanics to electronic orbits could be only approximate and provisional, since it disregarded radiative corrections to the Coulomb forces. This is why Bohr tried to isolate the assumptions of his theory that could be formulated without appeal to formal classical laws. After a period of hesitation he reached this aim in 1917. The resulting "postulates" were expressed in terms of purely quantum-theoretical concepts (like stationary state) or in terms of well-defined classical concepts, that is to say, concepts defined through an application of a classical theory within its range of validity. For this reason Bohr believed his theory to be solidly anchored; and he proved to be right, since Heisenberg's matrix mechanics of 1925 was based on exactly the same postulates as Bohr's 1918 essay "On the quantum theory of line spectra": the postulate of stationary states and the relation $\Delta E = h\nu$.

Conversely, Bohr often insisted on the provisional character of additional assumptions. In light of new empirical or formal developments he was always ready to reconsider his and others' preconceptions about the motion in stationary states. He successively considered strictly periodic motions obeying ordinary mechanics (1913–1916), motions of multiperiodic systems also subjected to ordinary mechanics (1917), multiperiodic motions of not necessarily multiperiodic systems still subjected to ordinary mechanics (1918–1922), multiperiodic motions eluding ordinary mechanics (1922–1925); and, finally, in spring 1925 he completely gave up the notion of definite electronic orbits.

Bohr's theory was deliberately incomplete and systematically open to revision. Around the stable pillars of the quantum postulates it needed metatheoretical "principles" that could direct constructive developments. The main principle was the correspondence principle, a procedure for deriving quantum analogues of relations between motion and radiation based on classical electrodynamics. In 1917 the initial successes of this adaptation convinced Bohr of the possibility of a "rational generalization" of classical electrodynamics based on the quantum postulates. However, the precise expression and the scope of the correspondence principle depended on the assumptions made about the electronic motion. Whenever this motion was a priori determined, the "correspondence" aided in *deducing* properties of emitted radiation. In the opposite case, characteristics of the electronic motion could be *induced* from the observed atomic spectra. This ambiguity made the correspondence principle a very flexible tool that was able to draw the most from the permanent inflow of empirical data.

In the gradual process of freeing atomic motion from classical preconceptions, the deductive side of the correspondence principle shrank, until nothing seemed to be left of it, at least in the eyes of Bohr's most open critic, Wolfgang Pauli. The heuristic power of this principle, however, was not yet exhausted. Even before the final collapse of the motion of electronic orbits, Bohr's closest disciples had started a symbolic translation of classical mechanical relations into purely quantum-theoretical ones, that is to say, a translation in terms of the basic quantities entering Bohr's postulates: energies of stationary states, quantum numbers, and transition probabilities. Heisenberg's matrix mechanics was, in fact, the ultimate result of this extended process, a symbolic system naturally and automatically integrating the formal analogy expressed in the correspondence principle.

Hence, tar trom being a naive or irrational extension of classical concepts, the correspondence principle allowed for the development of formal structures that could fill the conceptual void created by the breakdown of classical laws. The later interpretation of these structures within the framework of "complementarity" (which I will not describe here) fulfilled Bohr's early hope of a "rational generalization" of classical electrodynamics.

# CHAPTER V

# The Bohr Atom
# (1913–1916)

## HORRID ASSUMPTIONS

Bohr's leading role in the development of atomic theory began in 1913 with a fundamental memoir "On the constitution of atoms and molecules," published in three parts in the *Philosophical Magazine*. The first part of this trilogy introduced the concept of "stationary states" in atoms, the stability of which transcended mechanical explanation. As Heilbron and Kuhn have shown, the essential motivation for the introduction of this bold hypothesis was the impossibility of adapting the mechanical stability arguments of the Thomson atom to the new planetary models.[1]

In Thomson's model of 1904, electrons rotated within a sphere of uniformly distributed positive electricity, and they were arranged symmetrically on densely populated rings (their number being much higher than the atomic number). The first feature provided mechanical stability for certain types of electronic configurations, the second one approximate radiative stability, at least until $\beta$ and $\alpha$ scattering experiments required a reduction of the total electron number down to the atomic number. Most interestingly, the criterion of mechanical stability limited the relative numbers of electrons on the rings in a way which seemed to duplicate the periodic structure of Mendeleev's table, if only the electron numbers of successive chemical elements were assumed to differ by one unit.[2]

1. Bohr 1913; Heilbron and Kuhn 1969.
2. See Heilbron 1964, 1968.

In Rutherford's nuclear atom, mechanical and radiative stability were both lacking, so that it was only by Bohr's fiat that electrons remained on stationary orbits. Also, new clues for the electronic configuration of atoms were needed, and a new explanation of chemical periodicity would have to be found. Having been proposed in this context, Bohr's original notion of stationary state included only the "normal" (unexcited) state of atoms. However, in order to exploit spectral data, particularly the Balmer formula, Bohr introduced excited stationary states, with higher energies but similar indifference to mechanical and electrodynamic perturbations. Changes in atoms could occur only as sudden jumps between two such supposedly stable states, and electromagnetic radiation could be emitted only during these jumps.[3]

As a first step toward the determination of the stationary states and the emitted radiation, Bohr introduced two "principal assumptions":

I. "that the dynamic equilibrium [i.e., the motion] of the systems in the stationary states can be discussed by help of the ordinary mechanics, while the passing of systems between different stationary states cannot be treated on that basis," and

II. "that the latter process is followed by the emission of a *homogeneous* [i.e., monochromatic] radiation, for which the relation between the frequency and the amount of energy emitted is the one given by Planck's theory [of blackbody radiation]."

This second assumption was the boldest of all, for it generally gave for the frequency of the emitted radiation a frequency different from that of the motion in the original state of the emitting atom. Einstein had well anticipated, for instance in his theory of specific heats (1906), the idea of a discrete selection of mechanical states implied in the notion of stationary state; but he was "astonished" by Bohr's further departure from ordinary conceptions and judged it to be an "enormous achievement."[4]

In their first rough formulation, the "principal assumptions" were not sufficient to determine the stationary states, even in the simplest case of the hydrogen atom. Some other hypothesis was needed to select a discrete subset of motions among all those permitted by ordinary mechanics. I will not explain here how Bohr reached this hypothesis, the so-called quantum rule; rather, I will discuss the various formulations found in the 1913

3. See Heilbron and Kuhn 1969.
4. Bohr 1913, 7; Einstein's reaction (at the Vienna congress) is taken from Hevesy to Bohr, 23 Sept. 1913, *BCW* 2:[532].

trilogy. In the first formulation the quantum rule takes the form of a simple relation between the orbital frequency $\bar{v}$ in a given stationary state and the frequency $v$ of the monochromatic radiation emitted by an electron making a transition from rest at infinity to this stationary state.

Bohr's focus on the latter type of transition deserves a comment: his main interest was not in the theory of spectra but in the building of atoms, which could be imagined as being synthesized from originally separated electric charges. As we shall later see, in the early twenties Bohr based the construction of his second atomic theory on this idea of a correspondence between the properties of stationary states and the characteristics of the radiation emitted during their formation.

For the hydrogen atom, Bohr simply assumed $v = \bar{v}/2$, without any solid justification but success. This made, somewhat naturally, the frequency of the radiation emitted during the formation of a given stationary state the average of the initial (zero) and final ($\bar{v}$) orbital frequencies. He then applied assumption (II) to this process, giving to the energy emitted during the above process one of the values permitted by Planck's theory of radiation, that is to say, an integral multiple of the quantum $hv$. This yields the relation

$$E_\infty - E_n = nhv, \tag{1}$$

where $E_\infty$ is the energy in the original, unbound state, and $E_n$ the energy in the final stationary state. In purely orbital terms, one has the formal quantum rule

$$E_\infty - E_n = nh\bar{v}/2. \tag{2}$$

Assumption (I) includes the validity of ordinary mechanics for the motion of the charged particles in a given stationary state. In Rutherford's nuclear model (the one adopted by Bohr) this gives, for the hydrogen atom, a Keplerian elliptic motion of the electron around the nucleus. For an infinitely heavy nucleus, the total mechanical energy $E$ (taking $E_\infty = 0$) is related to the orbital frequency $\bar{v}$ through

$$(-E/\bar{v})^3 = \pi^2 \mu e^4/2\bar{v}, \tag{3}$$

where $\mu$ is the mass of the electron and $e$ its charge. Combined with the rule (2), this relation gives[5]

$$E_n = -Kh/n^2 \quad \text{with } K = 2\pi^2 \mu e^4/h^3. \tag{4,4'}$$

5. Bohr 1913, 4–5. The system of units (Gaussian) is such that the Coulomb energy is given by $-e^2/r$.

The stationary states being now completely determined, the energy spectrum of the hydrogen atom can be calculated by a direct application of assumption (II). For a transition from $n'$ to $n''$, the frequency $\nu_{n'n''}$ of the emitted radiation satisfies

$$E_{n'} - E_{n''} = h\nu_{n'n''} \tag{5}$$

if one-quantum processes are the only ones permitted. This gives

$$\nu_{n'n''} = K\left(\frac{1}{n''^2} - \frac{1}{n'^2}\right), \tag{6}$$

in conformity with Balmer's empirical formula.

Altogether, the above reasoning hardly provided a *deduction* of the hydrogen spectrum, since it relied on a loose, even inconsistent, analogy with Planck's radiation theory. For instance, the process of formation of the stationary state $n$ was taken to be a $n$-quantum radiation process, whereas in the derivation of the series spectrum the quantum transitions were taken to induce a one-quantum radiation process. Bohr, always excellent at self-criticism, immediately perceived the fault and mended it in the last part of the *same* paper. There he abandoned his tentative conception of the formation of atoms and introduced instead the idea of an asymptotic agreement between quantum theory and classical electrodynamics. With this new constraint he was able to re-derive the value of $K$ (Rydberg's constant) and also to discover the correct correspondence between orbital frequencies and radiation frequencies.

In general, the frequency of the radiation emitted in a transition between two successive stationary states is different from its orbital frequency in the original state (and from any harmonic of it), in sharp contrast with classical electrodynamics, as Bohr put it. However, in the limit of high quantum numbers, for which the orbital motion is very slow, one should expect an approximate equivalence between these two frequencies. In conjunction with relation (5), this implies

$$E_n - E_{n-1} \sim h\bar{\nu}_n \quad \text{or} \quad dE_n/dn \sim h\bar{\nu}_n. \tag{7,7'}$$

Assuming the form $E_n = -Kh/n^2$ (now taken from the empirical series), one has[6]

$$2Kh/n^3 \sim h\bar{\nu}_n, \quad \text{then } 2E_n/n \sim -h\bar{\nu}_n. \tag{8,8'}$$

The latter relation, being identical with Bohr's original quantum rule (2), leads to the same expression (4') of the Rydberg constant as before. More

---

6. Ibid., 8–9. As he later realized (Bohr 1914), Bohr could have *derived* the asymptotic form of $E_n$ by integrating the differential equation resulting from (3) and (7').

generally, for transitions in which $n$ decreases by a relatively small amount $\tau$, the following asymptotic relation holds:

$$E_n - E_{n-\tau} \sim \tau h \bar{v}_n, \qquad (9)$$

that is to say, the frequency of the emitted radiation nearly equals the harmonic $\tau$ of the orbital frequency. Bohr commented:

> The possibility of an emission of a radiation of such a frequency $[\tau\bar{v}_n]$ may also be interpreted from analogy with the ordinary electrodynamics, as [read "since" instead of "as"] an electron rotating round a nucleus in an elliptical orbit [the circular orbit would give only harmonic oscillations] will emit a radiation which according to Fourier's theorem can be resolved into homogeneous components, the frequencies of which are $[\tau\bar{v}]$, if $[\bar{v}]$ is the frequency of revolution of the electron.[7]

This was the first germ of the analogy later extended under the name of the correspondence principle. In 1913, however, the theory was still at an exploratory stage. The basic assumptions were too unsettled to allow for fruitful exploitation of formal analogies. A little after the publication of his trilogy, Bohr expressed in private great doubts about the "horrid assumptions" of his new theory and about the possibility of a generalization to systems more complex than the hydrogen atom or the harmonic oscillator:

> I tend to believe that in this problem there are buried very considerable difficulties, which can be avoided only by departing from the usual considerations to an even greater extent than has been necessary up to now, and that the preliminary success is due only to the simplicity of the systems considered.

This kind of utterance, a Kierkegaardian readiness to forecast great conceptual leaps, would recur whenever Bohr perceived a state of crisis—for instance in 1922–23, when the Bohr-Sommerfeld theory underwent difficulties, and in the late twenties, when the atomic nucleus seemed to disobey quantum-mechanical laws.[8]

## CONFIRMATIONS AND PERTURBATIONS

Although not mentioned in the above account of the basic assumptions of Bohr's new atomic theory, the main object of the 1913 trilogy was to

7. Bohr 1913, 14.
8. Bohr to McLaren, 1 Sept. 1913, BCW 5:[91]; Bohr to Oseen, 28 Sept. 1914, BCW 2:[557]–[560].

develop a new ring model, a substitute for Thomson's models of atoms and molecules (to which I will later return), in an endeavor to recover the properties of Mendeleev's table of elements. This part of Bohr's work was too speculative to receive unambiguous empirical confirmation. Nevertheless, in the years 1913–1914, various events improved Bohr's confidence in the basic truth of at least part of his "horrid assumptions." These assumptions appeared to give the first germs of a successful theory of atomic spectra when applied to the spectra of one-electron systems (H, He$^+$) and to the new field of X-ray spectra.[9]

Furthermore, Franck and Hertz unwittingly provided an independent confirmation of the concept of stationary state. In 1914 they observed an energy threshold in the electron-stopping power of mercury vapor. This threshold, Bohr explained the following year, corresponded to a transition between the normal state and another (first excited) stationary state of the mercury atom (and not, as Franck and Hertz had originally thought, to the ionization of the atom). The supramechanical stability implied by the assumption of stationary states was now empirically proved to comprehend stability with regard to electron impacts.[10]

In a paper published in 1914 Bohr examined the effect of electric and magnetic fields on the spectrum of the hydrogen atom. According to Stark's observations in the electric case, every spectral line split into a number of components; the separation between the components was proportional to the intensity of the applied field, with a proportionality coefficient depending on the line. The available method of quantization was too restricted to permit a complete explanation of this effect. Bohr therefore limited his investigation to large values of the quantum number $n$ and to strictly periodic motions. In this case he could derive the quantized energy levels $E_n$ from the relation (7'):

$$dE_n/dn \sim h\bar{v}_n,$$

which he named "correspondence principle" in one of his manuscripts. This is, as far as I know, the earliest occurrence of this expression in Bohr's writings; at this early stage it only meant the asymptotic agreement between the quantum-theoretical and the classical values of the frequency of the emitted radiation.[11]

9. See Heilbron and Kuhn 1969.
10. Bohr 1915b, 410–411. The relevant transition of the mercury atom was, in modern notations, 6 $^1S_0 \to$ 6 $^3P_1$ (resonance line at 2537 Å). For the history of this episode, see Bohr [1920a], [237]–[239]; Hon 1990.
11. Bohr 1914. For Stark's empirical discovery see Jammer 1966, 106–107. The mentioned manuscript is in BCW 2:[382].

For the hydrogen atom in an electric field $\mathbf{X}$, there are only two periodic motions of a given amplitude $2a$. The corresponding orbits are rectilinear and parallel to the field; the total energy $E$ and the frequency $\bar{\nu}$ are given by the formulae

$$E = -\frac{e^2}{2a} \pm 2aeX, \qquad \bar{\nu}^2 \sim \frac{e^2}{4\pi^2 \mu a^3} (1 \pm 3Xa^2/e). \qquad (10,10')$$

The elimination of $a$ and the "correspondence principle" $(7')$ give doubled energy levels:

$$E_n \sim -\frac{Rh}{n^2} \pm \frac{3Xh^2}{8\pi^2 \mu e} n^2. \qquad (11)$$

According to the relation $\Delta E = h\nu$, a double level-splitting should imply a quadruple line-splitting. However, Bohr retained only the doublet

$$\nu_{n'n''} = \nu^0_{n'n''} \pm \frac{\Delta\nu}{2}, \qquad \text{with } \Delta\nu \sim \frac{3}{4\pi^2} \frac{h}{e\mu} X(n'^2 - n''^2). \qquad (12,12')$$

His justification deserves special attention: "In order to obtain the continuity necessary for a connection with ordinary electrodynamics, we have assumed that the system can pass only between the different states in each series." Bohr meant that a transition between a level with the $+$ sign and a level with the $-$ sign would have corresponded to a *discontinuous* change in the orbit, even in the limit of high quantum numbers for which the classical principle of continuity ought to hold approximately. This is the first historical example of a derivation of a selection rule on the basis of an assumed correspondence between properties of the classical motion and quantum-theoretical radiation. Interestingly, the precise form taken here by this correspondence requirement, an identification of the connectivity (in the sense of the possibility of quantum transitions) of stationary states with the connectedness (in the topological sense) of classical motions in these states, was the one later used by Bohr in his application of the correspondence principle to the helium atom.[12]

Within the explicit high-$n$ limit, Bohr's theory of the Stark effect of hydrogen was a successful one; it gave the observed proportionality of the splitting to the field intensity, the approximative position of the two extreme components (the dominant ones) of the multiplets observed in the case of the lines $H_\beta$ and $H_\gamma$, for which the quantum numbers are not too

---

12. Bohr 1914, 515.

small ($n' = 4, 5$; $n'' = 2$), and the correct polarization of these two components (the latter again being determined by analogy with the classical motion).[13]

To Bohr the Zeeman effect was a different matter. In the case of hydrogen, Lorentz's classical model of an elastically bound electron or Rutherford's classical model led to the observed triple splitting with a linearly polarized unshifted central line and two circularly polarized lines shifted by

$$\delta v = \pm eB/4\pi\mu c, \tag{13}$$

where $B$ is the intensity of the magnetic field. This frequency shift was independent of the quantum number $n$, and the resulting spectrum violated Ritz's combination principle, according to which spectral lines are generated by taking all possible differences of series of spectral "terms" (as implied by Bohr's relation $v = \Delta E/h$). Furthermore, Bohr mistakenly believed that classically, the total energy of the atom remained unchanged during the application of a magnetic field. All these circumstances seemed to exclude an explanation of the Zeeman effect of the same nature as that given for the Stark effect.[14]

Consequently, Bohr interpreted the Zeeman effect as a violation of assumption (II), the rule $\Delta E = hv$. "In order to obtain the connection with the ordinary mechanics" this rule had to be modified into

$$\Delta E = h(v \pm \delta v). \tag{14}$$

More generally, until Sommerfeld inaugurated the quantum theory of multiperiodic systems, Bohr believed that the relation $\Delta E = hv$ applied only to strictly periodic motions. In other cases, he wrote in 1916, "we cannot, even in the limit [of large quantum numbers] obtain a relation between

---

13. See Bohr 1914, 516. The later energy formula based on the Bohr-Sommerfeld method was

$$E_n = -(Kh/n^2) - (3Xh^2/8\pi^2\mu)nn_e$$

with $n_e = 0, \pm 1, \pm 2, \ldots, \pm(n-1)$.
In the case of $H_\beta$ and $H_y$ the Stark components with maximal intensities correspond to $n_e = \pm(n-1)$, which provides approximate agreement with Bohr's first theory. Note that the orbits with $n_e = \pm n$—the only ones originally considered by Bohr—were precisely the ones excluded in the Bohr-Sommerfeld theory, on the ground that they implied a collision between the electron and the nucleus.
14. Bohr 1914, 518–520. Bohr believed that the increase of the kinetic energy due to Larmor's precession was compensated by a negative "magnetic potential energy" corresponding to the work of the electric field $E = -(1/c)\,\partial A/\partial t$ during the turning on of the magnetic field. Such is not the case: the total energy of a charged particle in a Coulomb potential $V(r)$ is always $\frac{1}{2}\mu v^2 + V(r)$, *even in the presence of a magnetic field.*

the frequency of the radiation and the motion of the system if we assume that the radiation is monochromatic."[15]

PERIODIC SYSTEMS

In 1915 Bohr undertook to consolidate the general foundations of the quantum theory. The proliferation of careless calculations and speculations based on Bohr's model had made such a task necessary. For instance, in early 1915 Debye and Sommerfeld had published calculations of the dispersion of light by the Bohr atom, in which they assumed classical electrodynamics to apply to the perturbation of stationary orbits. The results contradicted empirical data. Bohr and his friend Oseen were not at all surprised by this failure: classical electrodynamics, since it failed to explain the stability of stationary states, could a fortiori not be expected to represent correctly their reaction to external radiation. Observing that a theorist of Sommerfeld's stature could overlook such a central point, Bohr was compelled to clarify his positions.[16]

The results of this general clarification were ready for publication in early 1916, under the title "On the application of the quantum theory to periodic systems." On this occasion Bohr first isolated the "fundamental assumption" of his quantum theory, namely,

> that an atomic system can exist permanently only in a certain series of states corresponding with a discontinuous series of values of its energy, and that any change of the energy of the system including absorption and emission of electromagnetic radiation must take place by a transition between two such states. These states are termed "the stationary states" of the system.

From then on this would be the first, most unshaken postulate of Bohr's atomic theory, in spite of (or because of) its "entirely negative character." The other assumptions, those necessary to positively determine stationary states, emitted radiation, and statistical properties, could be specified only in particular cases, and were not well ascertained. In 1915–16, in conformity with his previous investigation of the Zeeman effect, Bohr did not even believe in the generality of the relation $\Delta E = h\nu$ (which would nevertheless become his second postulate a year later).[17]

When he wrote this essay, Bohr could formulate positive assumptions only for the quantization of strictly periodic systems. In this case he could

15. Bohr 1914, 520; Bohr [1916], [445].
16. Debye 1915; Sommerfeld 1915a; Oseen 1915; Bohr to Oseen, 20 Dec. 1915, BCW 2:[564]–[566]. See Hoyer's comments in BCW 2:[336]–[339].
17. Bohr [1916], [434], [445].

rely on Ehrenfest's adiabatic hypothesis (1913) for deriving a general quantum rule, that is, a prescription to derive the energy of the stationary
states of the system. Granted that stationary states are obtained through
a discrete selection among the classically possible motions of the system,
Ehrenfest's hypothesis further required "permitted" motions to transform
into other permitted motions during infinitely slow ("adiabatic") changes
of the forces acting in and on the system. Bohr found the assumption
most natural. Indeed, part of it, the continuous transformability of stationary states during slow variations of external conditions, could be viewed
as a "direct consequence of the necessary stability of these states," while
the rest of it, the applicability of ordinary mechanics to the slowly trans-
forming system, was hardly less plausible than the similar assumption for
the electronic motion of a given system in a stationary state. Bohr once
justified the latter point:

> If . . . the variation [of the external conditions] is performed at a constant or
> very slow changing rate, the forces to which the particles of the system will
> be exposed will not differ at any moment from those to which they would be
> exposed if we imagine that the external forces arise from a number of slowly
> moving additional particles which together with the original system form a
> system in a stationary state.[18]

Take the simplest possible case, a quantized harmonic oscillator. As
results from an old theorem of Boltzmann (to be later stated in detail),
an adiabatic change of the parameters (mass or elastic constant) does not
alter the ratio $E/\bar{v}$, so that the oscillator remains in a stationary state.
According to Ehrenfest's adiabatic hypothesis, this property admits a wide
generalization: under *any* (nonsingular) adiabatic deformation of the original oscillator—including ones that destroy the harmonicity—the system
remains in a stationary state.[19]

Consider now an arbitrary mechanical system with one degree of freedom allowing finite motions. Then any finite motion (which does not converge toward a fixed point) is periodic (as is shown on p. 107) and is
adiabatically connected to the motion of a harmonic oscillator. Consequently, the stationary states of the system are completely determined by

---

18. Ehrenfest 1914a. See Klein 1964b, 1970c, and Ehrenfest 1923b. Ehrenfest used the
word "adiabatic" because the theorem by Boltzmann on which his consideration was
based had been established in the context of formal mechanical analogies for the relation
$dS = \delta Q/T$. In these analogies a slow variation of a parameter of a simple periodic mechanical system represented an adiabatic transformation of a thermodynamic system. Quotations
from Bohr to Ehrenfest, 5 May 1918 (AHQP), and Bohr 1918a, 8, also in BCW 3:[74].
    19. Ehrenfest 1914a; Boltzmann's theorem is in Boltzmann 1904, 2:182.

the quantization of the latter oscillator. Even better, the quantum rule for the original system can be explicitly written as

$$\bar{T} = nh\bar{v}/2, \tag{15}$$

where $\bar{v}$ is the frequency of the motion and $\bar{T}$ is the average of the kinetic energy over one period. Indeed, $\bar{T}/\bar{v}$ is adiabatically invariant according to the above-mentioned theorem of Boltzmann, and in the case of a harmonic oscillator it is identical with $E/2\bar{v}$. Since this invariance also holds for any number of degrees of freedom, Bohr assumed the same form of the quantum rule to apply to any periodic system (i.e., systems for which all bounded motions are periodic).[20]

A simple proof of Boltzmann's theorem given by Bohr goes as follows.[21] The system is described by the canonical coordinates $q = (q_1, \ldots q_i, \ldots q_s)$ and $p = (p_1, \ldots p_i, \ldots p_s)$, which satisfy Hamilton's equations

$$\dot{q} = \partial H/\partial p, \qquad \dot{p} = -\partial H/\partial q, \tag{16}$$

where $H(q, p)$ is the Hamiltonian function, and the collective derivatives are defined as

$$\frac{\partial}{\partial p} = \left( \frac{\partial}{\partial p_1}, \frac{\partial}{\partial p_2}, \ldots \frac{\partial}{\partial p_i}, \ldots \frac{\partial}{\partial p_s} \right), \text{ etc.} \tag{17}$$

The kinetic energy $T$ being a quadratic function of $p$,

$$2T = p \cdot \frac{\partial T}{\partial p} = p \cdot \frac{\partial H}{\partial p} = p \cdot \dot{q}, \tag{18}$$

where the dot products are defined according to

$$p \cdot \dot{q} = \sum_i p_i \dot{q}_i, \text{ etc.} \tag{19}$$

This implies

$$2\bar{T}/\bar{v} = \oint 2T \, dt = \oint p \cdot dq. \tag{20}$$

I call the latter integral $I$. Its adiabatic invariance is proved as follows:

Consider periodic functions $q(t) + \delta q(t)$ and $p(t) + \delta p(t)$, which are infinitely close to solutions $q(t)$ and $p(t)$ of Hamilton's equations for $H$; these functions do not have to be solutions of the original equations of motion, and their common period does not have to be identical with the original period. Since the integral $I$ is independent of the choice of the

20. Bohr [1916], [435].
21. Ibid., [435]–[436]. A more explicit version of the same proof is in Bohr 1918a, 10–12, also in BCW 3:[76]–[78].

integration parameter, let us temporarily choose this parameter with a range independent of the period of $q(t)$ (for instance, $\bar{v}t$ would do). Then the variation of $I$ during the infinitesimal variation of $q$ and $p$ is simply given by

$$\delta I = \oint (\delta p \cdot dq + p \cdot d\delta q) = \oint (\delta p \cdot dq - \delta q \cdot dp), \qquad (21)$$

where the last expression has been obtained by partial integration of the second term. We now switch to the integration parameter $t$ (the time), in order to get

$$\delta I = \oint (\delta p \cdot \dot{q} - \delta q \cdot \dot{p}) \, dt \qquad (22)$$

and, with the help of Hamilton's equations,

$$\delta I = \oint \left( \frac{\partial H}{\partial p} \cdot \delta p + \frac{\partial H}{\partial q} \cdot \delta q \right) dt = \oint \delta H \, dt. \qquad (23)$$

We now further assume that $q + \delta q$ and $p + \delta p$ have been obtained from $q$ and $p$ through a Hamiltonian evolution from $t = -\theta$ to $t = 0$, with the Hamiltonian function

$$F = H + \lambda(t) \, \delta V, \qquad (24)$$

wherein $\lambda$ is a parameter varying slowly and smoothly from 0 to 1:

$$\lambda(t) = 0 \quad \text{for } t = -\theta, \qquad \lambda(t) = 1 \quad \text{for } t \geqslant 0, \qquad (25)$$

$$\theta \bar{v} \gg 1, \qquad \text{and} \qquad |\ddot{\lambda}/\dot{\lambda}\bar{v}| \ll 1 \quad \text{for } -\theta \leqslant t \leqslant 0. \qquad (26)$$

During this evolution the original Hamiltonian $H$ varies according to

$$\frac{dH}{dt} = \frac{d(F - \lambda \, \delta V)}{dt} = \frac{\partial F}{\partial t} - \frac{d}{dt} (\lambda \, \delta V) = -\lambda \frac{d}{dt} \delta V. \qquad (27)$$

Integration with respect to $t$ gives

$$\delta H(t) = H(t) - H(-\theta) = \int_{-\theta}^{t} -\lambda(t') \frac{d \, \delta V}{dt'} \, dt' \qquad (28)$$

or, after partial integration, and for $t \geqslant 0$,

$$\delta H(t) = -\delta V(t) + \int_{-\theta}^{0} \dot{\lambda}(t') \, \delta V[q(t')] \, dt'. \qquad (29)$$

In the latter integral, $q(t')$ may be replaced by its nonperturbed value (obtained by putting $F = H$), if only first-order contributions to $\delta H$ are retained. Then $\delta V[q(t')]$ is a periodic function of $t'$ with the period $1/\bar{v}$; since the variation of $\dot{\lambda}$ is negligible during such a period, $\delta V$ can be replaced in the integral by its average value $\overline{\delta V}$ over a period. This gives

$$\delta H(t) = -\delta V(t) + \overline{\delta V} \int_{-\theta}^{0} \lambda(t')\, dt' = \overline{\delta V} - \delta V(t). \tag{30}$$

The adiabatic invariance of $I$ results from combining (23) and (30):

$$\delta I = \oint \delta H(t)\, dt = \overline{\delta V - \delta V}/\overline{v} = 0. \tag{31}$$

In the above proof of the adiabatic theorem, Bohr accorded a special importance to the identity (23):

$$\delta I = \oint \delta H\, dt,$$

for it granted in the following way coherence between the quantum rule $I = nh$ and the postulate of stationary states. Consider two neighboring periodic motions of the same periodic system. Then $H$ is a constant for both motions, and the identity (23) reduces to

$$\delta H = \overline{v}\,\delta I, \tag{32}$$

which implies that $H$ depends only on $I$ (as long as the space of periodic motions of the system is connected in the topological sense). In this way the quantum rule completely determines the energy of the system. Bohr emphasized the importance of this result in the following terms:

> It will be seen that [the dependence of the energy on $I$ only] constitutes a necessary condition for the application of ordinary mechanics to the stationary states of periodic systems. Otherwise, we could by suitable variations of the external conditions make the systems emit or absorb energy, without a transition between stationary states corresponding to different values of $n$.[22]

This remark is very typical of Bohr's striving for mutual compatibility among his assumptions. Quantum theory, as based on the postulate of stationary states, was utterly anticlassical. Therefore, the importation of a whole piece of classical theory (for the determination of the motion in stationary states) could only be allowed insofar as the involved classical concepts did not conflict with the concept of stationary states. In judging this condition the adiabatic principle played an essential role, and so would, some time later, the correspondence principle. It was indeed part of Bohr's greatness to realize that successful heuristic principles also had something to say about the inner consistency of new theoretical schemes.

In the same text of 1916, Bohr generalized the idea of asymptotic correspondence, introduced in 1913, to any strictly periodic system. Here

22. Bohr [1916], [437].

again the identity (23) provided the necessary connection, when applied to two neighboring stationary states of a given system, $l = nh$ and $l = (n - \tau)h$. In this case too the relation (32) holds, for both motions derive from the same Hamiltonian $H$. This gives

$$E_n - E_{n-\tau} \sim \bar{\nu}_n[nh - (n - \tau)h] = h\tau\bar{\nu}_n, \qquad (33)$$

which means again an asymptotic agreement between the frequencies of the light emitted in the quantum transitions from a given stationary state, and the harmonics of the corresponding orbital frequency.[23]

The above essay never reached publication, because while he was holding proofs from the *Philosophical Magazine* Bohr became aware of two spectacular developments of his quantum theory: Sommerfeld and his followers had found quantum rules for nonperiodic systems, and Einstein had furthered the conceptualization of quantum radiation processes. Bohr immediately realized the possibility and necessity of extending the foundations of quantum theory.

## SUMMARY

Bohr's first atomic theory emerged in 1913 from an endeavor to explain the properties of chemical elements on the basis of Rutherford's new planetary model of atoms. While the most obvious property expected from real atoms was their stability with respect to external perturbations, Bohr found that Rutherford's model was unstable, both mechanically and electrodynamically. Not discouraged by this conflict, he proposed a supramechanical notion of stability that was embodied in his concept of "stationary state." By definition the stationary states were subjected to the following assumptions, which were mostly suggested by the quantum theory of Planck and Einstein, and the simple regularities of the hydrogen spectrum:

1. an atom can exist permanently only in a discontinuous series of stationary states.

1′. transitions between stationary states occur very suddenly, and therefore elude description in terms of ordinary mechanics and electrodynamics.

2. the motion of electrons in a stationary state is determined by applying ordinary mechanics to Rutherford's model.

3. the frequency of the radiation emitted or absorbed during a transition between two stationary states is given by the difference in energy

23. Ibid., [443]–[444].

between these states divided by Planck's constant (this is the frequency rule: $\Delta E = h\nu$).

Assumption (1) was obviously anticlassical but could be regarded as an extension of Einstein's ideas on quantization. Assumption (3) was more revolutionary, for it contradicted the hitherto unquestioned identity between the frequency of the motion in a source and the frequency of the radiation emitted by the source.

Assumptions (1) and (2) were not sufficient to determine the energy of stationary states. Bohr therefore introduced a "quantum rule" in order to select stationary motions from among the continuous manifold of classically possible ones. For the simple case of the hydrogen atom Bohr's 1913 memoir contained two conceptions of the quantum rule, both of which involved some analogy with classical electrodynamics. In the first, Bohr considered the spontaneous synthesis of the $n^{\text{th}}$ stationary state from an electron at rest at infinity and a hydrogen nucleus. Assuming (2) and an $n$-quantum version of (3) (namely, $\Delta E = nh\nu$), the energy of the $n^{\text{th}}$ stationary state of the hydrogen atom was obtained by setting the (homogeneous) frequency of the emitted radiation equal to half the final orbital frequency. In doing this Bohr renounced the exact classical relation between motion and emitted radiation but required an *analogous* relation to subsist at the quantum level. Such a requirement was the vaguest meaning of what Bohr would later call "correspondence principle."

A more precise anticipation of the correspondence idea appeared in Bohr's other conception of the quantum rule. Using his frequency rule, Bohr inferred from the empirically known properties of the hydrogen spectrum that the energy of the $n^{\text{th}}$ stationary state was proportional to $1/n^2$; and he derived the proportionality factor from the following condition: in the limit of high quantum numbers, for which stationary states are very close to one another, the frequency spectrum given by quantum theory must be almost identical with the spectrum which classical electrodynamics would yield (when applied to the motion in the initial state). More precisely, Bohr found that the transition between the stationary states $n$ and $n - \tau$ corresponded to the $\tau^{\text{th}}$ harmonic of the motion in the $n^{\text{th}}$ stationary state (in the following I will use quotation marks to imply this precise meaning of the verb "correspond"). This remark was essential to the later formulation of the correspondence principle.

Bohr's ambition was not limited to the hydrogen atom. His memoir of 1913 also pretended to give clues about more complex atoms, and even molecules. However, early confirmations of Bohr's ideas occurred only for one-electron systems, hydrogen and ionized helium, and for the general idea of stationary state, with Bohr's interpretation of collision experiments (Franck and Hertz). This was enough to persuade Bohr, if not all theorists, that his theory contained a kernel of truth.

In 1914 he achieved another success in explaining the Stark effect of the hydrogen atom (the splitting of spectral lines induced by an external electric field). His theory, albeit partial and approximate (limited to high quantum numbers and to periodic motions), is of special historical interest, for it provided a first glimpse of a very important aspect of the correspondence principle, the derivation of selection rules for quantum transitions. For each value of the quantum number $n$ of an unperturbed stationary state, Bohr's calculation gave two different orbits, say $+$ and $-$. The frequency rule should, therefore, have given a quadruple splitting of every line. But this was at variance with the observed dominance of doublets (only the most intense lines were retained). Bohr resolved the contradiction by noting that transitions between $+$ and $-$ orbits had no classical counterpart, because the $+$ orbits were qualitatively very different from the $-$ orbits, even in the limit of high quantum numbers. Asymptotic agreement with the classical spectrum could only be obtained by "selecting" transitions between orbits of the same kind.

Bohr was less successful with the Zeeman effect (the line-splitting induced by a magnetic field). Zeeman spectra did not seem to be compatible with the frequency rule (because of what would later turn out to be the selection rule $\Delta m = 0, \pm 1$); further, Bohr (wrongly) believed that the energy of the classical orbital motion was not affected by the magnetic field. Consequently, he excluded an explanation of the type employed in the Stark case and instead replaced the rule $\Delta E = h\nu$ with a modified rule $\Delta E = h(\nu + \delta\nu)$; then he had no better way to determine the frequency shift $\delta\nu$ than taking it from Lorentz's theory of the (normal) Zeeman effect.

In 1916 Bohr judged the time to be ripe for a clarification of the foundations of his theory. The only fundamental assumption of atomic theory, he now believed, was that of the (supramechanical) stability of stationary states (1 and 1'). All other assumptions, including the frequency rule, he thought, could hold only in the case of periodic systems. For the sake of definiteness, however, Bohr limited his considerations to this special case and applied ordinary mechanics to the motion in stationary states. For periodic systems (like a harmonic oscillator or the hydrogen atom), ordinary mechanics, together with Ehrenfest's adiabatic hypothesis, led to an explicit quantum rule.

According to Bohr's adaptation of Ehrenfest's hypothesis (1913), slow continuous ("adiabatic") deformations of an atomic system kept the system in a stationary state, and they were governed by ordinary mechanics. As a corollary, in order to determine the stationary states of a given system one just had to imagine a continuous deformation leading from the original system to a simpler system. If the quantum rule was already known for the simpler one, one had the rule for the original.

Bohr adopted the adiabatic principle not only on account of its calculational benefits but also because it blended harmoniously with his concept of stationary state. Since atomic systems could rest only in stationary states, it followed that slow deformations of them had to proceed along stationary states. Moreover, if ordinary mechanics applied to the motion in stationary states, it also had to apply to the slow deformations, because the system responsible for the forces inducing the deformation could be associated with the atomic system to form a larger, closed system in a strictly stationary state. Last but not least, the quantum rule deduced from the adiabatic principle automatically provided the necessary asymptotic agreement between quantum and classical spectrum. These very Bohrian comments tended to show a subtle harmony between two antagonistic aspects of the new atomic theory: the anticlassical notion of stationary state, and the limited recourse to ordinary mechanics.

Bohr had just finished writing down the above considerations when he learned of important developments of his theory that had taken place abroad. For this reason he decided not to publish anything until he could reach a, now necessary, synthesis.

# CHAPTER VI

# Postulates and Principles

After Sommerfeld's and Einstein's eminent contributions, it took Bohr a year of unusually solitary thinking (owing partly to the lack of communication which the war brought about) to produce the correspondence principle and the first parts of "On the quantum theory of line spectra," a masterly exposition of what would be known as the Bohr-Sommerfeld theory. Among the novelties that permitted these developments, the first I will recount is the generalization of Bohr's quantum rule to a much larger class of mechanical systems, the so-called multiperiodic systems.

## MULTIPERIODIC SYSTEMS

Since Michelson's measurements of 1891, the hydrogen spectrum was known to exhibit a fine structure: that is to say, most of its spectral lines could be resolved into narrow multiplets. Considering how small it was, this structure could hardly be held against Bohr's theory of 1913. On the contrary, in 1915 Bohr looked for an explanation based on the relativistic correction of the electron mass on a circular orbit of his model. Sommerfeld also tried to explain the fine structure within Bohr's theory, but without relativity. He originally believed that a new quantum condition, when added to Bohr's, would produce the observed splitting.[24]

In the case of circular orbits the quantum rule (2), $2E_n/\bar{\nu}_n = -nh$, could be rewritten in terms of the polar angle and the conjugate momentum

24. Bohr 1915a, 334–335; Sommerfeld 1915b. On the empirical discovery of the fine structure see Jammer 1966, 91.

(the angular momentum) as

$$\oint p_\theta \, d\theta = kh, \quad (k = n) \tag{34}$$

a form earlier given by Bohr himself.[25] Sommerfeld, unlike Bohr, applied the latter to elliptic motions, together with the extra quantum rule:

$$\oint p_r \, dr = n'h, \tag{35}$$

which was to Bohr's rule what the canonical pair $(r, p_r)$ was to $(\theta, p_\theta)$. The resulting expression for the energy,

$$E_{n',k} = -\frac{Kh}{(n' + k)^2}, \tag{36}$$

was quite disappointing, since it provided nothing but a relabeling of Bohr's terms.

In 1916 Sommerfeld combined his idea with Bohr's appeal to relativity. In this case the motion is no longer strictly periodic: a slow rotation of the main axis around the center of force is superposed upon the Kepler motion; and the energy of the precessing ellipses, quantized according to the above rules, becomes

$$E_{n',k} = \mu c^2 \left[ \left( 1 + \frac{\alpha^2}{(n' + \sqrt{k^2 - \alpha^2})^2} \right)^{-1/2} - 1 \right], \tag{37}$$

with $\alpha = 2\pi e^2/hc$ (a derivation of these results will be given after the introduction of the Hamilton-Jacobi method). Since $\alpha \sim \frac{1}{137} \ll 1$, a good approximation of this formula is

$$E_{n',k} = -Kh \left[ \frac{1}{n^2} + \frac{\alpha^2}{n^4} \left( \frac{n}{k} - \frac{3}{4} \right) \right] \tag{38}$$

with $n = n' + k$. The second term seemed to provide the expected splitting. A quantitative agreement was reached with later experiments, after Bohr and Kramers had derived the necessary selection rules and intensities.[26]

A few months after the publication of Sommerfeld's results, Schwarzschild and Epstein justified and widely generalized the new quantum rules in two fundamental papers on the Stark effect of the hydrogen atom. They used analytical methods from celestial mechanics to quantize not only the hydrogen atom in an electric field but any multiperiodic system.[27] Since

25. For instance in Bohr 1913, 24–25, and Bohr [1916], [451].
26. Sommerfeld 1915c, 1916a. See Nisio 1973. For the difficulties of an empirical confirmation of Sommerfeld's formula see Kragh 1985.
27. Schwarzschild 1916; Epstein 1916a, 1916b. See Jammer 1966, 103–104.

these methods played an essential role in the formulation and exploitation of the correspondence principle, I will now present them in some detail. For the sake of clarity I will rely principally on a purified version to be found in the appendices of Sommerfeld's *Atombau,* with some improvements borrowed from Bohr and Kramers.[28] The reader already familiar with Hamiltonian mechanics and action-angle variables need read only the paragraph on quantum rules (pp. 110–111) and that on Bohr's golden rule (pp. 115–116).

Consider a mechanical system with the configuration $q = (q_1, \ldots q_i, \ldots q_s)$, the Lagrangian function $L(q, \dot{q}, t)$, and the action integral

$$S = \int_{t_0}^{t_1} L(q, \dot{q}, t) \, dt. \tag{39}$$

For fixed values of $q(t_0)$ and $q(t_1)$, the motion between $t_0$ and $t_1$ is given by Hamilton's principle $\delta S = 0$, which is fulfilled if and only if Lagrange's equations

$$\frac{d}{dt} \frac{\partial L}{\partial \dot{q}} - \frac{\partial L}{\partial q} = 0 \tag{40}$$

are satisfied. Alternatively, one may use a Legendre transformation of $L$,

$$L \to H = p \cdot \dot{q} - L \qquad \left( p = \frac{\partial L}{\partial \dot{q}} \right). \tag{41}$$

This gives the Hamiltonian function $H(q, p, t)$ and the canonical equations of motion:

$$\dot{q} = \frac{\partial H}{\partial p}, \qquad \dot{p} = -\frac{\partial H}{\partial q}. \tag{42}$$

$H$ represents the energy of the system; if $L$ does not explicitly depend on time, $H$ is a constant of the motion.

For a fixed value of $q(t_0)$, to a given value of $q_1$ and $t_1$ corresponds (in general) one and only one motion for which $q(t_1) = q_1$; the corresponding value of $S$ is noted $S(q_1, t_1)$. An explicit expression of the differential of this function results from the following reasoning. We first consider $t_1$ to be fixed and $q_1$ to vary by $\delta q_1$, and denote by $\delta q(t)$ the corresponding

---

28. Sommerfeld 1919, appendices; Bohr 1918a, par. 3: "Conditionally periodic systems," 16–36; Kramers 1919, 289–294. Burgers 1917 and Born 1925 gave the most general and extensive treatment of the quantization of multiperiodic systems. In this context the most useful treatise of classical mechanics is Goldstein 1956.

variation of the motion $q(t)$ between $t_0$ and $t_1$. The resulting variation of $S$ is

$$\delta S = \int_{t_0}^{t_1} \left( \delta q \cdot \frac{\partial L}{\partial q} + \delta \dot{q} \cdot \frac{\partial L}{\partial \dot{q}} \right) dt = \delta q \cdot \frac{\partial L}{\partial \dot{q}} \Big|_{t_0}^{t_1} - \int_{t_0}^{t_1} \left( \frac{d}{dt} \frac{\partial L}{\partial \dot{q}} - \frac{\partial L}{\partial q} \right) \cdot \delta q \, dt.$$

(43)

The second term vanishes by virtue of the Lagrange equations; the first gives $\delta S = p_1 \cdot \delta q_1$, since $\delta q_0 = 0$. Consequently,

$$\frac{\partial S}{\partial q_1} = p_1.$$

(44)

We now consider a simultaneous variation of $t_1$ and $q_1$, but along a given motion. Using (39), the resulting variation of $S$ reads:

$$\delta S = L_1 \, \delta t_1.$$

(45)

Using (44), the same variation reads:

$$\delta S = \frac{\partial S}{\partial t_1} \delta t_1 + p_1 \cdot \delta q_1 = \left( \frac{\partial S}{\partial t_1} + p_1 \cdot \dot{q}_1 \right) \delta t_1.$$

(45')

Consequently, we have

$$\frac{\partial S}{\partial t_1} = L_1 - p_1 \cdot \dot{q}_1 = -H_1.$$

(46)

Finally, combining (44) and (46) and omitting the index 1, we get

$$dS = p \cdot dq - H \, dt.$$

(47)

We now suppose the system to be conservative (i.e., $L$ does not depend explicitly on $t$). Since $H$ takes a constant value $E$ during a given motion, one may advantageously introduce the Legendre transformation

$$S \to S'(q, E) = S + Et$$

(48)

through which $t$ is eliminated and $E$ and $q$ become the natural variables:

$$dS' = p \cdot dq + t \, dE.$$

(49)

As results from the latter differential expression, the function $S'$ obeys the so-called Hamilton-Jacobi partial differential equation:

$$H\left( q, \frac{\partial S'}{\partial q} \right) = E.$$

(50)

Suppose that the general integral of the above equation has been found under the form $S'(q, \alpha, E)$, where $\alpha = (\alpha_1, \ldots \alpha_i, \ldots \alpha_{s-1})$ are integration constants (I have omitted a trivial additive constant in $S'$). Then taking the derivative of the Hamilton-Jacobi equation with respect to $\alpha_i$ gives

$$0 = \frac{\partial H}{\partial p} \cdot \frac{\partial}{\partial \alpha_i} \frac{\partial S'}{\partial q} = \dot{q} \cdot \frac{\partial}{\partial q} \frac{\partial S'}{\partial \alpha_i} = \frac{d}{dt} \frac{\partial S'}{\partial \alpha_i}. \tag{51}$$

Consequently, the derivative $\partial S'/\partial \alpha_i$ is a constant of the motion. The $s - 1$ equations

$$\beta_i = \frac{\partial S'}{\partial \alpha_i} (q, \alpha, E) \tag{52}$$

determine a trajectory in $q$-space, and the equation

$$\partial S'/\partial E = t, \tag{53}$$

the so-called equation of time, specifies the motion along this trajectory. Thanks to this remarkable theorem of Jacobi, the complete solution of the mechanical problem results from simple differentiations, once the general integral of the Hamilton-Jacobi equation is known.

The practical importance of this theorem comes from the fact that most solvable mechanical problems fall into a category to which Jacobi's method is well adapted: namely, one for which the Hamilton-Jacobi equation is "separable," meaning that for a proper coordinate choice it can be split into $s$ independent equations of the type

$$\left(\frac{dS_i}{dq_i}\right)^2 = f_i(q_i) \tag{54}$$

with

$$S' = \sum_i S_i(q_i). \tag{55}$$

I will now discuss two simple examples of such problems, and some important properties of the resulting motions.

TWO EXAMPLES

First consider a nonrelativistic system with only one degree of freedom and with the Lagrangian $L = \frac{1}{2}m\dot{q}^2 - V(q)$. The Hamilton-Jacobi equation is trivially separated as

$$\left(\frac{dS'}{dq}\right)^2 = 2m[E - V(q)], \tag{56}$$

which gives

$$S' = \pm \int \sqrt{2m[E - V(q)]} \, dq. \tag{57}$$

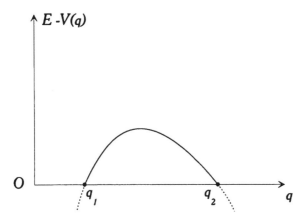

Figure 10. Form of the available kinetic energy leading to periodic motions.

The equation of time (53) then gives

$$t = \pm \int \sqrt{\frac{m}{2[E - V(q)]}} \, dq. \tag{58}$$

If we limit ourselves to motions capable of corresponding to stationary states, we have to exclude cases for which $q$ can reach infinity or can converge toward a fixed point. This supposes the existence of an interval $[q_1, q_2]$ in which $E - V(q)$ is positive and at the limits of which it vanishes, as in figure 10, and for which the integral

$$\int_{q_1}^{q_2} \frac{dq}{\sqrt{E - V(q)}} \tag{59}$$

has a finite value. Then $q$ is a monotonous function of time until it reaches either of the extremities of the above interval; at such a point the momentum $p = dS'/dq$ vanishes, and $q$ reverses its motion until it reaches the other extremity with zero velocity, and so forth. The resulting motion is periodic with the period

$$\frac{1}{\nu} = \int_{q_1}^{q_2} \sqrt{\frac{2m}{E - V(q)}} \, dq. \tag{60}$$

Our second example will be that of relativistic Kepler motion. In any relativistic motion the kinetic energy $T$ is related to the (rest) mass $m$ and to the momentum $p$ by

$$(T + mc^2)^2 = p^2 c^2 + m^2 c^4. \tag{61}$$

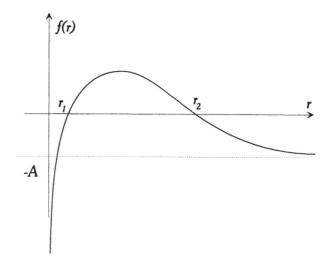

Figure 11.   Form of the radial kinetic energy in the Kepler problem.

This implies, for the (binding) energy $E = T - e^2/r$ in a Coulomb potential, the equation

$$\left(E + \frac{e^2}{r} + mc^2\right)^2 = p^2c^2 + m^2c^4.$$   (62)

Therefore, the Hamilton-Jacobi equation in polar coordinates $(r, \theta)$ in the plane of the trajectory reads:[29]

$$\left(E + \frac{e^2}{r} + mc^2\right)^2 = c^2\left(\frac{\partial S}{\partial r}\right)^2 + \frac{c^2}{r^2}\left(\frac{\partial S}{\partial \theta}\right)^2 + m^2c^4.$$   (63)

This equation is separable according to

$$S = S_\theta(\theta) + S_r(r),$$   (64)

with

$$dS_\theta/d\theta = p_\theta \qquad (\dot{p}_\theta = 0)$$   (65)

and

$$(dS_r/dr)^2 = f(r),$$   (66)

wherein $f(r) = -A + 2B/r - C/r^2$,

$$A = -\frac{(mc^2 + E)^2}{c^2} + m^2c^2,$$

$$B = \frac{e^2}{c^2}(E + mc^2), \qquad C = p_\theta^2 - \frac{e^4}{c^2}.$$   (67)

---

29. For the sake of brevity I assume the planarity of trajectories in central potentials to be already known.

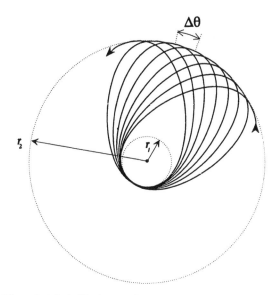

Figure 12.   The relativistic Kepler motion.

For bound motions the energy $E$ must be negative, which implies $A > 0$; for quantized motions, as will appear later, $p_\theta \geqslant h/2\pi$, which implies $C > 0$. Consequently the function $f(r)$ varies as indicated in figure 11.

Without recourse to Jacobi's theorem, the general aspect of the motion may be determined by the following simple consideration.

The component $p_r$ of the momentum has the form

$$p_r = \mu(r, E)\dot{r}, \tag{68}$$

where $\mu$ is the "relativistically increased mass," and it is related to the action $S_r$ by $p_r = dS_r/dr$. Combined with (66), this implies the differential equation

$$dt = \pm\mu(r, E)\, dr/\sqrt{f(r)} \tag{69}$$

for the time variation of $r$. By analogy with the case of the one-dimensional motion treated in the previous example, $r$ must be a periodic function oscillating between the limits $r_1$ and $r_2$ of the positive section of $f(r)$.

The constant $p_\theta = dS_\theta/d\theta$ represents the conjugate momentum of $\theta$, that is to say, the angular momentum, and is therefore given by

$$p_\theta = \mu(r, E)r^2\dot{\theta}. \tag{70}$$

From this equation it results that $\dot{\theta}$ is a periodic function of time, with the same frequency $\bar{\nu}_r$ as the function $r(t)$. Consequently, after each return of $r$ to its maximal value, the electron describes a portion of trajectory that is simply obtained by a global rotation of the previous portion. The resulting trajectory has the "rosette" shape given in figure 12.

Furthermore, if $\Delta\theta$ is the variation of $\theta$ during a period of $r(t)$ (which is called the advance of the perihelium), the angle

$$\theta' = \theta - \bar{v}_r \Delta\theta\, t \qquad (71)$$

is a periodic function of time with the same period (note that $\theta$ modulo $2\pi$ is *not* a periodic function). In the complex plane of the trajectory the position of the electron is therefore

$$re^{i\theta} = re^{i\theta'}e^{i\bar{v}_r\Delta\theta\, t}, \qquad (72)$$

the Fourier spectrum of which has only two fundamental frequencies, $\bar{v}_r$ and

$$\bar{v}_\theta = \bar{v}_r \Delta\theta/2\pi, \qquad (73)$$

with the harmonics $\tau\bar{v}_r \pm \bar{v}_\theta$, wherein $\tau$ is a positive integer. The motion is said to be biperiodic.


QUANTUM RULES

The following generalization of the above results holds for any separable Hamiltonian system: For coordinates $q$ that allow separation of the Hamilton-Jacobi equation, and for any motion in which none of these coordinates tends toward a fixed point (including infinity), each of the canonical couples $(q_i(t), p_i(t))$ repeatedly describes in the course of time a closed trajectory in the $(q_i, p_i)$-plane, provided that different values of $q_i$ leading to the same configuration of the system are identified (for example, $\theta$ and $\theta + 2\pi$, if $\theta$ is an angle). Then, even though the variation in time of these couples is in general not periodic, the motion is multiperiodic: that is to say, the configuration of the system may be expressed in terms of $s$ (or less) periodic functions of time (where $s$ is the number of degrees of freedom), as will be proved after the introduction of the action-angle variables.

For such multiperiodic motions a natural generalization of Bohr's quantum rule lies at hand. As we have seen, the rule (15) for a strictly periodic motion reads (using (20))

$$\oint p \cdot dq = \sum_i \oint p_i\, dq_i = nh. \qquad (74)$$

Since, in the separated multiperiodic case, $p_i = dS_i/dq_i$ is a function of $q_i$ only, it seems natural to split this rule into $s$ different rules

$$\oint p_i\, dq_i = n_i h, \qquad (75)$$

where the integrations are performed over the closed trajectories referred to in the above discussion of separable systems. In general, these conditions completely determine the energy of the system, since their number is equal to the number of parameters in the action function $S'$. It remains to prove that the resulting energy spectrum does not depend on the choice of the separating coordinates. This will be done later, after the introduction of the action-angle variables.

### QUANTIZATION OF THE RELATIVISTIC KEPLER MOTION

In this case the separating coordinates are the azimuth $\theta$ and the radius $r$. Accordingly, there are two quantum conditions. The azimuthal one reads

$$\oint p_\theta \, d\theta = 2\pi p_\theta = kh, \tag{76}$$

which expresses the quantization of angular momentum in terms of the "azimuthal quantum number" $k$. The radial condition reads

$$\oint p_r \, dr = n'h \tag{77}$$

or, with the notation introduced in (67),

$$J_r = \oint \sqrt{-A + 2\frac{B}{r} - \frac{C}{r^2}} \, dr = n'h. \tag{78}$$

The latter integral is easily computed through the method of residues. In the complex plane the radical has a "cut" along the real segment $[r_1, r_2]$ and two poles, at $z = 0$ and $z = \infty$. The integral $J_r$ is identical with the integral on the loop represented in figure 13, if only the square root is

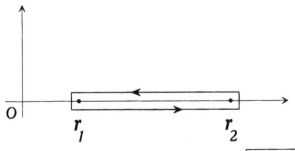

Figure 13.   The slit in the complex plane of the function $\sqrt{-A + 2\dfrac{B}{r} - \dfrac{C}{r^2}}$.

determined to be positive under the cut and negative above it. If this loop is considered to enclose the region outside the small rectangle, Cauchy's theorem gives

$$J_r = -2\pi i [\text{Res}(0) + \text{Res}(\infty)], \tag{79}$$

where

$$\text{Res}(0) = -i\sqrt{C} \tag{80}$$

and $\text{Res}(\infty)$ is the residue of $-(1/u^2)\sqrt{-A + 2Bu - Cu^2}$ for $u = 0$, that is,

$$\text{Res}(\infty) = iB/\sqrt{A} \tag{81}$$

The resulting expression for $J_r$ is

$$J_r = 2\pi\left(\frac{B}{\sqrt{A}} - \sqrt{C}\right). \tag{82}$$

Using the expressions (67) of $A$, $B$, $C$ in terms of $E$ and $p_\theta$, the two quantum rules imply the energy formula (37):

$$E_{n',k} = mc^2\left[\left(1 + \frac{\alpha^2}{(n' + \sqrt{k^2 - \alpha^2})^2}\right)^{-1/2} - 1\right] \tag{83}$$

with $\alpha = 2\pi e^2/hc$ for the "fine structure constant." This is, as Sommerfeld put it himself, "the royal road" to the Sommerfeld formula. Needless to say, his first derivation was more hesitating.[30]

### CANONICAL TRANSFORMATIONS

A little before his premature death, Schwarzschild found the method best suited to the determination of stationary states, namely, the introduction of the so-called action-angle variables. Unlike common users of analytical mechanics, astronomers like him sometimes favored this technique, for it provided direct access to the periods of celestial motions. The passage from the original canonical variables to the action-angle variables is a particular case of a more general type of transformation preserving the Hamiltonian structure of the equations of motion. I will first recall some general definitions and results about these transformations.

Since $q$ and $p$ play (anti)symmetrical roles in the equations of motion

$$\dot{q} = \frac{\partial H}{\partial p}, \qquad \dot{p} = -\frac{\partial H}{\partial q} \tag{84}$$

30. Sommerfeld 1919, 327–357, 520–522.

deriving from the Hamiltonian $H(q, p, t)$, a natural question is: What is the most general transformation from $(q, p, t)$ to $(Q, P, t)$ for which there exists a new Hamiltonian $K(Q, P, t)$ such that

$$\dot{Q} = \frac{\partial K}{\partial P}, \qquad \dot{P} = -\frac{\partial K}{\partial Q} \tag{85}$$

holds? The answer lies in the following theorem.

There exists a function $K$ if and only if the transformation $(q, p, t) \rightarrow (Q, P, t)$ is the result of a combination of the three following types of transformation. The first type simply involves re-scaling

$$(q, p) \rightarrow (\lambda q, \mu p) \tag{86}$$

and leads to $K = \lambda \mu H$. The second type involves a permutation

$$(q, p) \rightarrow (p, q) \tag{87}$$

and leads to $K = -H$. The third type consists of any transformation for which there exists a "generating function" $F(q, P, t)$ such that $p$ and $Q$ considered as functions of $q$ and $P$ are given by

$$p = \frac{\partial F}{\partial q} \quad \text{and} \quad Q = \frac{\partial F}{\partial P}. \tag{88}$$

The new Hamiltonian is then given by

$$K = H + \frac{\partial F}{\partial t}. \tag{89}$$

Combinations that do not involve a re-scaling are called *canonical transformations*. For an elementary proof of this theorem I refer the reader to Goldstein's textbook.[31]

ACTION-ANGLE VARIABLES

We now return to a conservative separable system. In a coordinate system for which the Hamilton-Jacobi equation and the action $S'$ are separated, the action variables are defined as

$$J_i = \oint p_i \, dq_i, \tag{90}$$

---

31. Goldstein 1956. The type of generating function defined here is not the one most commonly found in textbooks; but it is the most useful, since it includes transformations in the neighborhood of the identity.

where the integrations are performed over the cycles earlier introduced in the $(q_i, p_i)$-planes. Through the relations

$$p_i = \frac{dS_i}{dq_i}(q_i, \alpha, E) \qquad (91)$$

the $J$'s are in general in a one-to-one correspondence with the parameters $\alpha$ and $E$ and can therefore be taken as new parameters of the action, according to

$$S^*(q, J) = S'(q, \alpha, E). \qquad (92)$$

As results from $p = \partial S'/\partial q$, the function $S^*$ generates a canonical transformation from $(q, p)$ to $(w, J)$, with

$$w = \partial S^*/\partial J \qquad (93)$$

for the "angle variables."

Since $S^*$ does not explicitly depend on time, the new Hamiltonian is simply the old one expressed in terms of the new coordinates, or $E(J)$ (the energy of a given motion is completely determined by the action variables only). The new Hamiltonian equations are

$$\dot{J} = 0, \qquad \dot{w} = \partial E/\partial J. \qquad (94)$$

The second equation implies a linear variation in time of every angle variable.

The angle variables have another remarkable property. For a given choice of $J$, the partial variation $\Delta w_i$ of $w_i$ during a "full variation" of the coordinate $q_j$ (i.e., a variation for which the canonical couple $(q_j, p_j)$ completes a cycle) is

$$\Delta w_i = \oint \frac{\partial w_i}{\partial q_j} dq_j = \oint \frac{\partial^2 S^*}{\partial q_j \partial J_i} dq_j = \frac{\partial}{\partial J_i} \oint p_j \, dq_j = \delta_{ij}. \qquad (95)$$

Consequently, the configuration of the system is a periodic function of every $w_i$ with period unity. This is of course why the $w$'s are called angle variables, even though they generally are not angles in the geometric sense (as the reader will easily verify in the case of the relativistic Kepler motion). Furthermore, for a given motion the configuration is a multiperiodic function of time with frequencies

$$\bar{\nu}_i = \partial E/\partial J_i \qquad (96)$$

since, according to (94), the angle variables are linear functions of time with the rate $\partial E/\partial J_i$.

BOHR'S GOLDEN RULE

As can easily be proved, a system performing a multiperiodic motion returns as close as one wishes to its initial configuration after a sufficiently long time $T$. This is why a multiperiodic system is also called "conditionally periodic." Consider a nearly closed motion during the time $T$ and a neighboring motion of the same system. The relation (23) proved earlier for a strictly periodic system gives approximately

$$\delta I = \oint \delta H \, dt = T \, \delta H \tag{97}$$

The integral $I$ is related to the $J$'s by

$$I = \oint p \cdot dq = \sum_i N_i J_i, \tag{98}$$

where $N_i$ is the number of "cycles" of the couple $(q_i, p_i)$ during the time $T$. Therefore, relation (97) may be rewritten as

$$\delta H = \sum_i \frac{N_i}{T} \delta J_i, \tag{99}$$

where $T/N_i$ is the so-called "average period" of the coordinate $q_i$ (the variation of which is not periodical in general, as I repeatedly mentioned). In the case where the two neighboring motions are described in the same set of separating coordinates, another expression of $\delta H$ is obtained by taking the differential of the function $E(J)$ according to (96),

$$dE = \sum_i \bar{v}_i \, dJ_i. \tag{100}$$

Comparison with (99) gives $\bar{v}_i \sim N_i/T$ and thereby a more intuitive interpretation of the frequency $\bar{v}_i$ as the average number of cycles of the coordinate $q_i$ in a unit of time.

We now assume that the two neighboring motions are given in two different (but infinitely close) systems of separating coordinates. If, in spite of this change of coordinates, the corresponding $J$'s are given the same numerical values, the energy remains unchanged according to (99). Consequently, the energy spectrum obtained from the condition $J = nh$ does not depend on the choice of the separating variables, as long as all possible choices are connected continuously. This very elegant proof of the unambiguous character of the Bohr-Sommerfeld rules is due to Bohr.[32]

32. Bohr 1918a, 10–12, 22–23.

In the so-called nondegenerate case, for which the frequencies $\bar{v}_i$ are incommensurable, the arbitrariness in the choice of separating coordinates is limited; only transformations mixing each pair $(q_i, p_i)$ can be allowed, and the choice of the set of action variables is unique, as Schwarzschild proved. If there are, instead, $r$ (independent) relations

$$\tau_1 \cdot \bar{v} = \tau_2 \cdot \bar{v} = \cdots = \tau_r \cdot \bar{v} = 0 \tag{101}$$

with integral coefficients relating these frequencies, the following transformation is possible. First, the $w$'s can always be permuted in such a way that the $s - r$ last ones have incommensurable frequencies. Then the function

$$F(w, J') = \sum_{k=1}^{r} (\tau_k \cdot w)J'_k + \sum_{k=r+1}^{s} w_k J'_k \tag{102}$$

generates from $(w, J)$ new action-angle variables $(w', J')$ such that, for $1 \leqslant i \leqslant r$, $w'_i$ is a constant of any motion and $J'_i$ does not appear in the energy expression $E(J')$; for $i \geqslant r + 1$, $w'_i$ is identical with $w_i$. Consequently, the number of independent quantum conditions is always equal to the degree of periodicity of the system (that is, the number of independent frequencies).

To summarize, greatly benefiting the Bohr-Sommerfeld theory, the introduction of action-angle variables for separable Hamiltonian systems made it easy to derive several important properties: the multiperiodicity of all motions that do not converge toward a fixed point, the unambiguous character of the quantum rules, the degree of multiplicity of the resulting energy spectrum, and the relation (100),

$$dE = \sum_i \bar{v}_i \, dJ_i$$

which I will call "Bohr's golden rule" because it subsequently played a fundamental role in the formulation of the correspondence principle. Finally, as J. M. Burgers could show in his important dissertation (1918), action-angle variables were best suited to verify that quantum rules—or action variables—were adiabatically invariant, as required in Bohr's notion of stationary state. The following is a sketch of Burgers's reasoning, which can be omitted at first reading.[33]

ADIABATIC INVARIANCE OF THE ACTION VARIABLES

Suppose that the Hamiltonian of the system contains a parameter $\lambda$ and that the system is separable and multiperiodic for every value of $\lambda$. Then

33. Burgers 1917, 1918.

there exists a generating function $S^*(q, J, \lambda)$, defined as in (92), for new canonical variables $(w, J)$ depending on the parameter $\lambda$. Now assume that $\lambda$ is a function of time with zero value for $t \leqslant -\theta$, a very slow and smooth (in a sense to be later specified) increase for $-\theta \leqslant t \leqslant 0$ and the constant value $\varepsilon$ for $t \geqslant 0$. Before and after the variation of $\lambda$ the canonical variables generated by $S^*$ are action-angle variables. But during the variation of $\lambda$, their evolution is ruled by the new Hamiltonian (given by (89))

$$H' = H + \frac{\partial S^*}{\partial t} = E(J) + \dot{\lambda}\frac{\partial S^*}{\partial \lambda}, \tag{103}$$

and the $J$'s are no longer constants, as implied by the canonical equation

$$\dot{J}_r = -\frac{\partial H'}{\partial w_r} = -\dot{\lambda}\frac{\partial^2}{\partial \lambda\,\partial w_r} S^*[q(w, J, \lambda), J, \lambda]. \tag{104}$$

To first order in $\varepsilon$, $q(w, J, \lambda)$ and $S^*$ may be calculated as if $\lambda$ were constant. In this approximation $S^*$ increases by $J_r$ when $w_r$ increases by one unit, since for a cycle of the coordinate $q_r$

$$\oint dS^* = \oint p_r\,dq_r = J_r. \tag{105}$$

Consequently, $S^* - w \cdot J$ is a periodic function of each $w_r$ with period one. The same is true for the derivative

$$\partial S^*/\partial \lambda = \partial(S^* - w \cdot J)/\partial \lambda, \tag{106}$$

which therefore admits the Fourier development

$$\frac{\partial S^*}{\partial \lambda} = \sum_\tau a_\tau e^{2\pi i w \cdot \tau}. \tag{107}$$

The resulting Fourier series for the second derivative occurring in (104) is

$$-f_r(w) = \frac{\partial^2 S^*}{\partial \lambda\,\partial w_r} = \sum_\tau 2\pi i a_\tau \tau_r e^{2\pi i w \cdot \tau}. \tag{108}$$

After substitution of $w = \bar{v}t$, the time average of this expression over a long time (much longer than any period of the motion) is zero, unless there exists a sequence $\tau$ of integers for which $\bar{v} \cdot \tau = 0$ without $\tau_r$ being zero. Roughly, this singular case does not occur as long as we limit ourselves to transformations for which the degree of degeneracy of the system does not change.[34]

---

34. Unfortunately, this condition is never rigorously met, as realized by Burgers himself. Complete proofs were given as late as 1924 by Dirac and Laue (see part C, p. 306).

The total variation of $J_r$ during the adiabatic transformation is given by

$$\Delta J_r = \int_{-\theta}^{0} \dot{J}_r \, dt = \int_{-\theta}^{0} \dot{\lambda} f_r(\bar{v}t) \, dt. \tag{109}$$

If we take the variation of $\dot{\lambda}$ to be negligible during the periods of the motion, in the latter integral $f_r$ may be replaced by its average value over a large number of periods, which we just proved to be zero. This seals the proof of the invariance of the action variables $J$ for any adiabatic transformation that does not alter the degree of degeneracy of the system.[35]

The extension of Bohr's theory to multiperiodic systems raised a general wave of enthusiasm. As Sommerfeld and Born put it, the Hamiltonian formulation of classical mechanics almost seemed to have been created for the sake of quantum theory. The action variables of celestial mechanics permitted a strikingly simple expression of the quantum rules, and the theory of complex integration, a no less beautiful mathematical tool, appeared to be very well suited to the remaining calculations of the energy spectrum.[36]

In the following years, theoreticians of the Munich and Göttingen schools generally concentrated their attention on systematically carrying out the Bohr-Sommerfeld quantization procedure; they tended to neglect all aspects of quantum phenomena that did not fit into this well-defined mathematical framework (for instance, the intensities of spectral lines). As we shall presently see, Bohr reacted in a quite different way: in spite of his admiration for the concrete achievements of these schools, he emphasized the still provisional and incomplete character of the newly extended quantum theory; he insisted on a careful analysis of the degree of compatibility between the various physical concepts involved, and he concentrated his attention precisely on the questions to which the mathematical art of quantization by itself gave no answer.

## EINSTEIN'S TRANSITION PROBABILITIES

In addition to the quantization of multiperiodic systems, there was another important event that helped Bohr to formulate the correspondence principle: a new theory of thermal radiation proposed by Einstein in 1916.

35. Burgers generalized the notion of action-angle variables to nonnecessarily separable systems. In his new definition the canonical variables $(w, J)$ are action-angle variables if and only if they verify the properties enumerated on p. 114, and if they are generated by a function $S^*$ such that $S^* - w \cdot J$ is a periodic function of each $w_i$ (the latter condition preserves the proof of adiabatic invariance of the $J$ variables). Bohr and Born adopted this generalization, which enabled one to talk about action-angle variables without discussing Jacobi's problem.

36. Sommerfeld 1919, 520–522; Born 1925, v.

Most early proofs of Planck's blackbody law suffered from a fundamental inconsistency, which Einstein denounced as early as 1906.[37] On
the one hand the sources of radiation, generally Planck's resonators, were
quantized in a sharply anticlassical manner in order to reach the desired
formula for the resonator entropy; on the other hand, the same resonators
were assumed to interact classically with the electromagnetic radiation in
the derivation of the relation

$$u_v = \frac{8\pi v^2}{c^3} U \qquad (110)$$

between the spectral density $u_v$ and the average resonator energy $U$. In
1916 Einstein offered an in-depth resolution of this conflict by providing
a quantum-theoretical treatment of the interaction between the sources
of radiation—Bohr atoms in general—and the surrounding radiation.[38]

Like Bohr, Einstein assumed the existence of discrete stationary states
of atomic systems. Consider a homogeneous gas of such quantized atoms
at thermal equilibrium with radiation. According to statistical mechanics,
the canonical probability for a given atom to be in its stationary state $n$ is
proportional to $e^{-E_n/kT}$, if $E_n$ is the energy of this stationary state, and
$T$ the temperature.

In order to describe the interaction between the atoms and the surrounding radiation, Einstein relied on a natural analogy with classical electrodynamics and introduced two types of processes: *Ausstrahlung* (later
called "spontaneous emission" by Bohr), corresponding to the emission
of radiation by the oscillating atomic dipole by itself, and *Einstrahlung,*
corresponding to the interaction of this dipole with the radiation in which
the atom is immersed. The latter type of process is in turn decomposed
into a *negative Einstrahlung* (called "absorption" by Bohr) for which the
incoming radiation is in phase with the oscillation of the dipole, and a
*positive Einstrahlung* (called "stimulated emission" by Bohr) for which
the phases are opposed. Classically, the *Einstrahlung* probabilities are proportional to the density $\rho_v$ of the surrounding radiation at the natural
frequency $v$ of the atomic dipole.

In the absence of a detailed mechanism for these processes in the
quantum-theoretical case, Einstein, like Bohr, limited his consideration to
full, (almost) instantaneous, atomic transitions between pairs of stationary
states. According to the above analogy with classical electrodynamics, the
probability, per time unit and per atom, of the quantum jumps corresponding to the above-mentioned three types of processes had to take the

37. Einstein 1906.
38. Einstein 1916a, 1916b, 1917.

respective forms,

$$A_m^n, \; \rho_v B_n^m, \text{ and } \rho_v B_m^n,$$

$n$ and $m$ being the two stationary states involved, and $\rho_v$ the spectral density of the interacting radiation.

For the atoms and the surrounding radiation to be at thermal equilibrium, the number of quantum jumps from $m$ to $n$ must be equal to the number of reverse jumps, which gives[39]

$$B_n^m \rho_v e^{-E_m/kT} = (B_m^n \rho_v + A_m^n)e^{-E_n/kT}. \tag{111}$$

In the high temperature limit, for which $\rho_v \to \infty$, this condition degenerates into

$$B_m^n = B_n^m. \tag{112}$$

In general, taking into account the latter relation, equilibrium will be reached if and only if

$$u_v = \rho_v = \frac{A_m^n/B_m^n}{e^{(E_n - E_m)/kT} - 1}. \tag{113}$$

This is compatible with Wien's displacement law ($u_v = v^3 f(v/T)$, see p. 29) only if the energy difference $E_n - E_m$ is proportional to the frequency $v$ of the interacting radiation, which gives an independent confirmation of Bohr's frequency law. Furthermore, assuming $E_n - E_m = hv$, the Rayleigh-Jeans law ($u_v = 8\pi v^2 kT/c^3$) is recovered in the classical (low-frequency) limit if and only if the following relation between absorption and emission coefficients holds:

$$A_m^n/B_m^n = 8\pi hv^3/c^3. \tag{114}$$

With these two constraints Planck's blackbody law results from equation (113).

Einstein meant the second part of this study, a proof of the oriented character of the emission process,[40] to be his most essential contribution to quantum radiation theory. Instead, Bohr gave more importance to the new deduction of the blackbody law; for this deduction reinforced the basic assumptions of his atomic theory and completed them with a statistical description of radiation processes. He also emphasized, as I have done, the role of classical analogies in Einstein's demonstration. In ret-

---

39. The statistical weights (degrees of degeneracy) of the levels $m$ and $n$ are assumed to equal one.
40. Only in Einstein 1916b, and Einstein 1917, 124–128.

rospective comments he traced back the correspondence principle and Einstein's radiation theory to a common method, following which

> the attention is focused primarily on the emission and absorption processes and an attempt is made to draw certain general conclusions about these processes by comparing the assumptions underlying the quantum theory with the conceptions of classical electrodynamics to the extent to which the deductions from these laws have been essentially borne out of experience.[41]

## "ON THE QUANTUM THEORY OF LINE SPECTRA"

In late 1917 Bohr made up his mind to send "On the quantum theory of line spectra, part I: On the general theory" to the Proceedings of the Danish Academy, after, as he was wont to do, having written quite a few drafts.[42] In the introduction he explained how the new quantum rules and Einstein's radiation theory permitted an important extension of atomic theory. He also characterized his own contribution in the following terms: "On this state of the theory it might . . . be of interest to make an attempt to discuss the different applications [of the quantum theory] from a uniform point of view, and especially to consider the underlying assumptions in their relations to ordinary mechanics and electrodynamics." By the "relation with ordinary mechanics" Bohr essentially meant the adiabatic principle; by "the relation with electrodynamics" he meant the principle of analogy later called the "correspondence principle" and here hinted at in the following words: "It will be shown that it seems possible to throw some light on the outstanding difficulties by trying to trace the analogy between the quantum theory and the ordinary theory of radiation as closely as possible."[43]

In a previous draft of this introduction Bohr had been a little more specific about the significance of this analogy: "We shall see that the theory of line spectra based on $[\Delta E = h\nu]$ in a formal sense may be considered a natural generalization of the ordinary theories of radiation."

---

41. Bohr 1918a, 7; Bohr 1921b, 2.
42. "On the quantum theory of line spectra, part I: On the general theory" (Bohr 1918a) was ready for printing on 27 Apr. 1918. Bohr had originally planned four parts. Part II, "On the hydrogen spectrum" (Bohr 1918b), was ready for printing on 30 Dec. 1918. Part III, "On the spectra of elements of higher atomic number" (Bohr 1922b), although written in spring 1918, was printed only in November 1922, with an updating appendix of September 1922. Part IV, about the constitution of atoms and molecules, was never completed. See Rud Nielsen's introduction to BCW 3.
43. Bohr 1918a, 4.

Later on this idea of a "natural generalization" (or "rational generalization") became a leitmotiv of Bohr's atomic theory.[44]

No analogy can be constructive without a sufficiently sharp statement of the basic assumptions of the new theory, something that functions as the pivot of the analogy. In "On the quantum theory of line spectra" Bohr repeated almost word for word the assumption of stationary states of the withdrawn paper of 1916. However, he now added a second "fundamental assumption":

> that the radiation absorbed or emitted during a transition between two stationary states is "unifrequentic" [monochromatic] and possesses a frequency $v$, given by the relation $E' - E'' = hv$, where $h$ is the Planck constant and where $E'$ and $E''$ are the values of the energy of the two states under consideration.[45]

Bohr's original doubts about the generality of the second assumption had by now fallen away. Debye and Sommerfeld had been able to deduce the Zeeman effect in the hydrogen atom from the general quantization of multiperiodic systems. As will be presently recounted, Bohr himself had explained the apparent violation of Ritz's combination principle in this effect as a consequence of the correspondence principle. And Einstein's theory of radiation had given independent support to the relation $E' - E'' = hv$.

The other assumptions of the quantum theory, those concerning the motion in stationary states, the transition probabilities, the statistics of quantum states, and so on, were still regarded by Bohr as less fundamental. They were approximate, provisional, or incomplete, and would constantly need to be reexamined in the light of new experimental results. First of all, the application of ordinary mechanics to the motion of electrons in stationary states could be valid only in the approximation for which the interactions were given by Coulomb forces, and the coupling with the radiation field could be neglected, since, according to the quantum postulates, the latter coupling necessarily eluded classical theory. This explicit remark by Bohr was in fact very essential: on the one hand it made plausible a coherent *limited* use of classical mechanical concepts in the quantum theory; on the other hand it warned theoreticians against a blind application of classical electron orbits and even announced a necessary breakdown of this type of description as soon as finer details of atomic spectra would be considered.[46]

44. Unpublished manuscript, BCW 3:[48].
45. Bohr 1918a, 5.
46. Ibid., 6.

In the realm of atomic radiation Bohr regarded Einstein's assumptions about transition probabilities as consistent and even necessary. With his usual prudence, however, he did not entirely exclude the possibility of a finer description (necessarily beyond ordinary mechanics) of the process of transition. For instance, as late as 1922, he wrote: "*In the present state of the theory* the mode of occurrence of these transitions is considered to be a question of probability" (emphasis added). He even praised attempts like Whittaker's "to devise a mechanism which reproduces the characteristic features of the quantum theory." Nevertheless, he added, this type of consideration was "scarcely suited, from the nature of the case, to throw light on the actual applications in the present state of the theory." This comment reveals an important aspect of Bohr's general approach, an emphasis and concentration on the developments of the theory directed toward more fruitfully organizing and better encompassing of empirical results. As much as Einstein he struggled for clarity and consistency, but, unlike Einstein, he did not regard logical completeness as prior to empirical efficiency. He believed instead that progress was possible within a manifestly incomplete theory, under the guidance of organizing principles like the correspondence principle.[47]

## THE CORRESPONDENCE PRINCIPLE

The first systematic generalization of the correspondence idea is found in an early draft of "On the quantum theory of line spectra." There Bohr derived "selection rules" for the combination of spectral terms in the Zeeman and Stark effects and in the fine structure of the hydrogen atom. The empirical necessity of such rules had been immediately recognized by Sommerfeld in his treatment of the fine structure: if every variation of the quantum numbers $n$ and $k$ were possible during a quantum jump, the hydrogen spectrum would carry many more lines than it really does. In analogy with the condition $\Delta n \geqslant 0$ implied by the positivity of the radiated energy, Sommerfeld tentatively imposed a positive variation for every quantum number. Unfortunately, this *Auswahlprinzip* (principle of selection) soon proved to be violated for the quantum numbers introduced in the new theories of the Stark and Zeeman effects. While Epstein and Sommerfeld introduced *ad hoc* modifications of the original principle, Debye deplored the lack of a theory that would deduce correct selection rules, intensities, and polarization of spectral lines.[48]

47. Bohr 1923a, 279; Whittaker 1922. See also Bohr 1923b, trans., 20–21, 35.
48. Early draft of Bohr 1918a, in BCW 3:[48]–[52]; Sommerfeld 1915b; Sommerfeld 1916a, 23–26; Epstein 1916a, 511, 516; Sommerfeld 1916b, 494; Debye 1916, 511.

Take, for instance, the case of the Zeeman effect of the hydrogen atom, as calculated by Sommerfeld and Debye. The perturbation of the energy levels due to the presence of the magnetic field $\mathbf{B}$ is:

$$\delta E = m(eh/4\pi\mu c)B, \qquad (115)$$

where $\mu$ and $e$ are respectively the electron's mass and charge, and $m$ is the magnetic quantum number. The resulting line splitting is given by[49]

$$\Delta\nu = (m' - m'') \frac{e}{4\pi\mu c} B, \qquad (115')$$

where $m'$ and $m''$ refer to the initial and final stationary states. In order to obtain agreement with the observed triplet splitting one must impose the restriction

$$m' - m'' = 0, \pm 1. \qquad (116)$$

At this stage Bohr, perhaps remembering the consideration he had used in 1915 to reconcile a double line splitting with a double term splitting in his theory of the Stark effect (see p. 91), looked for a correspondence between the restriction (116) and the properties of the classical motion in stationary states. From his previous results in the case of strictly periodic motions, he knew that the harmonic components of the classical motion in a given stationary state corresponded to the various quantum transitions from this stationary state. Quite naturally he extended this correspondence to the Zeeman effect in the following manner: the possibility of a transition with, say,

$$n \to n - 1 \qquad \text{and} \qquad m' \to m''$$

had to correspond to the existence of a harmonic component in the quantized motion of the initial stationary state with the frequency

$$\bar{\nu} = \bar{\nu}_n^0 + (m' - m'')(eB/4\pi\mu c). \qquad (117)$$

Following a theorem by Larmor, the only effect of a weak magnetic field on motion in a central field with the frequency $\bar{\nu}_n^0$ is the superposition of a slow rotation around the magnetic axis, with the frequency $eB/4\pi\mu c$. Consequently, the harmonic components of the various projections of the motion can only have the frequencies

$$\bar{\nu}_n^0, \bar{\nu}_n^0 \pm eB/4\pi\mu c. \qquad (118)$$

Comparison with (117) immediately gives the selection rule (116).[50]

49. Sommerfeld 1916b; Debye 1916.
50. Early draft of Bohr 1918a, in BCW 3:[50].

Bohr promptly realized that this type of consideration could easily be extended to all multiperiodic systems and was therefore of a very general and profound nature. The simplest example was a pure harmonic oscillator, which had an empirical realization in the low-amplitude vibrations of molecules. On the one hand, in the observed spectrum only one frequency appears, as expected from a classical analysis of the oscillations; on the other hand the spectrum derived from energy quantization and the relation $\Delta E = h\nu$ contains all integral multiples of this frequency, since a quantum jump between the levels $n$ and $n - \tau$ leads to

$$E_n - E_{n-\tau} = nh\bar{\nu} - (n - \tau)h\bar{\nu} = \tau h\bar{\nu}. \tag{119}$$

Here also the "correspondence" between the quantum-theoretical spectrum and the spectrum of the classical motion reestablishes the agreement between theory and observation, since transitions $n \rightarrow n - \tau$ (with $\tau > 1$) would correspond to harmonics of the classical oscillation, which, by definition, do not exist in the case of a purely harmonic oscillator.[51]

In the general multiperiodic case the electric moment P of the system in the stationary state $n$, being a function of the configuration, is a multiperiodic function of time, with the Fourier expansion

$$P = \sum_{\tau} C_{\tau}(n)e^{2\pi i \tau \cdot \bar{\nu}(n)t}, \tag{120}$$

where $\bar{\nu}$ is the sequence of fundamental frequencies in the collective notation introduced in (19). According to ordinary electrodynamics, the spectrum of the emitted radiation would contain the frequencies

$$\sum_{i} \tau_i \bar{\nu}_i(n), \tag{121}$$

where the $\tau_i$ take all integral values with both signs; the corresponding intensities would be proportional to $|C_{\tau}(n)|^2$ and the polarization properties would be given by the orientation of the (complex) vector $C_{\tau}(n)$.

In the quantum theory the emitted spectrum has the frequencies

$$\nu_{n, n-\tau} = (E_n - E_{n-\tau})/h. \tag{122}$$

Bohr stated the following theorem: *In the limit of slow vibrations (high n) the quantum-theoretical spectrum is identical with the classical spectrum.* Indeed, according to Bohr's golden rule (100), we have[52]

$$E_n - E_{n-\tau} \sim \bar{\nu} \cdot \delta J = h\bar{\nu} \cdot \tau. \tag{123}$$

51. Ibid., [49].
52. Bohr 1918a, 14–15, 30.

Bohr introduced here the following important remark, which frequently occurred in his later writings as an affirmation of the contrast between quantum theory and classical electrodynamics, in spite of the above asymptotic agreement of the deduced spectra:

> It may be noticed, however, that, while on the first theory radiations of the different frequencies $[\tau \cdot \bar{v}]$ corresponding to different values of $\tau$ are emitted or absorbed at the same time, these frequencies will on the present theory, based on the fundamental assumptions I and II, be connected with entirely different processes of emission and absorption, corresponding to the transition of the system from a given state to different neighbouring states.

Considering this distinction there could be no agreement, even asymptotically, between the spectrum of the radiation emitted by a *single* atom and the one emitted by the corresponding classical system, since a single atom in a given state could emit only *one* line. But, in the spirit of Einstein's probabilistic treatment of radiation, one could still compare the spectrum of a statistical ensemble of such atoms with the classical spectrum. Even better: one could compare Einstein's $A$ coefficients with the classical intensities.[53]

In the limit of slow motions, Bohr expected classical electrodynamics to give correct values not only for the frequency of the spectral lines but also for their intensities and polarizations. Consequently, the following asymptotic proportionality had to hold:

$$A_{n-\tau}^{n} \propto |C_\tau(n)|^2. \tag{124}$$

At that stage of the quantum theory, this relation provided the only handle on the intensities of spectral lines. Bohr therefore suggested the following extrapolation, later named the "correspondence principle":

> Although, of course, we cannot without a detailed theory of the mechanism of transition obtain an exact calculation of the latter probabilities unless $n$ is large, we may expect that also for small values of $n$ the amplitude of the harmonic vibrations corresponding to a given value of $\tau$ will in some way give a measure for the probability of a transition between two states for which $n' - n''$ is equal to $\tau$.[54]

Bohr even assumed the "measure" to be exact in the case of a vanishing classical harmonic, which provided the selection rules. In this case the

---

53. Bohr 1918a, 15; see also Bohr 1923b, trans., 23: "Let it be once more recalled that in the limiting region of large quantum numbers there is in no wise a question of a gradual diminution of the difference between the description by the quantum theory of the phenomena of radiation and the ideas of classical electrodynamics, but only of an asymptotic agreement of the statistical results."
54. Bohr 1918a, 16.

analogy is between the presence or absence of a given harmonic, and the possibility or impossibility—not the probability—of a given transition. This is why in the earliest applications of the correspondence principle, which were limited to selection rules, Bohr had not needed Einstein's emission probabilities. He did so only in the later elaboration that required such probabilities as the quantum-theoretical concept "corresponding" to the (intensities of the) Fourier components of the classical electric moment.

As we shall presently see, Bohr generally used the correspondence principle in conjunction with a new method for quantizing perturbed systems, itself suggested by the correspondence principle. And he left the more sophisticated calculations to his gifted young associate Hendrik Kramers. In his dissertation (1919) Kramers derived the intensities and polarizations of the lines of the hydrogen spectrum, including fine structure and Zeeman and Stark effects, simply through Fourier analysis of the relevant classical motions in the stationary states and reinterpretation on the basis of the correspondence principle. As already mentioned, his results were indispensable for a proper comparison between theoretical and empirical spectra. Here I will limit myself to a very simple illustration of this type of consideration: how, in the case of the fine structure of hydrogen, the correspondence principle produces the selection rule $\Delta k = \pm 1$ for the azimuthal quantum number.[55]

In the solution of the relativistic Kepler problem (see p. 110), we have seen that the position $re^{i\theta}$ of the electron in the complex plane of its trajectory is a biperiodic function with fundamental frequencies $\bar{v}_r$ and $\bar{v}_\theta$, and the spectrum $\tau \bar{v}_r \pm \bar{v}_\theta$, where $\tau$ is any positive integer. The frequency $\bar{v}_\theta$ corresponds to the action variable $J_\theta$ and to the azimuthal quantum number $k$. According to the correspondence principle, a transition with a variation $\Delta k$ of this number can occur only if the harmonic $\bar{v}_\theta \Delta k$ is present in the spectrum of the classical motion in the initial stationary state. This implies the selection rule $\Delta k = \pm 1$.

The measured fine structure violated this rule and seemed thereby to threaten the correspondence principle. Fortunately, Bohr was well informed not only about experimental results but also about the detailed conditions under which they were obtained. He immediately observed that the degree of violation of his selection rule depended on the type of discharge tube used to excite the hydrogen atoms. This suggested that perturbing electric fields were responsible for the violation. Kramers's detailed calculations of the effect of such weak fields confirmed this intuition. If

---

55. Kramers 1919. On Kramers's biography and work see Dresden 1987.

we note that these calculations were also based on the correspondence principle, we have a typical example of the turning of an objection into a convincing confirmation.[56]

PERTURBATION THEORY

In the second part of "On the quantum theory of line spectra" published in December 1918, Bohr applied his point of view built around the correspondence principle to a first quantum theory of perturbations. Such a tool was needed not only to derive the effects of perturbing (static) fields on the hydrogen atom but also to investigate more complicated atoms, starting with the helium atom. From the original point of view of Sommerfeld, Epstein, and Schwarzschild, the quantization of the perturbed system seemed to require a *complete* solution of the corresponding (classical) dynamic problem through the method of separation. The correspondence principle, Bohr argued, suggested a more powerful alternative. The basic idea was first to investigate the character of the perturbed classical motion through *successive approximations* and then to deduce the quantum-theoretical spectrum from its "correspondence" with the classical spectrum.[57]

This program still necessitated a good deal of celestial mechanics, to be found in Charlier's and Poincaré's standard textbooks. But compared with Sommerfeld's, the method was more direct because it did not necessitate a nonperturbative solution of the mechanical problem. Also it was more general because the perturbed *system* did not have to be multiperiodic, only the perturbed *motion* had to be so (and only to a limited order of perturbation). Last but not least, Bohr's method provided not only the spectrum of the perturbed system but also, without much further effort, the corresponding (approximate) intensities, polarizations, and selection rules.[58]

The first mathematical expression of this program, the one found in the second part of "On the quantum theory of line spectra," was rather tentative, sometimes even awkward. But Kramers, in a systematic study of the effect of a weak electric field on the fine structure of the hydrogen atom, published in 1920, managed an elegant formulation of Bohr's ideas based on an extension of some aspects of Burgers's dissertation (and on the so-called Poincaré method of perturbation). In general, as Bohr had himself noted, the action-angle variables were extremely well suited not only to the determination of the quantum conditions but also to the study

56. Bohr 1918a, 69.
57. Bohr 1918b, par. 2: "The stationary states of a perturbed periodic system," 41–63.
58. Charlier 1907; Poincaré 1893.

of the periodicity properties of the motion, which played the most important role in the correspondence principle. Accordingly, the proper perturbation technique started with an infinitesimal canonical transformation
from the action-angle variables of the unperturbed system to those of the
perturbed system to first order. Then the correspondence principle directly
applied to the perturbed motion expressed in terms of the new action-angle
variables.[59]

To give an idea of the historical importance of this type of consideration, let us mention some later developments. In 1922 Born, Pauli, and
Heisenberg put their mathematical virtuosity in the service of the Bohr-
Kramers perturbation theory and managed to extend it to any order of
perturbation and to any type of degeneracy. They could even discuss the
convergence of the resulting series on the basis of older, sophisticated
theorems by Poincaré and Burns. In principle the quantization of atoms
had become, as Heilbron put it, "a problem for the nautical almanac."
Unfortunately, electrons would prove to be less docile than heavenly stars:
calculated spectra definitely departed from the observed ones.[60]

The following is a short account of the first-order perturbation theory,
in Kramers's canonical formulation. A perturbed system has, by definition,
a Hamiltonian of the form

$$H = H^0 + \varepsilon\Omega(q), \tag{125}$$

where $H^0$ is the Hamiltonian of the unperturbed system, $\varepsilon$ is a small parameter, and $\varepsilon\Omega$ is the perturbing potential. We assume that the mechanical
problem corresponding to the unperturbed motion has been solved in
terms of action-angle variables $J^0$, $w^0$. These variables are no longer
action-angle variables for the perturbed Hamiltonian $H$, but they remain
canonical (since the canonical character of a transformation is obviously
independent of the Hamiltonian of the system).

The new mechanical problem is solved as soon as the action-angle
variables $(w, J)$ of the perturbed system are known in terms of the unperturbed ones. Let us call $F$ the generating function of the canonical
transformation[61]

$$(w^0, J^0) \rightarrow (w, J). \tag{126}$$

Since $F = J \cdot w^0$ would generate the identity, $F$ has the general form

$$F(w^0, J) = J \cdot w^0 + \varepsilon\varphi(w^0, J). \tag{127}$$

59. Kramers 1920; Burgers 1917, 1918.
60. Born and Pauli 1922; Born and Heisenberg 1923a; Heilbron 1983, 288.
61. Here I use a generating function in the sense defined on p. 113.

This gives

$$w = w^0 + \varepsilon \frac{\partial \varphi}{\partial J}, \qquad J = J^0 - \varepsilon \frac{\partial \varphi}{\partial w^0}. \tag{128}$$

To first order in $\varepsilon$ the derivatives $\partial \varphi / \partial J$ and $\partial \varphi / \partial w^0$ may be replaced by $\partial f / \partial J^0$ and $\partial f / \partial w^0$, where $f$ is a function of $w^0$ and $J^0$. Therefore, the most general infinitesimal canonical transformation of $w^0$ and $J^0$ has the form

$$w = w^0 + \varepsilon \frac{\partial f}{\partial J^0}, \qquad J = J^0 - \varepsilon \frac{\partial f}{\partial w^0}. \tag{129}$$

Since the unperturbed problem has been solved, the Hamiltonian $H$ can be expressed as a function of $J^0$ and $w^0$. To first order in $\varepsilon$ the effect on this function of the above canonical transformation (recalling $\partial H / \partial J = \bar{v}$) is given by

$$H = H^0(J^0) + \varepsilon \Omega = H^0(J) + \varepsilon \bar{v} \cdot \frac{\partial f}{\partial w} + \varepsilon \Omega, \tag{130}$$

wherein the index $o$ has been dropped in the terms preceded by $\varepsilon$. Through the coordinates $q$ the potential $\Omega$ is a periodic function of the $w$'s with periods unity; we therefore look for a function $f$ with the same periodicity properties. Then the following Fourier developments hold:

$$f = \sum_\tau f_\tau(J) e^{2\pi i \tau \cdot w}, \qquad \Omega = \sum_\tau \Omega_\tau(J) e^{2\pi i \tau \cdot w}. \tag{131}$$

Substituting these series into (130) gives

$$H = H^0(J) + \varepsilon \Omega_0(J) + \varepsilon \sum_{\tau \neq 0} (2\pi i \bar{v} \cdot \tau f_\tau + \Omega_\tau) e^{2\pi i \tau \cdot w}. \tag{132}$$

For $J$ and $w$ to be action-angle variables for $H$ at first order in $\varepsilon$, the above expression must not contain $w$, which implies

$$2\pi i \bar{v} \cdot \tau f_\tau + \Omega_\tau = 0 \tag{133}$$

for any nonvanishing value of $\tau$. This is possible whenever the unperturbed system is nondegenerate, that is, when the frequency combinations $\bar{v} \cdot \tau$ never vanish. Through this condition $f$ is completely determined (up to an irrelevant constant) as

$$f = \sum_{\tau \neq 0} -\frac{1}{2\pi i} \frac{\Omega_\tau}{\bar{v} \cdot \tau} e^{2\pi i w \cdot \tau}. \tag{134}$$

As may easily be verified, the induced canonical variables satisfy all requirements to be action-angle variables for $H$.

From (132) and (133) results the expression for the first order perturbed energy:

$$H = H^0(J) + \varepsilon\Omega_0(J). \tag{135}$$

The correction $\varepsilon\Omega_0$ is simply interpreted as the time average (over a large number of periods) of the perturbing potential.

In the $r$-degenerate case, the function (102) can be used to induce new action-angle variables for the unperturbed system such that the first $r$ canonical couples $(\alpha, \beta)$ are constants of the motion which do not appear in $H^0$, and the $s - r$ remaining couples $(w, J)$ correspond to incommensurable nonvanishing frequencies. The transformation defined by (134) can then be applied, *mutatis mutandis,* to the latter canonical couples, which yields a new Hamiltonian:

$$H(J, \alpha, \beta) = H^0(J) + \varepsilon\Omega_0(J, \alpha, \beta). \tag{136}$$

The function $\varepsilon\Omega_0$ now plays the role of a Hamiltonian for the evolution of the $2r$ parameters $\alpha$ and $\beta$. If action-angle variables $w', J'$ can be found for this Hamiltonian, the mechanical problem will be entirely solved to first order in $\varepsilon$. In general the number of periods of the system will increase, as will the number of independent quantum conditions. The reader familiar with the perturbation theory of modern quantum mechanics will have noticed a striking similarity.

In the discussion of simple examples like the Stark and Zeeman effects, Bohr, instead of applying this powerful but exceedingly learned technique, simply exploited the specificities of the perturbation in a way that allowed for a more direct use of the correspondence principle. Consider for instance the Zeeman effect of the hydrogen atom (in the nonrelativistic approximation). As already mentioned, the main effect of the field on the original elliptic motion is the superposition of a slow uniform rotation around the field axis (Oz) with the frequency $\bar{v}_L = eB/4\pi\mu c$. If $\bar{v}_0$ is the frequency of the original motion, the resulting Fourier development of the various components of the electric moment are

$$P_z = \sum_{\tau = -\infty}^{+\infty} C_\tau^z e^{2\pi i \tau \cdot \bar{v}_0 t}, \qquad P_x \pm iP_y = \sum_{\tau = -\infty}^{+\infty} C_\tau^{\pm} e^{2\pi i(\tau \cdot \bar{v}_0 \pm \bar{v}_L)t}. \tag{137}$$

According to the correspondence principle, every Fourier component of the orbital motion corresponds to a possible transition from the corresponding stationary state. Therefore, $P_z$ gives, in a direction of observation perpendicular to the field, a spectrum identical with the original spectrum with linear polarization, while $P_x + iP_y$ and $P_x - iP_y$ give, in the directions of observation parallel to the field, a symmetrical splitting of every line of the original spectrum, the two components of each line having opposite circular polarizations.[62]

In order to quantitatively determine this splitting Bohr used his golden rule (100),

$$dE = \sum_i \bar{v}_i \, dJ_i,$$

which gives a connection between the energy of the system, the fundamental frequencies of the system, and the action variables to be quantized. In the present case the value of the frequency $\bar{v}_L$ is independent of the characteristics of the motion. Consequently, the relation $\partial E / \partial J_L = \bar{v}_L$ immediately gives $\delta E = \bar{v}_L J_L$ for the energy shift corresponding to the value $J_L$ of the action variable conjugated to the Larmor precession ($J_L$ is also the angular momentum around Oz). This variable is quantized as $J_L = mh$, where $m$ is called the magnetic quantum number, and the final Zeeman formula is

$$\delta E = mh\bar{v}_L. \tag{138}$$

From the "correspondence" between the variation $\Delta m$ during a transition and the harmonics $0, \pm 1$ of the frequency $\bar{v}_L$ in the classical motion results the selection rule

$$\Delta m = 0, \pm 1. \tag{139}$$

In contrast, Sommerfeld's and Debye's original calculations of the same effect did not provide the selection rule, and they requested an explicit solution of the mechanical problem in polar coordinates, with three quantum conditions instead of the two employed by Bohr.

THE PRINCIPLE OF MECHANICAL TRANSFORMABILITY

In 1918, at the time of the publication of the first two parts of "On the quantum theory of line spectra," the correspondence principle was not yet so named, and Bohr emphasized its heuristic power more than its "rational" character, which would be emphasized later. Above all, the correspondence principle was a means to grasp aspects of atomic entities otherwise inaccessible to the fundamentally incomplete quantum theory. At the same time, Bohr emphasized the role of Ehrenfest's adiabatic principle in the consolidation of the conceptual basis of the quantum theory. For this reason (and also, more prosaically, to avoid the thermodynamic

62. See, e.g., Bohr 1923a, 287–289.

consonance of the word "adiabatic") he renamed it the "principle of mechanical transformability."[63]

In a direct continuation of his unpublished remarks of 1916, Bohr lent a great importance to the adiabatic invariance of the extended quantum conditions, as proved by Burgers. He also used Ehrenfest's principle to derive the a priori statistical weights of quantum states, the ones to be used in entropy calculations à la Boltzmann.[64] Most fundamentally, he argued that the very *definition* of the energy concept in the quantum theory rested on the physical possibility of continuous deformations of atomic systems:

> In this connection it may be pointed out that the principle of the mechanical transformability of the stationary states allows us to overcome a fundamental difficulty which at first sight would seem to be involved in the definition of the energy difference between two stationary states which enters the relation [$\Delta E = h\nu$]. In fact we have assumed that the direct transition between two such states cannot be described by ordinary mechanics, while on the other hand we possess no means of defining an energy difference between two states if there exists no possibility for a continuous mechanical connection between them. It is clear, however, that such a connection is just afforded by Ehrenfest's principle which allows us to transform mechanically the stationary states of a given system into those of another, because for the latter system we may take one in which the forces which act on the particles are very small and where we may assume that the value of the energy in all stationary states will tend to coincide.[65]

Here we get a glimpse into a deep layer of Bohr's thought, one that anticipated an important aspect of complementarity. The definition of a concept, more generally of a word, demands continuity. The remark might seem obscure if not related to Bohr's early reflections on the inner working of language and thought. Unfortunately, Bohr does not seem to have ever made such considerations explicit before his last interview with Thomas Kuhn in 1962. The authenticity of his ultimate remembrance is nevertheless made very plausible by some allusions in the correspondence with his brother Harald, and by the well-known interest of the young Bohr in the psychology of cognition.[66]

63. Bohr 1918a, 8.

64. Burgers 1917, 1918, and also Ehrenfest 1916. For the derivation of statistical weights, Bohr relied on the adiabatic invariance of these weights proved in Ehrenfest 1914b. Bohr 1918a, 9, 25–27, also Bohr 1923b, trans., footnote on 16–17, where a most elegant proof of Ehrenfest's theorem is given.

65. Bohr 1918a, 9.

66. Bohr [1962], session 5 (17 Nov. 1962); Niels Bohr to Harald Bohr, 5 July 1910, BCW 1: "I must confess that I do not know what I am most happy about. . . . Probably the only answer is that feelings, like cognitions [Erkendelsen], must be arranged in planes [planer] that cannot be compared." See, e.g., Folse 1985.

As pointed out by Norton Wise, there is indirect but strong evidence that Bohr was inspired by Høffding's philosophical reflections on this theme. For instance both men, in different contexts, associated rationality with continuity and irrationality with discontinuity, and they regarded the tension between subject and object as a source of irrationality. More generally, some idiosyncratic aspects of Bohr's terminology, as well as the phrasing of his discussion of stationary states, seem to have been inspired by an underlying analogy between quantum phenomena and psychological processes. For more details I refer the reader to Wise's original study. My own discussion of Bohr's ideas on language will begin with the explicit confidence he made to Kuhn the day before he passed away.[67]

Every word we utter is defined only if it is immersed in a continuous context of meaning. Since the context is generally not unique, a word isolated from its context is essentially ambiguous. If language has to be globally coherent, there must be a way to connect the various meanings of a given word. Therefore, the various contexts of meaning have to belong to a single continuum.

Bohr illustrated this point of view or something similar through an analogy with the structure of Riemann's surfaces for functions of a complex variable (which were the subject of his brother's dissertation). This analogy may be illustrated by considering the Riemann surface of a logarithm, as drawn in figure 14. If a point in the complex plane represents a word, the various points $M_1, M_2, M_3, \ldots$ on the Riemann surface represent various meanings of this word, according to the "contextual" sheet in which it is immersed. As required above, the various meanings can be "compared" through a continuous path since the Riemann surface is arc-connected.

Bohr probably noticed that his first hint at the role of Ehrenfest's principle in the definition of the energy concept was not entirely satisfactory: a "shrunk" energy spectrum is still a discontinuous spectrum, no matter how close to each other the energy levels have become. He soon improved the argument by imagining cyclic adiabatic transformations connecting a given stationary state of a given system to any other stationary state of the *same* system. Such transformations provide precisely the required "continuous mechanical connection." At first glance their existence would seem to contradict the adiabatic invariance of the action-angle variables. But, as Bohr ingeniously noticed, there are exceptions to this invariance, when the degree of degeneracy of the system changes during the transformation; and these exceptions are of such a nature that they permit the jumps in

67. Wise 1987; Bohr [1962].

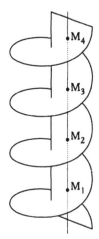

Figure 14.   A portion of the Riemann surface of a logarithm.

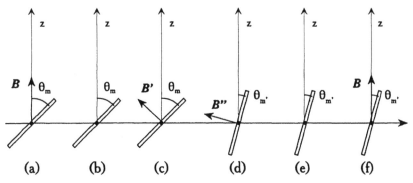

Figure 15.   Diagrams for the proof of the adiabatic connectivity of Zeemann sublevels. The thin rectangles represent the trace of the electronic orbit, its plane being assumed to be perpendicular to the plane of the drawing.

the action variables needed to connect different stationary states. This will be seen from a simple example.[68]

Consider a hydrogen atom originally immersed in a static homogeneous magnetic field **B** along the vertical axis Oz, as represented in figure 15:a. Initially, the angle of the (precessing) plane of the trajectory is in one of its quantized values $\theta_m$ (as imposed by the quantization of the component of the angular momentum along Oz). The adiabatic transformation starts

68. Bohr 1918a, 24–25; also Bohr 1923b, trans., 14–15.

with a slow turning off of the field while keeping it parallel to Oz. During this process the quantum numbers do not change, but the degree of degeneracy is increased by one unit (fig. 15:b). Next a magnetic field is slowly turned-on perpendicular to the plane of the trajectory (fig. 15:c); then the field is rotated within a vertical plane, at constant intensity (fig. 15:d). According to the adiabatic theorem, the plane of the trajectory "accompanies" the latter rotation of the field, that is, it remains perpendicular to the field. The rotation is stopped precisely when the angle made by the plane of the trajectory with the axis Oz is equal to $\theta_{m'}$, one of the quantized values of this angle in the original system. The field is now turned-off slowly with a constant orientation (fig. 15:e). Finally the original magnetic field B is slowly restored along Oz (fig. 15:f). We are now back to the original system, but in a different stationary state corresponding to the angle $\theta_{m'}$.[69]

In the spring of 1920 Bohr met Einstein in Berlin for the first time. As revealed by a subsequent letter of Einstein referring to "the way [Bohr] derive[d] quantum states from other quantum states ('in the manner of Riemann surfaces')," the mutual transformability of stationary states must have been a central argument in Bohr's defense of the rationality of his theory. Most interestingly, the analogy with Riemann's surfaces appears to have played a role in this context, which suggests a connection between the reflections on energy definition and the earlier speculations on language.[70]

In the context of atomic theory, the content of Bohr's allusion to Riemann's surfaces may be inferred from an early draft of "On the quantum theory of line spectra":

> The singular position of the degenerate systems in the general theory of conditionally periodic [i.e., multiperiodic] systems allows us to connect the different stationary states of a system of this kind with each other in a continuous way which gives the general theory a character which reminds us of the usual analytical theory of multiply valued functions.

To a point in the complex plane corresponded a choice of the potential function (from which the atomic forces derived); to the notion of multiply valued functions corresponded the set of stationary states as a function

---

69. I have preferred this example to that given in Bohr 1918a, 24–25, because of its greater simplicity. Strictly speaking, the proof of the adiabatic theorem given earlier does not apply to the case of a varying magnetic field, because there is always an electric field induced by this variation. Dirac gave an appropriate generalization in 1924; see part C, n. 70.

70. Einstein to Bohr, 2 May 1920, BCW 3:[634].

of the potential. Then, to a singularity of a type similar to that found in the function $\sqrt{z}$ corresponded a degenerate system: indeed, for $z = 0$ the two branches of $\sqrt{z}$ converge, just as different energy levels converge toward a single level in the case of degeneracy.[71]

Even though Bohr refrained from publishing such marginal considerations, they might help the understanding of his later insistence, in his notion of complementarity, on the necessity of having recourse to classical concepts in order to communicate physical results *unambiguously*. Classical theory, as the best possible expression of the ideal of continuity in physics, provided the best conceivable language to describe quantum phenomena, in spite of the intrinsically ambiguous behavior of quantum objects implied by the quantum postulate (in its later acceptation). Before he came to this judgment—that is, before the advent of complementarity—Bohr never quite declared classical concepts to be indispensable. Nevertheless, he already cared to show that their use in the new quantum theory, albeit provisional and approximate, was still coherent in the sense of his earlier reflections on language.

## THE MEANING OF THE CORRESPONDENCE PRINCIPLE: MAGIC OR REASON?

### A RATIONAL GENERALIZATION

With these fundamental insights into continuity and definition, the power of the adiabatic principle seemed to be exhausted. Instead, in the subsequent development of the Bohr theory the correspondence principle became more and more important as a "guide" toward a more definite quantum theory. In a lecture given in Berlin on 27 April 1920, Bohr established the technical acceptation of the word "correspondence":

> Although the process of radiation cannot be described on the basis of the ordinary theory of electrodynamics, according to which the nature of the radiation emitted by an atom is directly related to the harmonic components occurring in the motion of the system, there is found, nevertheless, to exist a far-reaching *correspondence* between the various types of possible transitions between the stationary states on the one hand and the various harmonic components of

71. Bohr, "On the principles of the quantum theory," fragments of an early draft (1917?) of Bohr 1918a, BMSS 7 (AHQP), section 1, on 24th sheet (the passage expressing the analogy with Riemann surfaces, starting with "which gives the general theory . . . ," is crossed out). The analogy should not be taken too literally: in the case of the Riemann surface for $\sqrt{z}$, one may travel continuously between the two determinations of the square root without crossing the origin $z = 0$.

the motion on the other hand. This correspondence is of such a nature, that the present theory of spectra is in a certain sense to be regarded as a rational generalization of the ordinary theory of radiation.[72]

By "a certain sense" Bohr provisionally meant "a formal sense," as appears from his earlier unpublished comments on the relation between the new quantum theory and classical electrodynamics. Some of the *formal* relations between the source and the emitted electromagnetic field could be saved and reinterpreted in terms of quantum-theoretical concepts like stationary states and transition probabilities. In this process a good deal of the formal harmony of classical electrodynamics was left intact, in spite of the irreducible "contrast" between classical and quantum theory introduced by the quantum postulates. Most important, the "correspondence" was not between classical and quantum theory—a common misinterpretation—but between quantum-theoretical concepts of motion and radiation. This is why Bohr, even in the lack of a quantitative expression of the correspondence, believed that he had in hand a fundamental "principle" of the quantum theory; he continued his lecture under the heading "The correspondence principle."[73]

Bohr opposed his idea of a "rational generalization" brought about by the correspondence principle to the "tendency of considering the quantum theory as a set of formal rules," a hardly dissimulated allusion to the Munich school of atomic theory. He was also afraid that critical observers would take his approach to the quantum problem, especially the correspondence principle, to be some opportunistic mismatch of classical and quantum concepts. A number of his readers, even some historians and philosophers, have done so. Yet Bohr could reveal an astonishing harmony between the rules for fixing the stationary states and the rules governing the transition between these states. In his words,

> If the correspondence principle cannot instruct us in a direct manner concerning the nature of the process of radiation and the cause of the stability of the stationary states, it does elucidate the application of the quantum theory in

72. Bohr 1920b; "Guide" is ibid., 60, *correspondence* (Bohr's emphasis), 23–24.

73. "Formal sense" is in the manuscript cited in n. 44; "The correspondence principle" is in Bohr 1920b, 27. In Bohr 1923b, trans., 22n, Bohr warned against a possible misunderstanding of the word "correspondence": "Such an expression might cause confusion since, in fact . . . the Correspondence Principle must be regarded as a law of the quantum theory, which can in no way diminish the contrast between the postulates [of the quantum theory] and electrodynamic theory."

such a way that one can anticipate an inner consistency for this theory of a kind similar to the formal consistency of the classical theory.[74]

For instance, the correspondence principle presupposed the identity between the degree of periodicity of the motion in stationary states and the number of quantum conditions necessary to fix these states. And this identity was indeed warranted by the Bohr-Sommerfeld rules for the quantization of multiperiodic systems. In his Göttingen lectures of June 1922 Bohr explained this type of harmony through a suggestive metaphor:

> Corresponding to the quantum orbits and the electrons, let us imagine a number of bowls into which we are to throw balls. If we were to depend on classical mechanics, it would not be easy to get a ball into a bowl. According to the quantum theory it seems that the ball must necessarily land in a bowl, and that is very strange. However, when we consider that the quantum states, *i.e.* the places where the balls are located, as well as the processes which cause the transitions, are determined by the same periodicity properties [respectively through the Bohr-Sommerfeld rules and through the correspondence principle], then we need no more wonder so much.[75]

## SOMMERFELD'S RETICENCE

Even in its most spectacular application, namely the derivation of selection rules for quantum transitions, the correspondence principle did not immediately convince Bohr's main competitor in the development of atomic theory, Arnold Sommerfeld. In early 1919 the Munich professor communicated to Bohr his impressions about the two published parts of "On the quantum theory of line spectra": "Your formal principle of analogy between classical theory and quantum theory is very interesting and fruitful. However, the hypothesis of Rubinowicz, although not nearly as far-reaching, seems for the present more satisfactory to me." Rubinowicz had indeed managed to derive some selection rules without the correspondence principle, through an extension of the conservation of angular momentum to radiation processes.[76]

The reasoning, published in 1918, proceeded along the following lines. The monochromatic radiation emitted by an atom during a quantum jump is assumed to be spherical and to propagate according to Maxwell's equations. Thus the total angular momentum **M** of such a wave (as derived

---

74. Bohr, English manuscript for the Solvay congress of 1921, BCW 3:[364]–[380], on [378], where Bohr defends "the tendency of considering the quantum theory not as a set of formal rules, but as a theory of radiation constituting a rational generalisation of the classical theory of electromagnetism"; Bohr 1923b, trans., 25.

75. Bohr [1922c], [386].

76. Sommerfeld to Bohr, 5 Feb. 1919, BCW 3:[687]–[688]; Rubinowicz 1918.

from Maxwell's stress tensor) must obey the inequality

$$|\mathbf{M}| \leqslant \varepsilon/2\pi v, \tag{140}$$

where $\varepsilon$ is the total energy of the wave and $v$ its frequency. The equality is reached for circularly polarized radiation, and the value zero corresponds to rectilinear polarization. If one further assumes that the energy and the total angular momentum of the system atom + radiation is conserved during the emission process, the following relations hold:

$$\varepsilon = hv, \qquad \mathbf{M} = \Delta\boldsymbol{\sigma}, \tag{141}$$

where $\Delta\boldsymbol{\sigma}$ is the variation of the angular momentum of the atom. The inequality (140) therefore implies

$$|\Delta\sigma| \leqslant h/2\pi. \tag{142}$$

For a Sommerfeld hydrogen atom, the value of $\sigma$ is given by the azimuthal quantum number as $\sigma = kh/2\pi$. Consequently,

$$|\Delta k| = \frac{2\pi}{h}|\Delta|\sigma|| \leqslant \frac{2\pi}{h}|\Delta\sigma| \leqslant 1. \tag{143}$$

Since $k$ is an integer, the following selection rule results:

$$\Delta k = 0, \pm 1. \tag{144}$$

The value 0 corresponds to rectilinear polarization, and the values $\pm 1$ to circular polarizations.

In his famous *Atombau,* first published in 1919, Sommerfeld commented on this reasoning as follows: "In this way, by a remarkably rigorous manner of deduction, reminiscent of the incontrovertible logic of numerical calculations, we have arrived from the principle of conservation of angular momentum at a *principle of selection* and a rule of polarization." A few pages later one may read: "On the other hand, Bohr has discovered in his *principle of correspondence* a magic wand (which he himself calls a formal principle), which allows us immediately to make use of the results of the classical wave theory in the quantum theory." No doubt that Sommerfeld preferred logic to magic.[77]

FOR AN OPEN THEORY

Bohr, instead, gave little importance to Rubinowicz's reasoning. He had himself independently published very similar considerations in the first

77. Sommerfeld 1919, English trans. (London, 1923) of the 3d. ed. (1922), 265–265, 275.

part of "On the quantum theory of line spectra," but as he later commented, the implications of the correspondence principle were far more extensive and accurate. For instance, in the above case of the azimuthal quantum number, the correspondence principle gave $\Delta k = \pm 1$, whereas Rubinowicz could not exclude $\Delta k = 0$. Furthermore, Rubinowicz's argument implicitly employed the correspondence principle in order to justify the otherwise arbitrary assumption of a spherical emission.[78]

Rubinowicz's ambition, however, was not limited to a derivation of selection rules. He aimed at a fully quantum-theoretical treatment of the coupling between atoms and radiation, what Bohr later termed *Koppelungsgesichtspunkt* (coupling viewpoint). His idea, first expressed in 1917 (and anticipated by W. Wilson in 1915), was to regard the electromagnetic field in a cavity with perfectly reflecting walls as a multiperiodic system and to submit it to the standard rules of quantization. As a consequence of the purely harmonic character of the oscillations of the various modes of the field, their energy had to be an integral multiple of $h$ times their frequency. In this way Bohr's relation $\Delta E = h\nu$ was "deduced" from the energy principle applied to the emission process (although not quite, since $\Delta E = nh\nu$ was also possible).[79]

In 1921 Rubinowicz managed to integrate his selection rules for angular momentum into this framework. For this purpose he had only to imagine a spherical cavity, the standing waves of which had a quantized angular momentum. Then conservation of the net angular momentum during the emission process provided both selection rules and polarizations. The latter paper, and also Rubinowicz's visit to Copenhagen, induced a public reply from Bohr.[80]

First of all, according to Bohr, Rubinowicz had overlooked the degeneracy of the multiperiodic system given by the electromagnetic cavity and therefore could not account for the possibility of elliptic polarization, which was implied by the correspondence principle (the most general spatial vibration of a vector being elliptical). More fundamentally, the "coupling viewpoint" was a *closed* one, meaning that "it may hardly be possible to extend [its realm of application] until we are closer to a solution of the enigma of the quantum theory." Instead, Bohr went on,

> the situation may be different in the case of the correspondence view-point, which so far proves to have been fruitful in ever new realms of application,

---

78. Bohr 1918a, 34–35; Bohr 1920b, 52.
79. Rubinowicz 1917; Wilson 1915.
80. Rubinowicz 1921; Bohr 1921b.

without thereby bringing us a step closer to a [complete] solution of the quantum enigma; we have progressed toward such a solution only insofar as with each extension of the application of the quantum theory we [better] perceive the nature of this enigma. This is connected with the circumstance that this view-point is not at all a closed formal one, but may rather be regarded as a description of certain general features of the radiation process.

As Bohr's emissary at the Solvay congress of 1921, Ehrenfest emphasized the need to maintain this adaptability of the correspondence principle: "It is not desirable that, with the most automatic application in view, one already casts in a rigid form the condition of correspondence, which up to now has been variable and groping." Thus Bohr and followers distinguished between the well-established and sharply formulated quantum "postulates" on the one hand and the more progressive and adaptive "principles" on the other.[81]

Rubinowicz's "coupling viewpoint" was not the only target of Bohr's criticism. Einstein's struggles to endow the light quanta with some kind of reality were received with a similar skepticism. Bohr certainly admired the light-quantum explanation of the photoelectric effect and similar phenomena, but as he declared in 1920, "Einstein's theory has hardly brought us closer to an understanding of the interaction between light and matter." Even though some "formal" validity could not be denied to the light quanta, they could not be brought to explain interference phenomena. "What it would mean to forgo an understanding of [these phenomena]," Bohr continued, "may perhaps be seen most clearly in the fact that the frequency which enters into Einstein's expression for the energy of a light quantum can only be determined with the aid of interference phenomena."[82]

All the same, Bohr was perfectly aware of the futility of more conservative substitutes to the light-quantum hypothesis. For instance, in the case of the photoelectric effect, he condemned Lenard's "triggering hypothesis," which compared the metal target to a set of loaded pistols (the trigger of which would respond only to *frequencies* of the impinging radiation above a certain threshold), because it was incompatible with the existence of phenomena that could be described as reverse photoelectric effects. Indeed, in the experiments by Franck and Hertz an electron could induce

---

81. Bohr 1921b, 8–9; Ehrenfest 1923a, 254.
82. Bohr [1920a], [234].

the emission of light from a hit atom, the frequency of this light being proportional to the variation of the kinetic energy of the electron, just as in Einstein's photoelectric relation. Bohr ironically commented: "The difficulty for our imagination of conceiving the reverse of what takes place when a pistol is fired is obvious to anyone."[83]

Quite generally Bohr excluded any cheap solution of the fundamental paradoxes brought about by the dual aspects of radiation phenomena:

> We must admit that, at the present time [1920], we are entirely without real understanding of the interaction between light and matter; in fact, in the opinion of many physicists [not so many], it is hardly possible to propose any picture which accounts, at the same time, for the interference phenomena and the photoelectric effect, without introducing profound changes in the view-points on the basis of which we have hitherto attempted to describe the natural phenomena.

In other words Bohr expected a conceptual revolution, one that would even alter the *epistemological* status of physical theory. However, he did not believe the time to be ripe for attempts at providing the missing "picture." He even refrained from publishing his own guesses about what characteristics of the old picture should be dropped (the contents of these guesses will be given later). Premature theoretical constructs were likely to interfere with a proper theoretical exploitation of the ever-growing ranges of experimental data on quantum phenomena. Again the correspondence principle, with its temporary renunciation of a coherent picture of radiation processes, was the only adaptable and improvable strategy for circumscribing the enigma of atomic constitution:

> At the same time as we consciously renounce the cohesiveness in our picture offered by such an edifice as the electromagnetic theory, we may attain just what that theory turned out to be unable to give, namely, the possibility of beginning to reach an understanding of the properties of the chemical elements.[84]

In the second (1921) and third (1922) editions of his *Atombau*, Sommerfeld gave an honorable place to the correspondence principle:

> Bohr's method is not only of greater consequence [than Rubinowicz's] in the question of intensity, but also leads to sharper and more definite results as regards the question of polarization. . . . In the matter of method the principle of correspondence has the great advantage that it postulates that Maxwell's

83. Ibid., [235].
84. Ibid.

theory be generally valid for long waves (Hertzian vibrations of wireless telegraphy), and that it does not throw overboard the many useful results, whch the classical theory gives for optical waves and Röntgen rays.[85]

In spite of these generous comments Sommerfeld remained insensitve to the "philosophy" of the correspondence principle: he disagreed wth the assertion that it was a fundamental part of the present—or a future— theory. In a letter of November 1920 he confessed to Bohr his leftover dissatisfaction: "The origin of your correspondence principle out of the quantum theory is still a source of distress to me, even though I am ready to admit that it reveals a most important relation between the quantum theory and classical electrodynamics." Clearly, Sommerfeld was not likely to accept Bohr's characterization of the correspondence principle as a "principle *of the quantum theory.*" Such an evaluation conflicted with his struggle for a mathematically closed theory of atoms. He certainly appreciated the clear mathematical recipes provided by Bohr to calculate intensities and polarizations, and even applied them himself with Heisenberg in 1922 to the derivation of the relative intensities of spectral multiplets and their Zeeman components; but he distrusted the pervasive adaptability of the correspondence principle and, as I will later show, denied the "sharpening of the correspondence principle" introduced by Bohr and Heisenberg in 1924–25.[86]

In 1922, in a congratulatory letter for the third edition of *Atombau,* Bohr confided to his friendly competitor:

> In the past few years I have often felt myself scientifically very lonely, under the impression that my efforts to develop the principles of the quantum theory systematically to the best of my ability have been received with very little understanding. For me it is not a matter of didactic trifles but of a serious attempt to achieve such an inner coherence that one can attain a secure basis for the further development. I understand quite well how little the matters are clarified as yet, and how helpless I am at expressing my thoughts in easily accessible form.[87]

Fortunately for Bohr (and his devoted Kramers), this period of relative "loneliness" was about to end. The consciousness of a crisis of the Munich approach to the quantum theory would rise, and two genial newcomers,

85. Sommerfeld (see n. 77), 276.
86. Sommerfeld to Bohr, 11 Nov. 1920, BCW 3:[690]. Bohr's subsequent claim that the correspondence principle was *a law of the quantum theory* (see n. 73) might well have been a reaction to Sommerfeld's comments.
87. Bohr to Sommerfeld, 30 Apr. 1922, BCW 3:[691]–[692].

Heisenberg and Pauli, would then perceive the superiority of Bohr's philosophy of correspondence.

## SUMMARY

In the initial solitary development of his theory, Bohr had found a quantum rule only for periodic systems and had not tried to extend it to a larger class of systems, since he had expected the frequency rule to break down for nonperiodic systems. In 1916 Sommerfeld, Schwarzschild, and Epstein, having no such prejudice, managed to quantize multiperiodic systems, that is, systems for which any bound motion is obtained by composing several periodic motions with different periods. This included, for instance, the relativistic Kepler motion, and the effect of electric or magnetic fields on the Kepler motion, leading respectively to explanations of the fine structure and of the Stark and Zeeman splittings of the hydrogen spectrum.

The new calculations involved sophisticated methods of analytical mechanics. The best-adapted to the evolving quantum theory was a method unknown to ordinary users of mechanics but familiar to astronomers like Schwarzschild. In this method the configuration of a multiperiodic system (its coordinates and momenta) is expressed in terms of "action" and "angle" variables, which have the following properties:

1.  The configuration is a periodic function of all angle variables with period unity

2.  The energy of the system is a function of the action variables only

3.  The action variables are constants of the motion

4.  The angle variables increase linearly in time

5.  The action variables are adiabatically invariant

6.  The energy variation during an infinitesimal increase of an action variable is obtained by multiplying this variation by the corresponding frequency.

Once these variables have been introduced, the quantization of the system becomes a trivial problem. As suggested by (2), (3), (5), and the analysis of a few simple cases, stationary states are determined by setting each action variable equal to an integer (the "quantum number") times Planck's constant. In general this gives as many quantum conditions as there are degrees of freedom. Moreover, the periodicity properties of the motion are immediately known (through (1) and (4)), which proved to be essential to the application of the correspondence principle. In this respect Bohr attached great importance to property (6), which relates energy variations with periodicity properties; I call it "Bohr's golden rule." Altogether, the method of action-angle variables seemed almost to have been invented for the sake of quantum theory. Bohr adopted it as the basic formal apparatus of his theory and even improved the relevant mathematical demonstrations on several points.

The year 1916 brought another spectacular extension of Bohr's theory, namely Einstein's theory of the emission and absorption of radiation. While Bohr perceived radiation as a means to glean information about atomic structure, Einstein used Bohr's theory to reach new insights into radiation processes. Since according to his and Bohr's ideas, changes in atomic systems could occur only through discontinuous transitions, radiation processes had to be described by transition probabilities. In classical electrodynamics an oscillating charge system may act on incident monochromatic waves in two different ways, positive and negative absorption; and, left to itself, it may also spontaneously emit radiation. Guided by the classical analogy, Einstein distinguished three transition probabilities, two for positive and negative absorption and one for spontaneous emission. Naturally the absorption probabilities must be proportional to the intensity of the incoming radiation. Now, if an assembly of identical atoms is in equilibrium with the surrounding radiation, for any pair of stationary states upward quantum jumps must balance downward quantum jumps. From this condition and Wien's displacement law Einstein could derive Bohr's frequency rule, Planck's blackbody law, and a relation between absorption and emission coefficients. Einstein also analyzed the momentum fluctuations of atoms immersed in thermal radiation and concluded that emission was a directed process (light quanta). Bohr rejected the latter conclusion but applauded the rest of Einstein's considerations, which fitted so closely with his theory and his growing taste for the classical analogy.

The first conclusion Bohr drew from Einstein's and Sommerfeld's advances was the general validity of the frequency rule. He observed that Sommerfeld and Debye had been able to derive the (normal) Zeeman effect without violating this rule, contrary to his earlier opinion. Moreover, Einstein's new derivation of the blackbody law *implied* this rule. Consequently, from the first part of "On the quantum theory of line spectra" (1918) until the beginnings of quantum mechanics (1925–26) Bohr based his theory on *two* postulates, one asserting the existence of stationary states, the other being the frequency rule. All other assumptions Bohr regarded as limited and provisional. Most important, the application of ordinary mechanics to the motion in stationary states presumed the separation of Coulomb and radiation forces, which could be meant only as an approximation. The price of such flexibility was a characteristic theoretical incompleteness. In compensation Bohr integrated in his theory two "principles" obtained from examining the relation between classical and quantum theory.

Bohr reached the correspondence principle in 1917 through a criticism of the new theories of the Stark and Zeeman effects. In general, combining the energy levels through the frequency rule gave far too many spectral lines, so that Sommerfeld had to introduce *ad hoc* "selection rules" in order to restrict the possible variations of the quantum numbers during an atomic transition. Bohr, inspired by the classical analogy, noted a very systematic "correspondence" between (improved) selection rules and the periodicity properties of the stationary orbits: a variation of a quantum number by $\tau$ was allowed if and only if the motion in the initial stationary state involved a $\tau^{th}$-order harmonic of the fundamental frequency associated with this quantum number. That such correspondence held in the case of high quantum numbers was a result of property (6) of action variables, Bohr's golden rule. That it held quite generally was *assumed* by Bohr and applied with great success.

Bohr understood this correspondence as a formal analogue of the relation between charge motion and emitted spectrum found in classical electrodynamics. However, he continually insisted on the contrast between classical and quantum theory. Even in the limit of high quantum numbers a *single* excited atom could emit only one line (excluding cascades), while the radiation classically emitted would generally contain harmonics of the line. Therefore, the (asymptotic) agreement between classical and quantum theory could be only statistical. Einstein's probability coefficients, Bohr continued, provided the conceptual tool necessary to express this agreement: the emission probability corresponding to a given quantum transition had to be proportional to the intensity of the "corresponding" harmonic component of the motion in the initial stationary state. Bohr further assumed this proportionality to hold approximately for moderate quantum numbers. This hypothesis, together with the sharper one concerning selection rules, constituted the hard core of what Bohr later named the correspondence principle (in 1920).

Among the first empirical fruits of the correspondence principle were various selection rules and Kramers's calculations of the intensities of hydrogen lines. This principle also suggested to Bohr an important theory

of perturbations. Bohr's method was to calculate approximations of the
perturbed motion through techniques adapted from celestial mechanics
and then to apply the correspondence principle in order to deduce the
perturbed spectrum from the periodicity properties of this motion. Unlike the calculations of the Stark and Zeeman effects earlier made by
Sommerfeld and others, Bohr's perturbative calculations did not require
an exact solution of the perturbed mechanical problem (which is generally
impossible); they gave the polarizations and intensities of spectral lines;
they did not require the perturbed system to be multiperiodic (which is
seldom the case). Later elaborations of this method by Kramers (1920) and
by Born and Pauli (1922) would play a crucial role in testing the validity
of the orbital model.

The other principle of Bohr's theory derived from Ehrenfest's adiabatic
hypothesis, which Bohr renamed in 1918 the "principle of mechanical
transformability," in order to emphasize its role in the definition of stationary states. Perhaps drawing on Høffding's philosophy, Bohr believed
that all definitions presumed continuity, for both ordinary language and
quantum theory. In the case of ordinary language, isolated words are necessarily ambiguous; they acquire a definite meaning only when immersed
in a precise context; various meanings of the same word can be compared
only if there is a continuous connection between the corresponding contexts. An atomic system is also ambiguous, since it can be found in different stationary states; the various energy values can be compared only if
one can imagine a continuous deformation of the system connecting any
two stationary states. Bohr regarded this condition as essential to the definiteness of his theory and managed to prove it for multiperiodic systems.
In this case as well as in the ordinary-language case, Bohr illustrated his
idea by an analogy with Riemann surfaces. There the ambiguity occurs
in the choice of the branch of a multiply valued function; the various
branches can, nevertheless, be compared thanks to their continuous connection via the Riemann surface.

Such images helped convince Bohr of the possibility of a harmonious
blending of continuity and discontinuity. The formal analogy between
quantum theory and classical electrodynamics reinforced this conviction,
so that he regarded quantum theory as a "rational generalization" of classical theory. Accordingly, he insisted that the correspondence principle
was a principle *of* the quantum theory, one connecting atomic motion
(whatever it might turn out to be) to the emitted radiation. And he explained that the jumping of atomic systems between discrete states was
not as bizarre as it seemed, for both the discrete states and the jumps

were determined by the periodicity properties of stationary motions (respectively through the Bohr-Sommerfeld rule and through the correspondence principle).

This is not to say that Bohr underestimated the paradoxes which quantum discontinuity brought about. Like Einstein, he believed that the coupling between continuous radiation and quantized atoms was highly problematic. But he rejected Einstein's light quantum, and other attempts at further specifying this coupling, as being premature and inconsistent. In his opinion a proper account of both interference and quantum phenomena would require a more drastic reform of the physicist's mode of describing natural phenomena. In the meantime the best strategy would seem to be to anchor the theory on its two postulates and to let it be guided by its principles, allowing it to evolve symbiotically with empirical progress.

To Bohr's disappointment the correspondence principle failed to attract much sympathy beyond Copenhagen. After a short period of total rejection, Sommerfeld came to admit its usefulness (in the derivation of selection rules and intensities) but refused to take it as a principle *of* the quantum theory. He was reluctant to regard the product of an unsharply formulated analogy as a constitutive part of a theory and strove instead for logically closed theories of well-defined models. Reciprocally, Bohr criticized the trend of the Munich school to limit the quantum theory to a set of formal rules, which, in his opinion, would necessarily lead to stagnation.

# CHAPTER VII

# Harmonic Interplay

## BEYOND MULTIPERIODIC SYSTEMS

Bohr's characterization of the correspondence principle as a guiding principle in an open theory was not wishful thinking. Very soon this principle proved to be useful even for systems without the original limitation of multiperiodicity. As already mentioned, the first extension appeared in Bohr's perturbation theory, in which the perturbed system did not have to be multiperiodic: only the original (unperturbed) system and the successive orders of perturbation had to be so. In 1920 Kramers published the first concrete application of this method, a determination of the effect of a small electric field on the fine structure of the hydrogen spectrum. Far from being academic, this study explained why even very small parasitic fields were able to break the selection rule $\Delta k = \pm 1$ in the experimenters' tubes. In contrast, Sommerfeld's method was here completely impotent, since the corresponding mechanical system, the relativistic Kepler system with an additional homogeneous electric field, was not separable.[88]

*A fortiori*, one did not expect the more complicated $n$-body systems corresponding to higher atoms to be multiperiodic. In general the mutual perturbation of the electrons in a given atom led to hopelessly intricate motions. Yet the discrete character of atomic spectra was known to be general; the combination principle continued to apply (with selection

88. Kramers 1920.

rules, of course); and part of the observed spectra, both in the optical and X-ray frequency ranges, exhibited a striking similarity with the hydrogen spectrum.

There was a simple explanation of this similarity: the emitting system in the relevant stationary states had to be analogous to the hydrogen atom. In the optical case and for high quantum numbers, the so-called "series spectra" were supposed to be emitted by a single electron revolving at a large distance from the rest of the atom, so that the only effect of the rest (or "core") was a screening of the charge of the nucleus. In the X-ray case the various emission lines were attributed to the transitions of an electron from an upper shell to an inner incomplete shell, in which the attraction from the nuclear charge far exceeded the perturbation by other electrons.[89]

At first sight these explanations were independent of the correspondence principle. They were, indeed, first proposed by theoreticians like Kossel and Sommerfeld without any reference to this principle. Bohr thought differently. In his opinion the very idea of inferring the characteristics of the motion from the observed spectra belonged to the strategy dictated by the correspondence principle. In other words, the fundamental relation stated by Bohr between atomic spectra and the periodicity properties of the underlying motion had both a deductive and an inductive side. This relation was deductive to the extent to which the allowed motions of the planetary model could be determined by a priori means; it was inductive whenever some characteristics of the observed spectra were used to help in the determination of orbits' properties.

The latter inductive side of the correspondence principle gave a first and fundamental piece of information: the actual motions of the electron systems in atoms had to be multiperiodic in order to be capable of yielding the observed discrete character of the spectrum. In Bohr's words, there had to be a "harmonic interplay" between the various electrons. Therefore, a selection had to be operated among the general motions of the non-multiperiodic systems corresponding to atoms with several electrons. Then, for a given continuous class of multiperiodic motions, the golden rule (100)

$$dE = \sum_i \bar{v}_i \, dJ_i$$

could be used to determine the variables $J$ to be quantized. In this way Bohr could in principle extend the quantum theory to the non-multiperiodic

89. For the history of studies on X-ray emission and absorption see Heilbron 1967, 1983.

orbital system, while Sommerfeld and his followers were confined to the search for alternative multiperiodic models.[90]

I will first illustrate Bohr's type of reasoning in the case of optical series, for large values of the principal quantum number $n$. In order to account for the approximate validity of the Rydberg formula, the orbit of the emitting electron, at least a large portion of it, must be similar to a Kepler ellipse. But whenever this electron approaches the atomic core, the orbit must depart from that of the Kepler motion. A priori, there is no reason why the resulting motion should remain multiperiodic. But it must be so according to the correspondence principle, and the effect of the core must therefore be limited, in a first approximation, to a precession of the ellipse in its plane (and also a precession of this plane if the core does not have a spherical symmetry). Following the golden rule (100) and the correspondence between harmonics and lines, to this precession corresponds an azimuthal quantum number $k$ subjected to the selection rule $\Delta k = \pm 1$. The resulting spectral pattern fits well the empirical series S, P, D, F if S corresponds to $k = 1$, P to $k = 2$, and so on.[91]

Well before these considerations by Bohr, Sommerfeld obtained this result in a less profound but more transparent way. He first replaced the several-electron system by a simple multiperiodic model obtained by substitution of a spherical potential for the atomic core. The resulting Hamilton-Jacobi equation has about the same form (for a potential departing only a little from the Coulomb one) as the one for the relativistic Kepler problem, so that the usual quantization gives a similar type of spectrum with two quantum numbers $n$ and $k$, albeit with a larger splitting of the $n$-levels. This method was less profound than Bohr's, for it relied on a simplified mechanical model, different from the more fundamental several-electron system (which consists of several negative point charges and a positive one interacting through Coulomb forces).[92]

With his method and without much further calculation, in 1921 Bohr could consider the case of "dipping" electrons, those that penetrate the core and feel a stronger attraction from the nucleus. Here the part of the orbit inside the core strongly departs from the elliptical shape of the outer loops. To comply with the correspondence principle, Bohr nevertheless assumed that no energy exchange took place between the outer electron and the core and that the consecutive outer loops were related to one

90. The expression "harmonic interplay" is found for instance in Bohr 1923b, trans., 16; another common expression is "harmony of motion," for instance in Bohr 1921a, 208.

91. Bohr 1920b, 47–48; Bohr [1921d], [108]–[112]; Bohr [1922c], [375]. Of course, this reasoning does not take into account the complex structure of spectra.

92. Sommerfeld 1915b; Sommerfeld 1916a, 131.

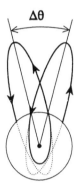

Figure 16.   A portion of a dipping orbit.

another by a constant rotation (fig. 16). The resulting periodicity properties being of the same type as before, the motion still had to be quantized in terms of the quantum numbers $n$ and $k$, the latter being subjected to its characteristic selection rule. On account of the large perturbing effect of the core, however, the Rydberg formula could no longer hold, even in a first approximation. Bohr corrected it in the following way.[93]

Assuming that the angular momentum of the outer electron is roughly conserved in the core, the radial action variable $J_r$ is the only one to be strongly affected by the penetration into the core. In a first approximation this modification, $\alpha_k$, depends only on the quantum number $k$, since the shape of the unperturbed ellipse inside the core (that is, near its focus) is roughly independent of $n$. Therefore the quantum conditions read:

$$J_r + \alpha_k = n'h, \qquad J_\theta = kh, \tag{145}$$

with $n = n' + k$. Since the largest portion of the orbit still belongs to unperturbed Kepler ellipses, the relation between energy and action variables is approximately the one used for the hydrogen atom,

$$E = -\frac{Kh^3}{(J_r + J_\theta)^2}. \tag{146}$$

Substituting (145) into this formula yields the modified Rydberg formula

$$E_n = -\frac{Kh}{(n - \alpha_k)^2}. \tag{147}$$

This type of expression fitted well the known series spectra of alkali atoms. The large observed values of the quantum defect $\alpha_k$, which of course motivated Bohr's above considerations, could be explained only if

93. Bohr [1921d], [141]–[144]; Bohr [1922c], [395]–[396]. See Kragh 1979a.

the penetration of the orbit into the core was deep enough. This implied a nonvanishing value of the eccentricity $n - k$ of the unperturbed ellipse. Consequently, the principal quantum number $n$ of the series electron had to exceed one in the fundamental state of alkali atoms. In early 1921 Bohr took $n = 2$ for all alkali metals.

Here again similar conclusions were independently reached on the basis of a definite multiperiodic model, provided this time by Erwin Schrödinger. In this model the atomic core was replaced by a thin uniform shell of negative electricity surrounding the nucleus, so that the potential had the Coulomb shape, both inside and outside the shell. Thus the action variables could be explicitly calculated, and a quantitative expression resulted for the quantum defect.[94]

In general, quantum physicists preferred explicit multiperiodic models, in Sommerfeld's style, to Bohr's subtle selection of multiperiodic motions from a more fundamental non-multiperiodic system. Nevertheless, even if the heuristic merits of the two methods could be compared, only Bohr's reference to the correspondence principle could show the essential and indispensable (model-independent) character of certain features of the theoretical description of series spectra: the existence of the quantum numbers $n$ and $k$, and the selection rule for $k$. A more thorough treatment of the interaction between core and outer electron would no doubt necessitate finer energy formulae and new quantum numbers; but, Bohr believed on the basis of the correspondence principle, the two quantum numbers $n$ and $k$ could not be contaminated.

As we shall later see, by late 1922 Bohr no longer believed that ordinary mechanics approximately applied to the mutual interaction between the electrons in a given atom. This did not affect in any manner his views about the alkali spectra and the essential connection of $n$ and $k$ with the periodicity properties of the outer loops of the series electron. The strange ability of the outer electron to cross the core without any energy loss instead confirmed their anticlassical behavior. As we shall presently see, Bohr based his subsequent construction of atoms on the two quantum numbers $n$ and $k$. And he would strongly oppose the opportunism with which the Munich physicists came to play with half-integral values of $k$.

## BUILDING ATOMS

In his trilogy of 1913 Bohr had proposed a model of all atoms (even molecules) in which electrons rotated on concentric coplanar circles with an

94. Schrödinger 1921.

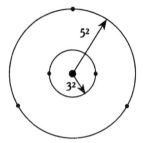

Figure 17. An example of a ring atom: nitrogen.

individual angular momentum $h/2\pi$ (see fig. 17). On a given circle or "ring" there were generally several electrons, arranged symmetrically in order to minimize their mutual repulsion. Superficially, this model resembled Thomson's old model, the only obvious differences being in the population scheme of the rings and in the distribution of the positive charge. However, the two models were radically opposed with regard to stability conditions. As mentioned above, Bohr's model lacked both the mechanical and the electrodynamic stability that Thomson had introduced. In principle the postulate of stationary states "took care" of both types of stability. In order to control—to some extent—the numbers of electrons that might occupy a ring, Bohr introduced a further criterion: one of energetic stability, namely, that the energy produced during the formation of the atom had to be the largest possible.[95]

As can be seen from his manuscripts, Bohr performed amazingly long numerical calculations based on the above stability criterion. Meanwhile, strong reasons accumulated to abandon the ring model and other related models. By 1920 Bohr himself had completely given up the planar rings. As he explained to Ladenburg, one could not build decent crystals or molecules with flat atoms. Nor could one account for band spectra and ionization potentials. Very much in the spirit of his subsequent atomic theory, to this wealth of empirical reasons Bohr added a theoretical one, far less definite but far more profound in his opinion: "This assumption [the ring atom] must be abandoned because of insufficient stability."[96]

Originally, by "stability" Bohr meant "stability in the sense of ordinary mechanics," as specified in his Berlin lecture of April 1920. In the ring model the only atom to provide such stability was the hydrogen atom; in

95. Bohr 1913, especially on 20–25 for the discussion of stability. See Heilbron and Kuhn 1969. For the history of Bohr's second atomic theory, see Kragh 1979a and Heilbron 1964.
96. Bohr to Ladenburg, 16 July 1920, *BCW* 4:[711].

other atoms a small perturbing force in the plane of the rings could lead to remote motions (the energetic stability implied mechanical stability only with regard to perturbations in the direction normal to the atomic plane). Stable motions, if there were any, had to be more complicated. But then the investigation of the effects of mechanical perturbations on such motions became practically impossible. Bohr therefore shifted toward a more constructive notion of stability, that the model should lead to a unique electronic motion for the normal state of a given atom.

Accordingly, in 1921 he gave the following retrospective justification of his rejection of the ring model: "The fundamental difficulty, involved in the assumption of a ring arrangement, consisted . . . therein that the picture offered no sufficient basis for an a priori fixation of a distribution of the electrons on the various rings." Indeed there was some freedom in the choice of the number of electrons on a given ring, since the effect of an inner ring on an outer one was limited to a screening of the nuclear potential. Stable atoms seemed, to Bohr, to require an intimate mutual coupling of all constituting electrons. The same criterion excluded a simple alternative to the ring model, the spatial shell models that were used by X-ray spectroscopists, because of the relative independence of successive shells.[97]

Bohr had another objection to the ring model, more generally to any model for which the configuration of the electrons was completely symmetrical at any instant (e.g., Sommerfeld's *Ellipsenverein*):

> We cannot expect [to find] in actual atoms, configurations of the type in which the electrons within each group are arranged in rings or in groups of polyhedral symmetry, because the formation of such configurations would claim that all the electrons within each group should be originally bound by the atom at the same time.

Bohr represented this mysterious argument as having been deduced from the correspondence principle, but without the detailed reasoning. His unpublished manuscripts nevertheless allow the following interpretation.[98]

Configurations with "polyhedral symmetry" have a zero electric moment, which makes them very singular with regard to the correspondence principle: they cannot emit or absorb any dipolar radiation, neither classically nor quantum-theoretically. Therefore they were not likely to be connected to other (asymmetrical) types of stationary states obtained by

97. Bohr 1920b, 60; Bohr 1921d, [105].
98. Bohr 1921a, 105.

adding electrons one by one to the bare nucleus. In general the correspondence principle suggested to Bohr the need to forbid transitions between states for which the electric moments (more specifically, their Fourier spectrum) were qualitatively different. The argument was of course far from being rigorous, but Bohr had convinced himself of its plausibility in the course of his and Kramers's study of the helium atom, as will presently be seen.[99]

Bohr had nothing to propose to replace the ring model before the fall of 1920. In July of that year he "confessed" to Ladenburg that he did "not consider any conception sufficiently assured as yet to make it possible to take a definite standpoint" about the constitution of atoms. Yet in December 1920 he was able to lecture on the main assets of a new theory of atomic structure, and in March 1921 he sent a letter to *Nature* summarizing his main results. There are two plausible origins to Bohr's sudden burst of inspiration: Landé's theory of the helium spectrum (1919) and Franck's measurement of the ionization potential of helium (1920), which will be discussed momentarily.[100]

The letter to *Nature* was in fact a reply to Norman Campbell, who had just denied in the same columns the possibility of a theory of atomic structure based on Bohr's orbits. According to the British philosopher-physicist the correspondence principle, with its characteristic abandonment of the classical relation between motion and field, had to lead one to consider the electronic orbits as "wholly fictitious." A more "real" model of atoms, he continued, would rather appeal to *static* electrons: following the views of Langmuir and Lewis, he argued that static models were more suited to explain chemical bonds and the building of crystals.[101]

Bohr strongly disagreed with this view. He admitted that "the correspondence principle, like all other notions of the quantum theory, [was] of a somewhat formal character"; but the success obtained from the application of this principle indicated a "reality of the assumptions of spectral theory" of a kind that allowed other physical and chemical properties of atoms to be explained on the same basis. In other words, orbiting electrons were not as real as a planetary system, but to Bohr they were more real than the competing static models, for they were supposed to provide a *universal* explanation of quite diverse phenomena, including spectra and chemical properties.[102]

99. Bohr [1921d], [135], [138]–[139].
100. Bohr to Ladenburg (see n. 96); Bohr 1920c, 1921a.
101. Campbell 1920.
102. Bohr 1921a.

Bohr went on to identify "a rational theoretical basis" for the construction of atoms: the correspondence principle, when applied to the radiation emitted during the formation of the normal state of atoms:

> [The correspondence principle] establishes an intimate connection between the character of the motion in the stationary states of an atomic system and the possibility of a transition between two of these states, and therefore, offers a basis for a theoretical examination of the process which may be expected to take place during the formation and re-organisation of the atom.[103]

Without much more detailed explanation Bohr gave the final result of his considerations, the electronic configuration of noble gases in terms of groups "$N_n$," wherein $N$ is the number of electrons in a group with the value $n$ of the principal quantum number:

| | | | |
|---|---|---|---|
| Helium | $(2_1)$ | Krypton | $(2_1\ 8_2\ 18_3\ 8_2)$ |
| Neon | $(2_1\ 8_2)$ | Xenon | $(2_1\ 8_2\ 18_3\ 18_3\ 8_2)$ |
| Argon | $(2_1\ 8_2\ 8_2)$ | Niton [radon] | $(2_1\ 8_2\ 18_3\ 32_4\ 18_3\ 8_2)$ |

This little table, and an improved version proposed in another letter to *Nature* (October 1921), roused a great excitement among other theoreticians of atoms. At the Solvay congress of 1921, in which the exhausted Bohr could not participate, everybody was very eager to hear Ehrenfest's report on this spectacular achievement. Sommerfeld did not wait long to assert, in the third edition (1922) of his *Atombau:*

> We have to recognize the complete superiority of the correspondence principle in the matter of atomic models. For here Bohr seems to have succeeded, by using classical mechanics and electrodynamics, in arriving at definite statements about the periodic system and the atomic shells which would have been inaccessible by any other route.[104]

Perhaps the mystery surrounding Bohr's considerations was not unconnected with the great enthusiasm they brought about. The letters to *Nature* were too vague and too concise to allow one to judge to what extent the correspondence principle had guided Bohr in his construction of the periodic table. At the Solvay congress information came only indirectly, through Ehrenfest, who had only partially been initiated into the realm of his friend's methods. The first more detailed account of Bohr's

103. Ibid.
104. Ibid.; Sommerfeld (see n. 77), 276. See Kragh 1979a.

ideas appeared in early 1922, but in Danish, while German and English readers had to wait a few more months. By June 1922, however, the Göttingen physicists had the privilege of hearing and freely interrogating the elusive Bohr.[105]

From the manuscript of these Göttingen lectures, and from an unpublished second part of Bohr's Solvay report, one may reconstruct Bohr's route to the so-called "second atomic theory" and the part that the correspondence principle played in obtaining it. In fact, Bohr appears to have made simultaneous and intricate use of two types of considerations. The first type was mostly inductive and relied on empirical information provided by series spectra, X-ray spectra, chemical properties, and so on. The second type, which Bohr hoped would be self-sufficient, pointed to a possible deductive use of the general assumptions of the quantum theory. The correspondence principle played a central role in both cases, respectively as a means of connecting empirical spectra and electronic orbits, and as an a priori procedure for deriving "exclusion rules" forbidding certain electronic configurations. Here again, we encounter the above-mentioned two-way use of this principle.[106]

The mysterious appearance of the deductive endeavors found in Bohr's second atomic theory disappears as soon as one considers that they were entirely inspired by an analogy with the Bohr-Kramers theory of the helium atom. I will now summarize the previous history of this theory on the basis of the abundant unpublished manuscripts found in the Bohr archive.

HELIUM

In the fall of 1916 Bohr and Kramers started extensive perturbative calculations of the classical orbits of the helium atom's two electrons. The spectrum of this atom had long been known to consist in two non-combining series spectra named ortho- and para-helium spectra. Transitions between the two types of terms never occurred, even in the presence of strong electric fields, so that physicists believed for a while that o-He and p-He were two different elements.[107]

In his first calculations with Kramers, Bohr investigated both coplanar and perpendicularly oriented orbits for the two electrons, and found, in the coplanar case, two simple classes of periodic motions that seemed

105. Bohr 1922a. On Ehrenfest's friendship with Bohr see Klein 1986.
106. Bohr [1921d], [1922c].
107. See Rud Nielsen's introduction to BCW 4, on [36]–[39], and Mehra and Rechenberg 1982a, 398. In modern quantum mechanics, the o-He and p-He spectra respectively correspond to the values $S = 1$ and $S = 0$ of the total spin, and the noncombination of the two spectra corresponds to the selection rule $\Delta S = 0$.

likely to correspond to the o- and p-helium spectra.[108] Nevertheless, a quantitative determination of the terms and of the ionization potential appeared to be far out of reach. The subsequent extension of the Bohr theory to multiperiodic systems did not ease the task: the classical three-body system was not multiperiodic, and the well-known difficulty of the corresponding celestial problem suggested that the helium problem would be at least as difficult. In 1919, however, Alfred Landé, an ambitious student of Sommerfeld's, jumped over these obstacles and proposed a simple multiperiodic model to which he could apply Sommerfeld's quantum rules.[109]

According to Landé, the inner orbiting electron could be replaced (for given values of the principal quantum numbers) with a rigid rotator making a constant angle with the angular momentum of the outer electron. This was a plausible approximation if the outer orbit was much larger than the inner, which should at least be the case for highly excited states. Then the quantization of the total angular momentum $j$, assuming unit angular momentum of the inner electron and a momentum $k$ for the outer electron, gave (in units $h/2\pi$):

$$j = k - 1, k, k + 1, \tag{148}$$

the two extreme values corresponding to coplanar orbits, and the middle one to almost perpendicular orbital planes (see fig. 18).[110]

Having excluded (for no good reason) the case $j = k - 1$, Landé could show that the quantized coplanar orbits fitted well with the o-He spectrum, while the perpendicular orbits fitted well with the p-He spectrum. He also identified the normal state as the stationary state corresponding to the lowest term in the o-He spectrum, since this state had the lowest energy. Even for this normal state Landé and Sommerfeld assumed the distinction between inner and outer orbits to be valid, which gave two concentric one-quantum ($n = 1$) orbits, in opposition to Bohr's earlier ring model (see fig. 19).

In the same year 1919 Franck and Knipping (also Horton and Davies) applied the Franck-Hertz technique of accelerated electrons to the first precise determination of the ionization potential of the helium atom (this

108. In the first class of motion the shape of the inner orbit was invariable, while in the second class the eccentricity of the inner orbit changed periodically under the perturbing influence of the outer electron. See Bohr 1922b (written in 1918), 105; Bohr to Landé, 26 June 1919, AHQP; and numerous helium calculations by Bohr or Kramers in BMSS.

109. Landé 1919a. See Forman 1970.

110. Landé developed his rather intricated procedures in Landé 1919b, 1920. For a clear account of Landé's method, see Bohr 1921d, [123]–[124].

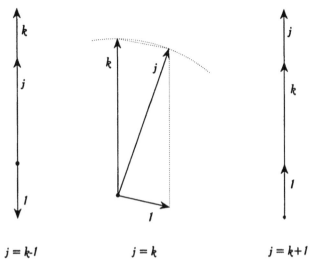

$$j = k\text{-}l \qquad\qquad j = k \qquad\qquad j = k\text{+}l$$

Figure 18.   Diagram for Landé's composition of a unit angular momentum with the momentum $k$.

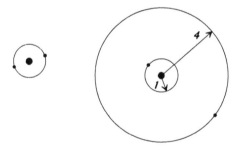

Figure 19.   Two planar models for helium, with $n = 1$ for both electrons: the ring model (left), and the Landé-Sommerfeld model (right).

energy was not accessible to optical measurements since the corresponding frequency lies in the far UV region). They found 25.5 V instead of the 30 V predicted by Landé and Sommerfeld (the ring model also gave about 30 V). This brought them to suspect that the normal helium state belonged to the p-He terms and that the state corresponding to the lowest o-term was only metastable. The two conjectures were soon supported by the following reasoning, due to Franck and Reiche.[111]

In 1914 Paschen had observed that the 10830 Å line of the o-He spectrum was a resonance line of the absorption spectrum obtained after the helium sample had been subjected to an electric discharge. On the one

111. Franck and Knipping 1919, 1920; Horton and Davies 1919.

hand, in Landé's interpretation of the helium spectrum, this line had to correspond to a transition from a two-quantum ($n = 2$) state of the outer electron. On the other hand, in Bohr's theory the large absorption and scattering (the frequency of the scattered light being the same as that of the absorbed light) characteristic of resonance lines were interpreted as due to transitions between a nondecaying stationary state (generally the normal state) and another excited state. Consequently, the two-quantum ($n = 2$) state of the o-He spectrum had to be metastable, and the one-quantum ($n = 1$) state of o-He had to be forbidden for some unknown reason. The normal state could now only belong to the p-He terms (see fig. 20).[112]

Bohr approved Franck and Reiche's clever analysis. He also congratulated Landé for his successful classification of the helium terms but criticized the roughness of the underlying model. He believed that his and Kramers's approach, being more true to the general principles of quantum theory, would be the only one able to provide quantitative results and a proper identification of the normal state. The strategy to be followed was clear but not easily executable. One first had to extract continuous classes of multiperiodic motions from among the motions of the exact orbital model; then one had to quantize the motions in each class with the help of Bohr's golden rule (100); finally, selection rules had to be established on the basis of the correspondence principle.[113]

Concerning the possible continuous classes of multiperiodic motions, Bohr and Kramers somewhat revised their original opinion (perhaps in reaction to Landé's connection of the p-He spectrum with perpendicular orbits). In 1920 they believed, until Born and Heisenberg proved the contrary, that there were only two such classes (once the class of symmetrical ring motions had been excluded), one corresponding to mutually perturbed concentric, circular, and coplanar orbits, the other corresponding to mutually perturbed perpendicular, circular orbits (the angle between the two orbits being affected by the perturbation). They also believed, until Born and Heisenberg proved the contrary of this as well, that these two classes were not connected by a continuous set of multiperiodic solutions. In this way they could explain very simply the existence of two separate series spectra and why they did not combine. Indeed, in the generalization to non-multiperiodic systems, the correspondence principle suggested a

112. See Franck and Reiche 1920. With his new vacuum spectrograph (for the extreme ultraviolet) T. Lyman soon detected the p-He spectral line (585 Å) corresponding to the transition from $n = 2$ to $n = 1$ (Lyman 1922).

113. Bohr to Franck, 18 Oct. 1920, BCW 3:[644]–[645]; Bohr to Landé, 24 Feb. 1920, BCW 4:[719]–[720]; Bohr [1921d], [132]–[133]. This manuscript contains, in its second paragraph ("Helium," [122]–[139]), the most detailed description of Bohr's and Kramers's ideas on helium by late 1921. Bohr 1922b, appendix (September 1922), [180]–[181].

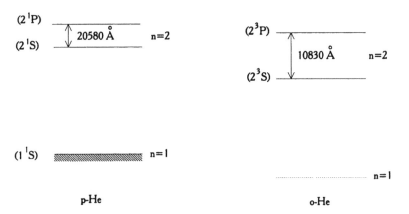

p-He                                                o-He

Figure 20.   Helium levels according to Franck and Reiche (the modern notation is given in parentheses).

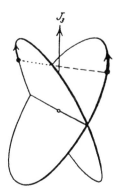

Figure 21.   The Bohr-Kemble model of the helium atom (from Born 1925, 331). $J_3$ denotes the total angular momentum.

correspondence between the topological structure (connectedness) of the space of classical motions and the connectivity (defined by the possibility of quantum transitions) of stationary states. In this respect, Bohr commented, Landé's model was necessarily wrong, since all resulting motions belonged to a unique continuous multiperiodic class.[114]

Among the multiperiodic motions of the orbital model Bohr still had to determine the one corresponding to the normal state, which was of course the most important task from the point of view of atomic building. There were three candidates for which the two electrons were in a one-quantum ($n = 1$) state: the old ring state, the Landé-Sommerfeld planar concentric state, and an intersecting state belonging to the p-He class (fig. 21). Bohr

114. Bohr [1921d], [122]–[139]; Bohr [1922c], [379]–[387]; Bohr [1921d], [124].

and Kramers first proved that the second candidate was mechanically excluded. Indeed, according to classical mechanics, the radial action variable corresponding to the motion of the outer electron in a Coulomb potential modified by the inner electron always exceeds the quantum of action if the inner electron is on a one-quantum orbit. This makes the one-quantum quantization of the motion of the outer electron impossible. It might be worth noticing that this mechanical type of exclusion already had an antecedent, namely, the exclusion of the value $k = 0$ in the Sommerfeld atom, on the ground that the corresponding orbit would cross the nucleus.[115]

To exclude the old ring state Bohr appealed to the correspondence principle, exactly in the same way as in the exclusion of transitions between o- and p-He states. The ring motion was not continuously related to the class of coplanar concentric motions and could therefore not be obtained by the radiative decay of stationary states corresponding to the latter motions. As Bohr put it in his Göttingen lectures, in order to form a ring atom from an o-He state, "one must so to speak demand a readiness on the part of the outer electron to come to an understanding with the inner electron." And such understanding presupposed some kind of continuity, as in Bohr's early meditations on language.[116]

At the very best, the one-quantum ring motion could have resulted from a simultaneous binding of the two electrons through a succession of larger ring motions with decreasing radius. But this was a very unrealistic way to form helium atoms. Here we have the key of Bohr's above-discussed exclusion of "configurations of polyhedral symmetry": such configurations could not be continuously related to earlier dissymmetric steps in the formation of the atom, at least if one could believe in the analogy with helium, the simplest atom with more than one electron.[117]

Having excluded the two planar candidates for the normal state of helium, one was left with the intersecting motion, obtained from the simple model of two intersecting one-quantum circular orbits by mutual perturbation. For the total angular momentum to be $h/2\pi$, the angle between the planes of the orbits has to be 120°. On the basis of this model (which was also proposed by Kemble on different grounds) Kramers calculated the ionization potential of the helium atom, taking the inverse of the charge number of the nucleus, $1/Z$, as a perturbation parameter (for large $Z$ the mutual perturbation of the two electrons can be neglected). To first order in $1/Z$ the result was 4V below the experimental value. When Bohr lectured in Göttingen in June 1922, he still hoped that the dis-

115. Bohr [1920c], [64]; Bohr [1921d], [133]–[134].
116. Bohr [1922c], [386].
117. Bohr [1921d], [135], [137]–[138].

crepancy resulted from the large value $\frac{1}{2}$ of the perturbation parameter. If this were true, the quantum theory, with the help of the correspondence principle, would account for all known empirical properties of helium.[118]

## TO HAFNIUM

Bohr and Kramers had reached the core of their theory of helium before the end of 1920. Bohr immediately turned to bigger atoms. The fundamental question (according to the *Aufbauprinzip*) was: "How may an atom be formed by the successive capture and binding of the electrons one by one in the field of the nucleus?" Bohr expected the ionic structure obtained after the addition of $z$ electrons to a given nucleus of charge $Z$ to be similar to the structure of the neutral atom with charge number $z$, as suggested by an (imaginary) adiabatic change of the charge of the nucleus from $Z$ to $z$. He also assumed that the quantum numbers $n$ and $k$, to which the correspondence principle gave a precise meaning for the last bound electron, were not modified by the further addition of electrons.[119]

In the case of alkali atoms the observed series spectrum implied, as explained above, the inequality $k < n$ for the outer electron, a necessary condition for the "dipping" character of the corresponding orbit. This excluded one-quantum outer orbits ($n = 1$). Consequently, in his letter to *Nature* of March 1921 Bohr took $n = 2$ for all alkali atoms; in the one of October 1921 he took for $n$ the number of the period in Mendeleev's table, which gave a better fit of the observed quantum defects. In both cases the lithium atom had the structure $1_1\ 1_1\ 2_1$ in Bohr's "$n_k$" notation. This differed of course from the old one-quantum ring arrangement.[120]

There were other convincing empirical reasons to introduce multi-quantum ($n > 1$) orbits in the building of higher elements. In 1915 Kossel had introduced the K, L . . . electronic shells in his interpretation of X-ray spectra; and the number of the shell, Sommerfeld could show, corresponded to the value $n$ of the quantum number $n$ appearing in spectral formulae of the Rydberg type. In 1919–20 Born and Landé needed elliptic orbits to explain the density of atom-packing in crystals, which also implied values of $n$ larger than unity, at least for the outer electrons.[121]

However, there was, according to Bohr, a far more fundamental reason to opt for increasing values of $n$ in successive electronic groups: the correspondence principle excluded other types of configurations, even without reference to empirical data. Although he may sometimes have given

118. Kemble 1921; Bohr [1922c], [387].
119. Bohr 1922a; Bohr 1922d, 75.
120. Bohr 1921a, 1921c.
121. See Heilbron 1967, 1983, and Kragh 1979a.

the opposite impression, Bohr did not have a general proof of this exclusion law. In fact, he relied entirely on an analogy with the reasonings made in the case of the helium atom. For instance, he forbade the $1_1$ $1_1$ $1_1$ configuration of lithium for the following reasons. In such a configuration the third one-quantum electron would be bound either in an orbit larger than those of the two first electrons, or in an "equivalent" orbit of the same type as the two previous ones. Bohr excluded the first alternative n analogy with the exclusion of the $1_1$ $1_1$ o-He (Landé-Sommerfeld) state, and the second in analogy with the exclusion of the ring state of helium. That is to say, the first exclusion was mechanical, while the second resulted from the correspondence principle. More generally, Bohr declared electron groups with a given $n$ (or subgroups with given $n_k$) to be closed whenever "the inclusion of a further electron would not show any resemblance with a process of transition between two stationary states of a multiple periodic motion."[122]

This statement expressed a hope rather than a deduction. In order effectively to determine the number of electrons in a closed group or subgroup Bohr had to rely on a questionable symmetry argument: no more than four equivalent orbits could exhibit a *spatial* symmetry allowing the "harmonic interplay" necessary for multiperiodic motion.[123] When no such a priori consideration was available, Bohr occasionally relied on empirical data. Such a complex mixture of deductive arguments, partial calculations, analogies, and empirical input is sometimes called physical intuition. Bohr certainly excelled in this type of reasoning. His periodic table of elements already resembled the modern one in several essential aspects, particularly the association of the principal quantum number with electronic groups. The greatest success came in 1922, when Coster and Hevesy discovered the missing element 72 (hafnium), after Bohr had told them where to find it, in zirconium ores (and not in rare earths, against a popular tradition).[124]

This spectacular achievement did not lead Bohr to overestimate the firmness of the ground on which his reasonings were based. At the end

---

122. Bohr [1921d], [141]–[143]; Bohr [1922c], [390]; quotation from Bohr [1921d], [15.].
123. See, e.g., Bohr 1922d, 91–92, 94. In his notion of harmonic interplay Bohr excluded a simultaneous visit of the core by the outer electrons, for reasons similar to those advanced in his earlier rejection of configurations of polyhedral symmetry. In opposition to his first atomic theory, he now assumed maximal stability to result from an "*intimate coupling* [by means of dipping orbits] *between the motions of the electrons in the various groups* characterized by different quantum numbers, as well as the *greater independence of the mode of binding within one and the same group of electrons* the orbits of which are characterized by the same quantum number" (ibid., 92).
124. See Kragh 1979a, 1980, and Rud Nielsen's introduction to BCW 4, on [30]–[32]

of his Nobel address of 11 December 1922 he announced Coster and Hevesy's findings and prudently concluded with the following general remarks:

> By a theoretical explanation of natural phenomena we understand in general a classification of the observations of a certain domain with the help of analogies pertaining to other domains of observation, where one presumably has to do with simpler phenomena. The most that one can demand of a theory is that this classification can be pushed so far that it can contribute to the development of the field of observation by the prediction of new phenomena. When we consider the atomic theory, we are, however, in the peculiar position that there can be no question of an explanation in this last sense, since here we have to do with phenomena which from the very nature of the case are simpler than in any other field of observation, where the phenomena are always conditioned by the combined action of a large number of atoms. We are therefore obliged to be modest in our demands and content ourselves with concepts which are formal in the sense that they do not provide a visual picture [Anskuelighed] of the sort one is accustomed to in the explanations with which natural philosophy deals.[125]

Not only had Bohr been aware, at least since his reply to Campbell, of the formal nature of his atomic orbits, but he was ready to modify their configurations in the face of new empirical information. In May 1924, after such modifications had proved necessary, he wrote:

> In fact the present state of the quantum theory hardly provides an unambiguous basis for conclusions as to the distribution of the electrons among the different subgroups of a completed or partially completed electron group and for testing such conclusions by comparison with experiment.[126]

The correspondence principle, even in the hands of its creator, failed to provide unambiguous guidance in the building of atoms. However, Bohr believed it to elucidate the character of the exclusions observed by nature.

---

125. Bohr 1923c, trans., 44. In the same vein, but with a humorous touch, Bohr had already written, in his preface (May 1922) to Bohr 1922d, vii: "It may be useful once more to emphasize, that—although the word 'explanation' has been used more liberally [in the Copenhagen lecture of 18 Oct. 1921]—we are not concerned with a description of the phenomena based on a well-defined physical picture. It may rather be said that hitherto every progress in the problem of atomic structure has tended to emphasize the well-known 'mysteries' [a probable allusion to Sommerfeld's "Zahlenmysterium," NW 8 (1920):61–64] of the quantum theory more and more. I hope the exposition in these essays is sufficiently clear, nevertheless, to give the reader an impression of the particular charm which the study of atomic physics possesses just on this account."
126. Bohr 1922d, 2d ed. (1924), appendix (May 1924), 127. The discussion of the spectrum of the carbon ion was one of the considerations that prompted Bohr, in February 1924, to revise the population of electronic subgroups: see Heilbron 1983, 286.

## SYSTEMATIC CALCULATIONS

Since Bohr's intuitive use of the correspondence principle seemed to perform miracles in the construction of atoms, the more mathematically inclined theoreticians in Munich and Göttingen engaged in rigorous calculations in order to check the validity of his reasonings or to reconstruct them when they were not publicly known. In a first period most of Bohr's guesses seemed to be confirmed. Sommerfeld's wonder-student, Wolfgang Pauli, calculated the $H_2^+$ ion, which could be reduced to a separable mechanical problem (Jacobi's problem of the two centers) in the approximation of fixed protons. He found three disconnected classes of motion, in conformity with Bohr's intuition of a general multiple connectedness of atomic motions. In order to select from among these motions those able to represent stationary states, he introduced four conditions, three of which had counterparts in Bohr's second atomic theory.[127]

Motions for which the electrons would collide with the nucleus had to be excluded, just like the orbits with $k = 0$ in Sommerfeld's atom; the temporal average of the total force acting on each proton had to be zero (this was automatically satisfied in the center-symmetric case of isolated atoms); according to the correspondence principle, the normal state had to belong to one of the continuous classes of motion, so that it could be reached by radiative decay from excited states. The latter condition was particularly faithful to Bohr's considerations. As Pauli explained, it excluded the old one-quantum ring model of the $H_2^+$ ion, for which the electron revolves on a circle at equal distances from the two protons, exactly in the same way as Bohr excluded the ring state of the helium atom: "Regarding energy and stability, the real normal state of $H_2^+$ is to the one-quantum circular orbit in the middle plane what the real normal state of He is to Bohr's earlier helium model."[128]

There was a fourth condition imposed by Pauli to the stationary motions: they had to be stable in the sense of ordinary mechanics. Here also Pauli seemed to believe that he was being faithful to Bohr. In the Berlin lecture of April 1920, Bohr had indeed excluded the ring atoms in the name of this type of stability; and there was, in Bohr's published writings prior to Pauli's work, no explicit renunciation of this type of criterion. We have seen, however, that mechanical stability did not play a role in Bohr's second atomic theory. The exclusion laws were believed to derive solely

127. Pauli 1922. See Jensen 1984. At that time there were no sufficiently accurate empirical data to be compared with the main quantitative prediction of Pauli's theory, the ionization energy of the $H_2^+$ ion.
128. Pauli 1922, 236, underlined by Pauli.

from the correspondence principle. The motion in the normal state did not according to Bohr necessarily have to be mechanically stable, since the postulate of stationary states could in principle take care of all kinds of mechanical instabilities. In fact the configuration proposed by Bohr and Kramers for the normal state of helium was *not* mechanically stable (although this might have not been known before mid-1922).

Nevertheless, in the case of $H_2^+$, Pauli was certainly right to demand mechanical stability. He gave the following fundamental reason for this: according to the general assumptions of Bohr's theory, ordinary mechanics had to apply to *slow* mechanical perturbations of $H_2^+$, since the corresponding mechanical system was a multiperiodic one, within the scope of the adiabatic principle.[129]

At that point Pauli introduced the following digression, which shows well his early sympathy with Bohr's way of thinking. Stability with respect to slow perturbations was necessary but not sufficient: one also had to consider the case of small fast perturbations. In this case, ordinary mechanics could not apply, as Bohr had repeatedly emphasized in his discussion of electron-impact experiments. Pauli therefore proposed a "mechanical correspondence principle" that would imply an agreement between "the really observable averages" given by classical and quantum theory for the collisions between a simple particle and a target system in a state with large quantum numbers. Some of this agreement had to survive in the case of moderate quantum numbers, in the spirit of Bohr's correspondence principle.[130]

To express this condition, averages had to be taken over any uncontrollable parameter like the impact parameter or the phase of the target system. Pauli insisted on this statistical character of the agreement between classical and quantum theory, as Bohr had done in the case of radiation. He even doubted that the deterministic character of the classical description of collisions would survive in the expected future quantum theory: "Ordinary mechanics unambiguously [zwangläufig] gives the course of the collision as a function of the initial conditions; one may doubt that this unambiguous relation corresponds to reality."[131]

Bohr approved the substance of Pauli's suggestion but not the terminology. The term "mechanical correspondence principle," he argued, ought to be reserved to a principle truly analogous to the correspondence principle. Just as the latter principle was a law of the quantum theory ruling

129. Ibid., 181–182.
130. Ibid., 184–189.
131. Ibid., 186.

quantum radiation processes, a mechanical version of it had to be a law of a yet unknown "quantum kinetics" ruling collisions between atomic particles. Instead Pauli's tentative principle, as helpful as it could be, seemed only to assert the approximate applicability of classical kinetics and not to help in the construction of its quantum version.[132]

Pauli finished this work in early 1922 in Göttingen and then worked on Born's program of systematic exploration of the formal apparatus of the quantum theory. As I have already mentioned, this collaboration led to a powerful extension of the Bohr-Kramers perturbation theory to any order of perturbation.[133] The resulting method, however, was not well adapted to the case of "accidental degeneracy," the situation in which the fundamental frequencies become commensurable *for a given value of the action variables*. While this type of degeneracy rarely occurs in celestial mechanics, it is systematically found in the case of the unperturbed charge system obtained by "switching off" the Coulomb repulsion between electrons in the nuclear atom, since the frequencies of quantized (unperturbed) Kepler motions are all commensurable.

A few months later Born and Heisenberg found in Poincaré's *Mécanique céleste* the proper way (Bohlin's method) to handle this most interesting case. In principle they could now determine, order by order, the stationary states of all atoms. They had not yet treated realistic examples, but results obtained on slightly simplified systems (with only one accidentally degenerate variable) were very encouraging. As Heisenberg reported to Pauli in December 1922:

> I am working with Born on an extension of your perturbation theory to include the case of "accidental" degeneracy. . . . The results are very exciting and quite remarkable. As a first result we obtain the *phase relation* between electrons; as a second result, the complete elimination of some periods. At this moment [and also in the final publication], I believe that every atom in its norma: state performs a *strictly* periodic [motion]; nevertheless, all degrees of freedom must be quantized [whereas there is only one quantum condition in the case of strictly periodic *systems*].[134]

Once again these results seemed to confirm Bohr's intuition. The phase relations granted the multiperiodicity of motions demanded by the correspondence principle; the strict periodicity of the motion in the normal state, if it was true, was likely to be related to the spatial symmetry requested by Bohr in electronic subgroups. Heisenberg further wrote: "Born

132. Bohr 1923b, trans., 12n.
133. See Born and Pauli 1922.
134. Heisenberg to Pauli, 12 Dec. 1922, *PB*, no. 30.

is very enthusiastic about these results, because we might now have a simple mathematical method to determine the relations of symmetry and the length of periods in the system of elements." Unfortunately, even before the publication of this beautiful theorem in March 1923, the adequacy of the formal apparatus of the Bohr-Sommerfeld quantum theory had become very questionable, as we shall now see.[135]

## SUMMARY

Although the first precise statement of the correspondence principle appeared in the context of multiperiodic systems, Bohr soon fruitfully applied it to more general cases. Systems with more than one electron were never multiperiodic, and their motion could not be calculated exactly. In such cases Bohr managed to organize knowledge and guide reasoning about atomic spectra and structure by combining two uses of the correspondence principle: a deductive one in which characteristics of the emitted spectrum were derived from a priori known properties of the motion, and an inductive one in which the reverse was done. The inductive use gave a first essential piece of information: the motion in stationary states had to be multiperiodic because of the discreteness of observed spectra. Therefore, from among all possible motions of a non-multiperiodic system one had to extract the multiperiodic ones. The correspondence principle was then applied a second time, deductively, in order to find quantum rules and selection rules and to make the implied predictions about the emitted spectrum.

With this kind of reasoning Bohr explained numerous features of atomic spectra, for instance the various series (S, P, D, F) of alkali spectra (1920) and their large quantum defects (departures from hydrogenlike series). The relevant multiperiodic motion was a planar precession of approximately elliptic loops of the outer electron (similar to the relativistic precession, but larger); the quantum number $k$ associated with this precession defined the various series, and the penetration of the loops into the atomic core accounted for the quantum defects. For sure, Sommerfeld had explained the S, P, D, F series well before Bohr, and Schrödinger dipped orbits into atomic cores at the same time as Bohr did. However, the method of Sommerfeld and his followers was fundamentally different. They first replaced the orbital system with a simplified multiperiodic model, for instance an electron coupled to a solid charged sphere, and then applied Sommerfeld's

135. Ibid.; Born and Heisenberg 1923a.

rules of quantization to the model. This procedure was more easily understood and therefore more popular than Bohr's, but did not allow one to
decide which features of the description were independent of the simplification introduced in the model. Instead, Bohr convinced himself that the
two quantum numbers $n$ and $k$ (principal and azimuthal) were incor-
ruptible attributes of atomic electrons, ones corresponding to well-defined
periodicity properties of the true orbital motion. This explains why he
used $n$ and $k$ to classify electrons in his subsequent atomic theory, and
why he disapproved of opportunistic plays with half-integral $k$.

From the beginning, Bohr's principal aim had been to explain the
properties of chemical elements and to deduce the length of the periods
in Mendeleev's table. His 1913 attempt was based on a "ring model" of
atoms in which electrons were arranged symmetrically on concentric circular rings. In the following years it became clear that no piling up of such
flat atoms would produce decent crystals and molecules. Bohr rejected not
only his own model but also some static models produced by chemists, because they were incompatible with the orbital picture; and he criticized the
"shell model" of X-ray spectroscopists on the grounds that the assumed
mutual independence of successive shells (save for the screening of the
nucleus's attraction) precluded a viable means of constraining the population of the various shells. Instead, he argued, the stability of atomic
systems would require a "harmonic interplay" of all electrons, the outer
electrons regularly coming close to the inner ones.

In 1921 Bohr elaborated this idea and reached his "second atomic
theory." The main result was a new table of elements, based on electron
"groups" and "subgroups," which were defined by the quantum numbers
$n$ and $k$. This achievement (a durable one, as we now know) raised great
excitement among physicists and chemists, the more so because it was
published in a very concise form, in two letters to *Nature*. Exhausted by
the cumulation of scientific and administrative duties (as director of the
newly built Institute for Theoretical Physics in Copenhagen), Bohr left
most of his secrets hidden for some time.

In his first letter Bohr just declared the correspondence principle to
be the "rational foundation" of his new theory, although the adiabatic
principle also played a role, especially in justifying the permanence of the
quantum numbers $n$ and $k$ during the building up of atoms. From his unpublished manuscript and from his later writings we may verify that most
of his arguments involved, again, subtle combinations of the inductive and
deductive aspects of the correspondence principle. On the inductive side,
atomic spectra provided information about the formation of atoms from

a bare nucleus through successive capture of electrons, and X-ray spectra about the reorganization of the electrons after perturbation of the inner structure. On the deductive side certain "exclusion rules". were obtained by applying selection rules to the theoretical formation process of the normal state. The mystery about these exclusion rules is largely dissipated as soon as one understands that Bohr's only evidence for them rested on an analogy with his and Kramers's extensive study of the helium case.

Bohr and Kramers had been working on helium since 1916, trying, among other things, to explain why the helium spectrum had two strictly noncombining sets of lines (o-helium and p-helium spectra). They progressed rather slowly until, in 1919, Landé found a successful multiperiodic model for the helium spectrum, and Franck and Reiche determined that the normal state of helium belonged to the p-helium terms. Bohr appreciated Landé's classification of the helium terms but rejected his detailed model, for to him it was not a legitimate approximation of the real two-electron system, and, most important, because it did not explain why o-helium and p-helium terms never combined. With Kramers's help he extracted two distinct classes of multiperiodic motions from the general motions of the two-electron system and identified their quantized states with the stationary states of o-helium and p-helium. He believed these two classes of motion to be topologically disconnected, which explained, through the correspondence principle, why o-helium and p-helium terms never combined. He could also show that the normal state had to correspond to the lowest stationary state of p-helium. Indeed, the other mechanically possible candidate, the old ring configuration, was excluded because a selection rule (resulting from the correspondence principle) forbade all transitions from excited helium states to the ring state. In other words, there was no possible history for the formation of a ring state.

For atoms with more than two electrons, Bohr's "deductions" largely depended on an analogy with the helium case. But this tenuous reasoning generally proceeded in parallel with induction from empirical data. For instance, the increase of the principal quantum number between successive groups was presented as derivable from the general assumptions of the theory but was also inferred from the quantum defects of alkali spectra. However, Bohr's presentations tended to emphasize the deductive side, which made his theory appear more profound and more predictive. His views seemed to be spectacularly confirmed when, in 1922, Coster and Hevesy found that the properties of element 72 agreed with his predictions. Yet Bohr was, at least in part, aware of the limits of his considerations. He regarded the atomic orbits as essentially formal, since they did not interact with light according to ordinary laws of electrodynamics. They were real only insofar as they provided a universal explanation of very different types of phenomena, both physical and chemical. Bohr also knew that his determination of electron groups and subgroups involved some questionable symmetry arguments (some symmetry being necessary for the harmonic interplay of electrons), reasoning which he had no qualms about modifying (e.g., in 1924).

In 1922–23 rigorous German calculations confirmed some of Bohr's intuitions. Pauli's exact treatment of the $H_2^+$ ion confirmed the possibility of disconnected classes of motion and thus the pertinence of exclusion rules similar to Bohr's. The improved perturbation theory of Born, Pauli, and Heisenberg also confirmed the general idea of a harmonic interplay. But very soon these wonderfully sophisticated methods brought more trouble than satisfaction.

# CHAPTER VIII

# A Crisis

## THE CATASTROPHE OF HELIUM

In November 1922 the American physicist John van Vleck published a new calculation of the ionization potential for the Bohr-Kemble model of helium. His method differed from Kramers's (the perturbation parameter being the sine of the angle between the two orbital planes), but his result, 20.71V, was strikingly close to Kramers's 20.65V. Although neither of the two perturbation techniques was a priori accurate, this convergence of results suggested the reality of the discrepancy with the improved empirical value, 24.6V. Before the end of the year, Kramers confirmed the clash on the basis of a more precise method of calculation.[136]

As we have seen in the preceding chapter, helium was the (half-hidden) paradigm of Bohr's second theory of atomic structure. Consequently, Bohr avoided interpreting Kramers's new calculation as leading to a complete breakdown of the orbital model. In the description of the interaction between the two electrons he admitted only a limited failure of classical mechanics. This mechanics, he hoped, still provided a proper classification of the types of orbital motions, to which the correspondence principle could then still be safely applied. But it could no longer be used to calculate the binding energy, or more generally the energy of the stationary state of atoms with more than one electron. The latter type of failure, Bohr now argued, was almost necessary, for the following fundamental reason.[137]

---

136. van Vleck 1922, 1923; Kramers 1923a.
137. Bohr 1923b, trans., 15–16; Kramers 1923a, 340–341.

Remember that Bohr believed the definition of energy to depend on the possibility of continuously deforming the system in a way that would connect the various stationary states. This property, established for the case of multiperiodic systems, could not be valid in the case of the helium model, for at least one reason: the motion in the normal state was mechanically unstable (as proved by Kramers). The energy of the helium model (and of all further orbital models) therefore suffered, Bohr concluded, an "indeterminacy [Unbestimmtheit] of a rather peculiar nature."[138]

In short, Bohr did not yet abandon orbits, the visual support of which he may have felt necessary, but he emancipated them from the rules of classical mechanics. After all, in his "On the quantum theory of line spectra" (1918) he had anticipated such a step, when noting that ordinary mechanics would apply to the Coulomb interaction of electrons only to the extent to which Coulomb forces could be separated from radiative forces. As we shall later observe (p. 180), at this early date he had also excluded (in the case of atoms beyond helium) an explanation of the complex structure and anomalous Zeeman effect on the basis of ordinary mechanics.

In his "Fundamental postulates" of November 1922, Bohr gave additional reasons to abandon the application of classical mechanics to electron orbits in atoms with more than one electron. One of them was a perceived analogy between the electron-electron interaction in a given atom, and the electron-atom interaction in electron impact experiments like those of Franck and Hertz. In these experiments occurred an utterly nonclassical, discontinuous exchange of energy above, and no exchange at all below, a certain threshold. The similarity was most obvious in the case of the interaction between atomic core and outer electrons, where a strange "transparency" of the core was required in Bohr's theory of alkali spectra. In general, for the description of encounters between atomic particles, Bohr pleaded for a new "quantum *kinetics*" that would provide a statistical connection between ingoing and outgoing stationary states and renounce a complete description of individual processes.[139]

But he maintained the concept of definite orbits in stationary states, in conformity with his general views on the necessity of classical concepts, as expressed in "The fundamental postulates":

The quantum theory presents a sharp departure from the ideas of classical electrodynamics in the introduction of discontinuities into the laws of nature.

138. Bohr, according to Kramers 1923a, 340.
139. Bohr 1923b, trans., 15, 11–12.

From the present point of view of physics, however, every description of natural processes must be based on ideas which have been introduced and defined in the classical theory. The question therefore arises, whether it is possible to present the principles of the quantum theory in such a way that their application appears free from contradiction.[140]

Five years later, Bohr dropped the prudent restriction: "from the present point of view." And he maintained the essential idea that the departure from classical laws imposed by the quantum postulate(s) still allowed for a consistent, limited application of classical concepts.

While Bohr was philosophizing on quantum riddles, in Göttingen Born and Heisenberg were striving to develop the applications of their grand perturbation theory. With this powerful mathematical machinery they soon struck the deductive side of Bohr's second atomic theory a fatal blow. In early 1923 they achieved a systematic derivation of the quantized orbits of excited helium, with devastating results. First of all, the derived spectrum departed quantitatively and qualitatively from the empirical one, in spite of a rough and partial agreement in the case of the lowest spectral terms (this had been provided by Landé's theory). More fundamentally, Bohr's two classes of orbits proved to belong to a larger continuous class of multiperiodic motions, and to correspond to two quantized choices of the total angular momentum, $j = k + 1$ and $j = k$, as prefigured by Landé's model. This ruined Bohr's correspondence argument for the non-combination of the o- and p-He terms.[141]

Heisenberg commented to Pauli: "Es ist ein Jammer." Born announced to Bohr a "Katastrophe." After Kramers and Pauli had checked the calculations in Copenhagen, Bohr wrote to Born: "In fact the result is very important as evidence of the inadequacy of the present basis of the quantum theory, as far as systems with several electrons are concerned." In their paper Heisenberg and Born concluded that some fundamental assumption of the Bohr-Sommerfeld theory had to be abandoned: either the expression of the quantum conditions (for instance, the quantum numbers would no longer be integers) or the application of ordinary mechanics to the electronic motion in stationary states.[142]

---

140. Ibid., 1.
141. Born and Heisenberg 1923b.
142. Heisenberg to Pauli, 19 Feb. 1923, *PB*, no. 31; Born to Bohr, 4 Mar. 1923, *BCW* 4:[669]–[670]; Bohr to Born, 2 May 1923, ibid., [673]; Born and Heisenberg 1923b, 243. In relation to his considerations on the anomalous Zeeman effect and after van Vleck's helium calculations, Heisenberg had tried half-integral quantum numbers in the helium model, with a certain success (see Heisenberg to Pauli, 12 Dec. 1922, *PB*).

In a subsequent essay, Bohr manifested his preference for the second alternative: "Born and Heisenberg's investigation may be particularly well suited to provide evidence of the fundamental failure of the laws of mechanics in describing the finer details of the motion of systems with several electrons." Unlike Born, however, he did not conclude that the quantum theory had been completely wrecked. The applicability of classical mechanics to the electronic motion in stationary states had never been, as has repeatedly been mentioned, a fundamental element of his quantum theory. To the contrary, in the light of his "Fundamental postulates" of November 1922, he wrote to Born: "It is possible to give a unified conception [einheitliche Auffassung] of the quantum theory in which the failure of mechanics for the stationary states fits naturally."[143]

Indeed, the two fundamental postulates, the one about stationary states and the relation $\Delta E = h\nu$, were independent of any assumption about the type of motion in the stationary state. So was the correspondence principle, once understood as a correspondence between spectrum and harmonics of motion. Finally, the adiabatic principle could be formulated without reference to classical mechanics, as "the principle of the existence and permanence of the quantum numbers" that would serve the same purpose in atom-building. The new type of "permanence" was considered to exist under both slow and fast (but small) external perturbations and therefore secured the desired stability of atoms. Not only was the latter principle expressed in purely quantum-theoretical concepts; it comprehended certain types of violations of ordinary mechanics, for instance the absence of atomic excitations in electron-impact experiments below the energy threshold, and the anomalous transparency of atoms to slow electrons that Ramsauer had just observed (and to which I will later come back).[144]

In an unpublished sequel to "The fundamental postulates" and in a subsequent paper on atomic structure (in both of which Pauli participated), Bohr explained how, thanks to the above-described formulation of the principles of the quantum theory without classical dynamics, he could save his construction of the periodic table of elements. The existence of the quantum numbers $n$ and $k$, and sometimes their value (in the case of alkali spectra), resulted from the interpretation of series spectra through the correspondence principle; their stability during the building process of

143. Bohr 1923e, 271n; Bohr to Born, 2 May 1923, BCW 4: [673].
144. Bohr 1923b, trans., 16; Bohr [1923f], [507]–[508].

atoms reflected the new principle of permanence. However, Bohr was now forced to eliminate any reference to the a priori determination of the value of $n$. Indeed, his original argument, as we saw, rested on an analogy with the existence of disconnected classes of multiperiodic motions in the case of helium. Once Born and Heisenberg had deprived this analogy of its basis, the correspondence principle lost all the deductive power Bohr had dreamed of in his second atomic theory.[145]

In June 1923, from Copenhagen Pauli reported to Sommerfeld that Bohr's theory could not explain the length of periods in Mendeleev's table. The next month, for the tenth anniversary of the Bohr atom, Kramers commented publicly on this (hopefully) provisional failure of the correspondence principle: "For the moment we have been unable to deepen this type of consideration on the closing of groups; this is mainly made difficult by the failure of classical mechanical laws in the description of [orbital] motion." At the very best, Kramers pertinently noticed, the correspondence principle would forbid the addition of an electron to a closed group through a radiative transition but not through a process of collision. Gone was what Sommerfeld had earlier called "the complete superiority of the correspondence principle in the matter of atomic building."[146]

## ORBIT-KILLERS IN THE ZEEMAN JUNGLE

In 1918, while writing the third part of "On the quantum theory of line spectra," Bohr gave some thought to a notorious anomaly of the atomic theory. According to the correspondence principle, the only possible effect of a weak magnetic field on a spectral line was a triplet splitting, as a result of the correspondence of the quantum spectrum with the Fourier spectrum of the classical electronic motion, which would be subjected to Larmor's precession in such a field. This triplet structure was observed in the case of hydrogen and helium (although a better spectral resolution would have given different results). Alas, in the case of other atoms, many lines were known to split into more complex patterns. More specifically, an anomalous pattern appeared whenever the unperturbed lines belonged to narrow doublets or triplets (forming the so-called "complex structure" of spectra). And it disappeared—that is to say, the normal Zeeman triplets reappeared—as soon as the magnetic splitting became much larger than

---

145. Bohr [1923f], [507]–[509]. See also Bohr 1923e; BCW 4:[576]–[578].
146. Pauli to Sommerfeld, 6 June 1923, PB, no. 37; Kramers 1923b, 557.

the complex structure, as F. Paschen and E. Back had observed for the first time in 1912.[147]

Bohr concluded: "These [anomalous Zeeman] effects which clearly have intimate connection with the unknown mechanism responsible for the doubling of the lines can obviously not be explained on the basis of the general considerations mentioned above [Larmor theorem and correspondence principle]." He further alluded to a possible connection of this anomaly with the absence of paramagnetism for many elements despite a nonvanishing magnetic moment (in the ring model). The natural way out was to admit a violation of Larmor's theorem, and a violation of ordinary mechanics for the interactions responsible for the complex structure and the anomalous magnetic effects.[148]

In the same text of 1918 Bohr attributed the complex structure to an interaction between the atomic core and outer electrons, and gave an a priori reason to expect a failure of ordinary mechanics in this interaction: the corresponding perturbation of the core could not be calculated on the basis of ordinary mechanics because a core made of electron rings was mechanically unstable. When, in September 1922, Bohr finally made up his mind to publish this text (as the third part of "On the quantum theory of line spectra"), he mentioned in an appendix that the latter argument remained valid after the abandonment of the ring model, because the atomic core was still mechanically unstable (as explicitly proved by Kramers in the case of a heliumlike core).[149]

As a corollary of the above discussion, from Bohr's point of view (at least) the complex structure and anomalous Zeeman effect could not be the source of the crisis of the quantum theory which arose in the winter 1922–23. Since 1918 he had been aware of the central difficulty connected with anomalous Zeeman splitting: if the correspondence principle were true, there could be no question of an explanation of this effect as long as the validity of ordinary mechanics in stationary states was maintained.

147. See Forman 1968; also Mehra and Rechenberg 1982a. In the modern terminology the complex structure is generally called fine structure, since there is no essential difference between the fine-structure splitting in hydrogen and the complex structure in higher elements: both are relativistic effects mostly due to spin-orbit (and spin-spin) coupling. The spin-orbit coupling energy is a rapidly increasing function of the charge number (about $(\alpha Z)^2 E_0$, if $\alpha$ is the fine-structure constant ($\alpha = e^2/hc$), and $E_0$ the fundamental energy of the outer electron). This explains why the fine structures of hydrogen and helium were the only ones which were not enough resolved to give an observable (anomalous) Zeeman splitting of their components (for a history of the latter point see Jensen 1985). As we shall presently see, in the old quantum theory the fine structure of hydrogen and the complex structure of higher elements were thought to be of different origins.
148. Bohr 1922b (written in 1918), 111.
149. Ibid., 105; ibid., appendix (September 1922), 114.

In his eyes, the real source of the crisis of 1922–23 was the failure of the helium model, for this failure occurred precisely where he expected a major success, one able to serve as a paradigm of his second atomic theory.

Bohr's initial belief that the helium orbits were immune to a breakdown of ordinary mechanics was supported by two considerations: an empirical one, namely, that the corresponding Zeeman patterns were normal at the available precision; and a theoretical one, namely, that in excited helium, the inner system, a single electron on a quasi-circular orbit, was obviously stable. Therefore, Bohr expected troubles to start only with lithium, and from his point of view that would have affected only the finer details of the orbital motion, ones irrelevant to atomic building.[150]

For some other specialists, like Sommerfeld and Landé, the anomalous Zeeman effect seems to have played a more important part in the recognition of a crisis in the quantum theory. However, they did not conclude that the Bohr-Sommerfeld theory was impotent in the realm of such problems until, in 1921–22, explicit, quantitative models of the Zeeman effect showed a necessary departure from Larmor's theorem. Bohr's earlier proof of a necessary breakdown of Larmor's theorem in higher atoms had little chance to convince them, since it rested on a not fully understood correspondence principle. They had no taste for Bohr's "philosophy" and were more attracted by the "number mysteries" (Sommerfeld) of atomic spectra, especially by the striking regularities of anomalous Zeeman patterns.[151]

By late 1922, the gap between what Bohr could prove on the basis of the general assumptions of his quantum theory and what model-builders could simulate had considerably widened. The successful but eccentric models of the anomalous multiplets contributed to the feeling of crisis. On the positive side, Bohr and Pauli hoped that these models could perhaps serve as indicators of the extent and nature of the breakdown of classical concepts in the quantum theory. In 1925, while a new quantum mechanics was about to be born, Pauli judged this intuition to have been largely confirmed:

How deep the failure of known theoretical principles is, appears most clearly in the multiplet structure of spectra. While during the past few years physicists were able to derive empirical regularities of an astonishing simplicity and beauty, and of a very considerable generality, one cannot do justice to the simplicity of these regularities within the framework of the usual principles of the quantum

150. See Bohr 1922b, 105, 117n.
151. Sommerfeld 1920a. See Forman 1970.

theory. It even seems that one must renounce the practice of attributing to the electrons in the stationary states trajectories that are uniquely defined in the sense of ordinary kinematics.[152]

In the following I will retrace the events that led Pauli to this questioning of ordinary kinematics.

In his Berlin lecture of April 1920, Bohr expounded his ideas on the nature of the magnetic anomalies. He emphasized the role of the correspondence principle in his conclusion about the failure of Larmor's theorem and manifested his belief that the complex structure of spectra and the anomalous Zeeman patterns had a common origin, to be found in a nonmechanical coupling between atomic core and outer electrons. As a result of this coupling, the motion of the outer electrons had to "possess a somewhat more complicated character than that of a simple central motion." Bohr also recalled Voigt's old theory of the Zeeman and Paschen-Back effects of the yellow sodium doublet, a simple model based on three coupled oscillators. This model, Bohr suggested, could be to the final quantum theory of the anomalous Zeeman effect what Lorentz's model had been to the Debye-Sommerfeld theory of the normal Zeeman effect.[153]

Meanwhile, Sommerfeld was trying to express the empirical data on complex structure and Zeeman multiplets in quantum-theoretical language. More specifically, he wanted to find a system of spectral *terms* and selection rules from which these data would derive. In the spring of 1920, this led him, in the case of the complex structure, to the "inner quantum number" $i$. To a given $n_k$ term in the Rydberg scheme ($n$ for the principal quantum number, $k$ for the azimuthal one) corresponded a multiplet, whose terms were labeled $i$, according to the following table:

| multiplicity | singlet | doublet | triplet |
|---|---|---|---|
| value of $i$ | $k$ | $k, k-1$ | $k, k-1, k-2$ |

The selection rules were

$$\Delta k = \pm 1 \quad \text{and} \quad \Delta i = 0, \pm 1. \tag{149}$$

152. Pauli 1926a, 167.
153. Bohr 1920b, 58–59.

Even though he alluded to a "hidden rotation," Sommerfeld left the precise origin of the inner quantum number open.[154]

A few months later Landé gave a similar treatment of the anomalous Zeeman patterns. In the presence of a (weak) magnetic field, a given $(n, k, i)$-term with the energy $E_0$ had to give rise to a multiplet with the energies

$$E(n, k, i, m) = E_0(n, k, i) + g(k, i)mh\bar{v}_L, \tag{150}$$

where $\bar{v}_L$ is the Larmor frequency, $m$ is the "magnetic quantum number," and $g$ is the so-called Landé factor, which was always given as a simple rational fraction composed of $k$ and $i$. For instance, $g = 1$ in the singlet case, and $g = 2i/(2k - 1)$ in the doublet case. Any departure of $g$ from unity directly implied a violation of Larmor's theorem. In analogy with the normal Zeeman effect Landé submitted the magnetic quantum number $m$ to the selection rule

$$\Delta m = 0, \pm 1. \tag{151}$$

In the singlet and triplet case $m$ was an integer restricted by $|m| \leqslant i$. But, very strangely, it took the following *half-integral* values in the doublet case:

$$m = \pm\tfrac{1}{2}, \pm\tfrac{3}{2}, \ldots, \pm(i - \tfrac{1}{2}). \tag{152}$$

This normalization was the only one compatible with the selection rule $\Delta m = 0, \pm 1$ and the symmetry of all Zeeman multiplets with respect to the unperturbed term.[155]

If one believed that the magnetic quantum number had something to do with a precession around the magnetic axis, Landé commented, the selection rule $\Delta m = 0, \pm 1$ was the only one compatible with the correspondence principle (however, Sommerfeld preferred the rule $\Delta m = 0, \pm 2$,

154. Sommerfeld 1920b, 232. In the modern theory of $L \cdot S$ coupling, Sommerfeld's singlets correspond either to $S = 0$ (first kind) or to $S = \tfrac{1}{2}$ with $L = 0$ (second kind), his doublets correspond to $S = \tfrac{1}{2}$ with $L \geqslant 1$, and his triplets correspond to $S = 1$ with $L \geqslant 1$. The correspondence with today's quantum numbers is $n \to n$, $k \to L + 1$; $i \to J + 1$ for singlets of the first kind, $i \to J + \tfrac{1}{2}$ for singlets of the second kind and doublets, $i \to J$ for triplets. Taking into account the selection rule $\Delta S = 0$, Sommerfeld's selection rules correspond to $\Delta L = \pm 1$ and $\Delta J = 0, \pm 1$.

155. Landé 1921a, 1921b. Except for singlets, Landé's $m$ is identical with the modern $m_J$, which takes the values $\pm J$, $\pm(J - 1), \ldots$, the last one being either 0 or $\tfrac{1}{2}$, according to the parity of $2J$. In the case of singlets of the first kind (see n. 154), the true range of $m$ should be $|m| \leqslant i - 1$. In the case of singlets of the second kind, $m = 0, \pm 1$, while $m_J = \pm\tfrac{1}{2}$, which is compensated by Landé's exclusion of $m = 0$ in this case, and by his choice of $g = 1$ (instead of 2 in the modern theory for $S = \tfrac{1}{2}$ and $L = 0$).

the values of $m$ being twice Landé's). The same principle, or Rubinowicz's considerations on the angular momentum of radiation, suggested to Landé an interpretation of $i$ as the total angular momentum of the atom. Bohr of course approved the reasoning and specified the type of motion to be associated with $i$ (or $j$ in Bohr's notation): a uniform precession of the plane of the outer electron orbit around the vector i.[156]

In the meantime, Sommerfeld followed Bohr's suggestion for a quantum-theoretical version (*Umdeutung*) of Voigt's spectrum, which promised to give, in the case of doublets, a generalization of Landé's formula that would apply for arbitrary magnetic fields. That is to say, he found a system of term energies and selection rules corresponding to this spectrum. For the splitting of a $k$-term in an arbitrary (small or large) magnetic field B, he derived

$$\delta E = h\bar{v}_L \left( m \pm \frac{1}{2} \sqrt{1 + 2v \frac{m}{k - \frac{1}{2}} + v^2} \right). \tag{153}$$

In this expression $m$ takes the half-integral values $\pm\frac{1}{2}, \pm\frac{3}{2}, \ldots, \pm(i - \frac{1}{2})$, $\bar{v}_L$ is the Larmor frequency $eB/4\pi\mu c$, and $v = \delta/h\bar{v}_L$, wherein $\delta$ is the doublet width. The selection rules are $\Delta m = 0, \pm 1$ and $\Delta(\pm) = 0$. For small fields ($\bar{v}_L \to 0, v \to \infty$), this gives

$$\delta E = \pm \delta/2 + m \left( 1 \pm \frac{1}{2k - 1} \right) h\bar{v}_L. \tag{154}$$

The first term yields the doublet in absence of field, and the second one reproduces Landé's expression (150) for the anomalous Zeeman effect of a doublet (since $1 \pm 1/(2k - 1) = 2i/(2k - 1)$, with $i = k, k - 1$). In the large field limit ($v \to 0$), the general formula (153) reduces to

$$\delta E = h\bar{v}_L(m \pm \frac{1}{2}), \tag{155}$$

where $m \pm \frac{1}{2}$ takes all integral values between $-k$ and $k$. This reproduces the normal Zeeman pattern, in conformity with the Paschen-Back effect.[157]

### HEISENBERG'S *RUMPF* MODEL

In late 1921, a young student of Sommerfeld in Munich, Werner Heisenberg, managed to find a simple quantum-theoretical model that repro-

---

156. Landé 1921a, 240. See, e.g., Bohr [1923f], [546]–[549].
157. Sommerfeld 1922.

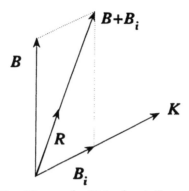

Figure 22.   Diagram for Heisenberg's *Rumpf* model.

duced all regularities expressed in Landé's and Sommerfeld's formulae and even predicted new ones for the case of triplets. In the case of doublets in alkali atoms, to which I will limit my account, Heisenberg's model was based on the following assumptions:[158]

1.   The atomic core borrows (on average) from the outer electron an angular momentum $R = \frac{1}{2}$ (in units $h/2\pi$), leaving a momentum $K = k - \frac{1}{2}$ the outer electron.[159]

2.   The rotating core (or *Rumpf*) orients itself in a direction parallel or antiparallel to the total magnetic field $\mathbf{B} + \mathbf{B}_i$, wherein $\mathbf{B}$ is the external field, and $\mathbf{B}_i$ the magnetic field created (on average) by the outer electron at the place of the core (see fig. 22).[160]

3.   The angle between the angular momentum $\mathbf{K}$ of the outer electron and the external field (if any) is quantized according to $\mathbf{K} \cdot \mathbf{B} = mB$, where $m$ takes half-integral values (in units $h/2\pi$).[161]

158. Heisenberg 1921. See Cassidy 1979.
159. In a manuscript entitled "Memorabilia" (BMSS, 1917 or 1918) Bohr has written, "possibility of an explanation of anomalous Zeeman effect by assuming that in a magnetic field the total angular momentum round axis (equal to entire multiple of $h/2\pi$) is divided in fractions between outer and inner electron," as probably suggested by the appearance of simple rational fractions (with the so-called "Runge denominator") in the expression of magnetic energy shifts.
160. Heisenberg assumed that K and $\mathbf{B}_i$ shared the same direction, even though the orbital picture of the outer electron leads to opposite directions.
161. Heisenberg left the origin of the range of $m$ ($|m| \leqslant i$ instead of $|m| \leqslant K$) open to further studies (Heisenberg 1921, 284).

4. The term shift of the atomic system is given by the model's magnetic energy, as determined by the ordinary theory of magnetic interactions.

This magnetic energy is

$$\delta E = \frac{e}{2\mu c} \, (\mathbf{K} \cdot \mathbf{B} + \mathbf{R} \cdot \mathbf{B} + \mathbf{R} \cdot \mathbf{B}_i), \qquad (156)$$

if $\mu$ is the mass of the electron. Assumptions (1) and (2) lead to

$$\mathbf{R} \cdot (\mathbf{B} + \mathbf{B}_i) = \pm R|\mathbf{B} + \mathbf{B}_i| = \pm \tfrac{1}{2}B(1 + 2v \cos (\mathbf{B}, \mathbf{B}_i) + v^2)^{1/2} \quad (157)$$

with $v = B_i/B$. The field $\mathbf{B}_i$, created by the orbital motion of the outer electron, is parallel to the orbital momentum $\mathbf{K}$, which implies

$$\cos (\mathbf{B}, \mathbf{B}_i) = \cos (\mathbf{B}, \mathbf{K}) = \frac{m}{K} = \frac{m}{k - \tfrac{1}{2}}. \qquad (158)$$

Assumption (3) and the formulae (156) and (157) finally give

$$\delta E = h\bar{v}_L \left( m \pm \frac{1}{2} \sqrt{1 + 2v \, \frac{m}{k - \tfrac{1}{2}} + v^2} \right), \qquad (159)$$

in agreement with the Voigt-Sommerfeld formula (153).

The success was great, but the method questionable, as Heisenberg himself commented in a first private report of his result to Pauli: "Success sanctifies means." From the point of view of the quantum theory of multiperiodic systems, the assumption (4) was the only conservative one: it maintained the classical expression for magnetic energies and the classical value $(e/2\mu c)$ of the gyromagnetic factor for the core and the outer electron. The assumptions (1) and (3) departed from the usual quantization rules since they both involved *half-integral* quantum numbers, and since assumption (3) deliberately omitted the contribution of the core in the spatial quantization of the *total* angular momentum ("passivity of the *Rumpf*"). Further, the empirically known range of $m$, given by $|m| < i$, received no explanation.[162]

Another major offense to general principles was contained in assumption (2): if classical mechanics applied to the motion in the stationary states of the *Rumpf* model, the equations of motions had to be

$$\dot{\mathbf{j}} = -\frac{e}{2\mu c} \, \mathbf{J} \times \mathbf{B} \qquad (160)$$

162. Heisenberg to Pauli, 19 Nov. 1921, *PB*, no. 16.

for the total angular momentum $J = R + K$, and

$$\dot{R} = -\frac{e}{2\mu c} R \times (B + B_i) \tag{161}$$

for the angular momentum of the core. The first equation implies a precession of $J$ around $B$, while the second, if assumption (2) is to be true, implies the permanence of $R$. These two conditions are clearly incompatible.

In spite of general admiration for Heisenberg's ingenuity, no major theoretician could accept the transgressions of general principles contained in the work, save for Sommerfeld, the leader of what Heisenberg called the seminar of *Atommystik*. In his Göttingen lectures, Bohr judged Heisenberg's paper "very promising" but found the involved assumptions "difficult to justify," which for Bohr meant perfectly intolerable. Bohr certainly admitted that any successful model of the anomalous Zeeman effect had to depart from the general assumptions of the Bohr-Sommerfeld theory. But he could not accept violations of rules that pertained to the corroborated part of the quantum theory. Accordingly, he judged Heisenberg's major sin to be his recourse to half-integral values of the azimuthal quantum number $k$ ($K$ in Heisenberg's notation). This conflicted both with the correspondence principle, according to which $k$ had to correspond to the advance of the outer loops of the outer electron, and with the permanence of $k$, which was part of the foundation on which the building (*Aufbau*) of atoms took place.[163]

To Pauli and Landé, Heisenberg's worst offenses were the "passivity of the core" implied by assumption (3) and the violation of the theorem concerning the motion of angular momentum. In general, Pauli judged severely the opportunistic trend of Heisenberg's works in this period: "He [Heisenberg] is very unphilosophical, he does not care about a clear elaboration of the fundamental assumptions and of their relation to the previous theories."[164]

STURM UND ZWANG IN COPENHAGEN

From September 1922, Pauli spent a year in Copenhagen, and helped Bohr to explore, among other things, the mysteries of the anomalous Zeeman

163. "Atommystik" is in Heisenberg to Pauli, 19 Nov. 1921, *PB*, no. 16; Bohr [1922c], [391]; Bohr 1923e; *BCW* 4: [647]n.
164. Pauli's position was inferred from Heisenberg to Pauli, 25 Nov. 1921, *PB*, no. 17, and 17 Dec. 1921, *PB*, no. 18; Landé 1922: in this article Landé managed to do without the passivity of the core; Pauli to Bohr, 11 Feb. 1924, *PB*, no. 73.

effect. He first tried his best to find a multiperiodic model of this effect: one that would necessarily involve some extramechanical property of the core but that would nevertheless retain the regular quantum theory of multiperiodic systems with integral quantum numbers. As Bohr put it in March 1923, after the failure of Pauli's attempt, "It was a desperate attempt to remain true to the integral quantum numbers; we hoped to find in the very paradoxes an indication of the path along which one should search for a solution of the anomalous Zeeman effect."[165]

Confronted with this failure, Bohr was pressed to locate the precise type of departure from ordinary mechanics needed to account for the success of Heisenberg's model. To this end he proposed the notion of *unmechanischer Zwang,* a form of nonmechanical stress occurring in the interaction between the atomic core and outer electrons. In order to be fundamental, this notion had to be independent of particular models, and of the specific labeling of multiplet terms favored by Sommerfeld and Landé. Thus Bohr reasoned in terms of the a priori statistical weights of the $n_k$ states, for they had a direct empirical meaning, as the total number of terms in a magnetic field corresponding to a given value of $n$ and $k$.[166]

Consider the case of alkali doublets. On the one hand, the multiplicity associated with $n_k$ had to be $2(2k - 1)$, because, according to Landé, there were $2k$ choices of $m$ corresponding to $i = k$, and $2k - 2$ choices of $m$ corresponding to $i = k - 1$. On the other hand, for a vanishing coupling between core and outer electron, the statistical weight of the core had to be *one* in order to account for the diamagnetism of the corresponding noble gas;[167] and the multiplicity of the outer electron would be that given by the Sommerfeld atom: $2k$ (Bohr and Sommerfeld excluded the value $m = 0$ on the grounds that the corresponding orbit is adiabatically connected to an orbit passing through the nucleus).[168] The resulting total

165. Bohr to Landé, 3 Mar. 1923, AHQP. At the end of 1922 Pauli sent a manuscript to Pauli containing an analysis of the anomalous Zeeman effect with integral quantum numbers only (see Heisenberg to Sommerfeld, 4 Jan. 1923, AHQP), but he soon surrendered to Heisenberg's objections (see Heisenberg to Pauli, 21 Feb. 1923, PB) and admitted with Bohr a half-integral $j$.

166. See the excellent account in Serwer 1977.

167. Indeed, the absence of paramagnetism implies a vanishing net magnetic moment of the atom, and therefore no magnetic splitting of its energy. This result seemed to be at variance with Bohr's second atomic theory, which gave a unit angular momentum to all noble gases (due to the heliumlike K-shell, the other shells being saturated and symmetrical). Bohr avoided the contradiction by admitting a violation of ordinary mechanics already in the diamagnetic behavior of noble gases (cf. n. 148).

168. Interestingly, the exclusion of $m = 0$ led to the correct number of Zeeman components for the levels of the hydrogen atom: to a $n_k$ level in the Sommerfeld atom corresponds a definite value of $J = k - \frac{1}{2}$ in the modern theory, which gives $2J + 1 = 2k$ magnetic sublevels.

multiplicity was thus $1 \times 2k = 2k$. This result was incompatible with the existence of a multiperiodic model of the interaction between core and electron, since in such a model an adiabatic variation of the coupling strength would have conserved the total statistical weight of an $n_k$ term.[169]

In March 1923 Bohr concluded:

> The coupling of the series electron to the atomic core is subject to a *stress* [*Zwang*] which is not analogous to the effect of an external field, but which forces the atomic core to adopt two different orientations in the atom, instead of the single orientation possible in a constant external field, while, at the same time, as a result of the same stress, in the atomic assemblage the outer electron can only assume $2k - 1$ orientations in an external field instead of $2k$.[170]

This way of splitting the multiplicity $2(2k - 1)$ into two factors was of course suggested by Heisenberg's *Rumpf* model. However, as Bohr and Pauli noticed, it explained why the S-states ($k = 1$) of alkali atoms were singlets instead of doublets, a fact for which Heisenberg had no satisfactory explanation. Indeed, in this case Heisenberg's model still gives two orientations for the *Rumpf* in the outer electron's field, and therefore a doublet (in absence of external field). Instead Bohr's *Zwang* gives the multiplicity $2(2k - 1) = 2$, to be attributed to a singlet with double magnetic splitting.[171]

More fundamentally, Bohr's *Zwang* was in perfect harmony with his previous analysis of the origins of the failure of the helium atom. In both cases the stability of the relevant dynamic structure eluded ordinary mechanics and requested a nonmechanical stress. This lack of mechanical stability implied a violation of the adiabatic theorem, which in turn created a gap in the definition of energy (in the helium case) or in the definition of statistical weights (in the alkali doublet case).

Altogether, Bohr did not think that the anomalous Zeeman effect made the situation worse than it already was as a result of the failure of the helium atom. He concluded without any sign of a disturbance: "Under these circumstances [the necessity of an unmechanical *Zwang*], we must presume that the coupling between the series electrons and the atomic core cannot be directly described according to the quantization rules of multiperiodic systems." Pauli reacted quite differently, identifying this failure of general principles as a personal failure, as appears in one of his letters to Landé (May 1923): "I am very depressed that I have not been able to

169. Bohr 1923e, 274–277; Bohr [1923f], [525]–[530].
170. Bohr 1923e, 276.
171. Ibid., 279.

find a satisfactory explanation of these dumbfoundingly simple regularities [of the anomalous Zeeman effect] in terms of a model."[172]

## DOUBLE MAGNETISM AND VECTOR MODEL

In reality, in spite of the lack of a proper model, Pauli had made great progress toward the identification of the necessary nonmechanical features of a possible model of the anomalous Zeeman effect, as he reported in the same letter to Landé. His approach started from the Paschen-Back effect. In this case the correspondence between the observed spectrum and a mechanical model was likely to be more transparent, for at least two reasons: the *normal* Zeeman splitting was observed, and the coupling between the core and the outer electron was negligible. For the Paschen-Back splitting of doublets, Sommerfeld and Heisenberg had obtained (formula 155)

$$\delta E = h\bar{v}_L(m + m_r) \quad \text{with} \quad m_r = \pm\tfrac{1}{2}, \tag{162}$$

and the selection rules $\Delta m = 0, \pm 1$; $\Delta m_r = 0$. The half-integral $m$ and $m_r$ respectively gave the angular momentum of the outer electron and of the *Rumpf* along $\mathbf{B}$ (in this approximation $\mathbf{B}$ and $\mathbf{B} + \mathbf{B}_i$ are used interchangeably).

In order to comply with Bohr's requirement of integral $k$, Pauli introduced a quantum number $m_k$ representing the component of the orbital momentum along $\mathbf{B}$, and taking the $2k - 1$ values $0, \pm 1, \pm 2, \ldots,$ $\pm(k - 1)$ (in conformity with Bohr's *Zwang*), and wrote

$$\delta E = h\bar{v}_L(m_k + 2m_r), \tag{163}$$

which gives the same spectrum as Heisenberg's formula, provided that the selection rules are $\Delta m_r = 0$, and $\Delta m_k = 0, \pm 1$, in conformity with the orbital interpretation of $m_k$. According to Heisenberg's *Rumpf* model, more generally according to Landé's interpretation of $m = m_k + m_r$ as the projection of the total angular momentum along $\mathbf{B}$, $m_r$ had to measure the angular momentum of the core (along $\mathbf{B}$). Consequently, the factor 2 in front of $m_r$ in Pauli's formula meant a *double magnetism* of the core.[173]

Pauli found another indication of this double magnetism in a study by Landé of the Zeeman effects of the higher multiplets (quadruplets,

---

172. Ibid., 276; Pauli to Landé, 23 May 1923, *PB*, no. 35. For the reasons and effects of Pauli's depressed mood, see Heilbron 1983.

173. Pauli to Landé, 23 May 1923, *PB*, no. 35. To $m_k$ and $m_r$ correspond, respectively, $m_L$ and $m_S$ in the modern spectroscopic notation, the electronic spin playing the role of the atomic core.

quintets, etc.) discovered by Catalán in 1922. Quite remarkably, Landé discovered a synthetic formula for a "Landé factor" applicable to the magnetic splitting of any multiplet:

$$g(k, j) = \frac{3}{2} - \frac{1}{2} \frac{(k - \frac{1}{2})^2 - (r + \frac{1}{2})^2}{j(j - 1)}, \tag{164}$$

here written with the Bohr-Pauli normalization: $k = 1, 2, \ldots, n$; $r = 0$ (singlet), $\frac{1}{2}$ (doublet), $1$ (triplet)$\ldots$; and (for $k > r$) $j = k - r$, $k - r + 1, \ldots, k + r - 1, k + r$ (as would be suitable for an angular momentum obtained by composition of $k$ and $r$).[174]

In his paper Landé also pointed to the similarity of this formula to the one resulting from a simple model, the so-called "vector model," in which the angular momentum **K** of the outer electron was coupled with the angular momentum, **R**, of the bare core, and with an angular momentum **R'** borrowed by the core from the outer electron in the manner of Heisenberg. Whereas Heisenberg always took $R' = \frac{1}{2}$ (with $R = 0$ in the case of doublets, and $R = \frac{1}{2}$ in the case of triplets), Landé had to take $R' = R$ for any multiplet.[175]

From Pauli's viewpoint, the latter identity between intrinsic core momentum and borrowed momentum meant nothing but a double magnetism of the core. His version of the vector model[176] was based on this double magnetism. Unlike Heisenberg's *Rumpf* model, the new model saved the equations of mechanics, including

$$\dot{\mathbf{j}} = \mathbf{M} \times \mathbf{B} \tag{165}$$

and

$$\dot{\mathbf{R}} = \mathbf{M'} \times (\mathbf{B} + \mathbf{B}_i), \tag{166}$$

174. Landé 1923a. What I called the Bohr-Pauli normalization is related to the modern one by $r \to S$, $k \to L + 1$, $j \to J + 1$. Landé's g-factor is now given by:

$$g = \frac{3}{2} - \frac{1}{2} \frac{L(L + 1) - S(S + 1)}{J(J + 1)}.$$

Landé's original normalization ($R \to S + \frac{1}{2}$, $K \to L + \frac{1}{2}$, $J \to J + \frac{1}{2}$) gave

$$g = \frac{3}{2} - \frac{1}{2} \frac{K^2 - R^2}{J^2 - \frac{1}{4}},$$

which is closest to the formula derived from the vector model. Pauli 1923a used Landé's R (which he called $i$) instead of $r$. The latter quantum number appeared for the first time in Pauli 1924, 372n, where it is called $\bar{r}$.

175. Landé 1923a, 197–205.

176. Pauli first described his version of the vector model in Pauli to Landé, 23 May 1923, *PB*, no. 35. A more detailed formulation is in Landé 1923b, also in Pauli 1924. Landé 1923b also gave a formula for multiplet splitting in the absence of an external field. I will not discuss this formula since it played a less important role in the identification of the necessary departures from the quantum theory of multiperiodic systems. See, e.g., Pauli 1926a, 229–231.

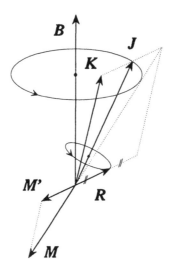

Figure 23.    Diagram for the Landé-Pauli vector model.

where M and M′ are respectively the magnetic moments of the atom and of the core (see fig. 23). However, the latter moment is given, according to Pauli's assumption of double magnetism, by

$$\mathbf{M}' = 2(-e/2\mu c)\mathbf{R}, \tag{167}$$

and the total magnetic moment by

$$\mathbf{M} = -\frac{e}{2\mu c}\,(\mathbf{K} + 2\mathbf{R}) = -\frac{e}{2\mu c}\,(\mathbf{J} + \mathbf{R}). \tag{168}$$

The part of the magnetic energy depending on the external field B is

$$W = -\mathbf{M}\cdot\mathbf{B} = \frac{e}{2\mu c}\,(\mathbf{J}\cdot\mathbf{B} + \mathbf{R}\cdot\mathbf{B}). \tag{169}$$

According to equation (165) the scalar product $\mathbf{J}\cdot\mathbf{B}$ is stationary. The product $\mathbf{R}\cdot\mathbf{B}$ is not, but, according to (166), if the field B is small ($B \ll B_i$), R rotates around J much faster than J rotates around B. Consequently the following identities hold approximately for the temporal average $\overline{\mathbf{R}\cdot\mathbf{B}}$:

$$\overline{\mathbf{R}\cdot\mathbf{B}} = RB\,\frac{\mathbf{R}\cdot\mathbf{J}}{RJ}\,\frac{\mathbf{J}\cdot\mathbf{B}}{JB} = \frac{(\mathbf{R}\cdot\mathbf{J})(\mathbf{J}\cdot\mathbf{B})}{J^2} = \mathbf{J}\cdot\mathbf{B}\,\frac{J^2 + R^2 - K^2}{2J^2}. \tag{170}$$

Using this expression, the temporal average $\overline{W}$ of the perturbing energy (169) becomes

$$\overline{W} = \frac{e}{2\mu c}\,\mathbf{J}\cdot\mathbf{B}\left(1 + \frac{J^2 + R^2 - K^2}{2J^2}\right). \tag{171}$$

The quantization of this model gives integral values to $J$, $R$, and $K$, and also to the projection $m$ of $J$ along $B$ (in units $h/2\pi$). The quantum-theoretical value of the magnetic energy shift is therefore (according to the first-order perturbation theory)

$$\delta_B E = mh\bar{\nu}_L\left(1 + \frac{J^2 + R^2 - K^2}{2J^2}\right),\qquad(172)$$

which gives the Landé factor

$$g = \frac{3}{2} - \frac{1}{2}\frac{K^2 - R^2}{J^2},\qquad(173)$$

to be compared to Landé's empirical formula (164),

$$g = \frac{3}{2} - \frac{1}{2}\frac{(k - \frac{1}{2})^2 - (r + \frac{1}{2})^2}{j(j - 1)}.$$

Even a translation of the quantum numbers by $\frac{1}{2}$ could not reestablish the agreement. One also had to perform the substitution

$$\frac{1}{j^2} \rightarrow \frac{1}{j(j - 1)}.\qquad(174)$$

The theoretical $g$ was related to the empirical one, Pauli commented, by the "substitution of a derivative to a difference." Indeed,

$$\frac{1}{j(j - 1)} = \frac{1}{j - 1} - \frac{1}{j}\quad\text{while}\quad\frac{1}{j^2} = -\frac{d}{dj}\left(\frac{1}{j}\right).\qquad(175)$$

Pauli repeated this consideration in a letter to Sommerfeld of July 1923, with the comment: "This seems to point to something unmechanical." To Landé he also wrote: "I am convinced that in the anomalous Zeeman effect there is no multiperiodic model and that something essentially new must be done."[177]

Being in a depressed mood, Pauli did not want to publish the above considerations, which were of a mostly negative character. Under Bohr's pressure, however, he consented to the publishing of the part of his reasonings which could be formulated without reference to any model. This included the formula (162) for the Paschen-Back effect, now introduced as the simplest possible empirical formula employing the selection rules $\Delta m_k = 0, \pm 1$ (with the corresponding rule of polarization) and $\Delta m_r = 0$. He also provided a new and ingenious derivation of the Landé factors in weak fields on the basis of the following assumption: "The sum of the

177. Pauli to Landé, 23 May 1923, *PB*, no. 35; Pauli to Sommerfeld, 19 July 1923, *PB*, no. 40; Pauli to Landé, 17 Aug. 1923, *PB*, no. 42.

energy values of all stationary states that belong to given values of $m$ and $k$ [and $n$ of course] remains a linear function of the field intensity during the whole transition from weak field to strong field."[178]

This "sum rule" had first been introduced by Heisenberg in order to justify his dual quantization of the *Rumpf*. Not only did it lack a fundamental proof, but it was incompatible with the vector model, a further sign of "something unmechanical." In a letter sent along with this paper, Pauli commented to Sommerfeld: "As you will see, I was so intimidated that I have carefully avoided even the term 'angular momentum'." However, he promptly confessed: "I would never have reached the given representation of the spectral terms in strong fields if I had not been guided by model representations."[179]

AMBIGUOUS MOMENTA

In the following months Pauli kept thinking about the extent of the breakdown of classical concepts. In a letter to Landé of October 1923, he rewrote Landé's empirical formula for the g-factor as

$$g = \frac{3}{2} - \frac{1}{2} \frac{k(k-1) - r'(r'-1)}{j(j-1)} \tag{176}$$

with $r' = r + 1$. "You see," he commented, "the formula could hardly be more symmetrical." Of course, this form depended on the normalization of the quantum numbers, which differed from the one originally favored by Landé on semiempirical grounds. Nevertheless, Pauli believed his choice to be the more fundamental, for it still made $k$ an integer and maintained the symmetry of the distribution of the values of $j$ around $k$. This led him to the following suggestion: "You see that every momentum is represented not by a single number but by a *pair* of numbers. In a certain sense the momenta seem to be *double-valued [zweideutig]*."[180]

Pauli further related the ambiguity of the angular momenta $r$ and $j$ to Bohr's unmechanical *Zwang*: "The violation of the conservation of statistical weights [as expressed in Bohr's *Zwang*] during the coupling results

---

178. Pauli 1923a, 155.
179. Pauli to Sommerfeld, 19 July 1923, *PB*, no. 40. In 1924 Pauli published a detailed account of his vector model, with the following motivation: being multiperiodic, this model is subject to the adiabatic theorem and therefore allows a derivation of the correspondence between the quantum numbers in a weak field and those in a strong field (Pauli 1924).
180. Pauli to Landé, 23 Sept. 1923, *PB*, no. 46. In Pauli 1924, 372n, Pauli redefined the quantum numbers in the manner adopted in Sommerfeld 1923a, 1923b, 1924 (with different notations and purposes). The new numbers are obtained by subtracting one unit from $k$, $r'$, $j$, and they are exactly the ones now in use ($L$, $S$, and $J$).

from the atomic core reacting either with one momentum value or with the other." The ambiguity of $k$ was far more puzzling, since it contradicted Bohr's use of a well-defined $k$ as a foundation of his second atomic theory. Pauli nevertheless found further support for this heresy in a remarkable property of X-ray spectra, brought to light by Coster and Wentzel in 1921.[181]

The complex structure of these spectra appeared to be made of two alternating types of doublets, a relativistic one given by Sommerfeld's formula (with proper re-scaling) and a screening type calculated by replacing, in the hydrogenlike system, the Coulomb potential by a different spherically symmetric potential. Globally, the terms of the X-ray spectrum were well represented by the Bohr-Coster formula:

$$E_{nk_1k_2}/Kh = \left(\frac{Z - a_{nk_1}}{n}\right)^2 + \alpha^2\left(\frac{Z - b_{nk_1k_2}}{n}\right)^4\left(\frac{n}{k_2} - \frac{3}{4}\right). \quad (177)$$

Strangely enough, two different quantum numbers, $k_1$ and $k_2$, were needed to express the screening parameters ($a$ and $b$) and the relativistic correction (compare with formula (38)). To Pauli this suggested an ambiguity of the azimuthal quantum number and of its orbital origin. None of the kinematic characteristics of Bohr's theory seemed to survive in atoms with several electrons.[182]

A NEW QUANTUM PRINCIPLE IN GÖTTINGEN

In conformity with his *Rumpf* model, Heisenberg preferred a half-integral $k$ to an ambiguous $k$. However, he welcomed Pauli's suggestion of an ambiguous $j$ and managed to integrate it in a revolutionary program just started by his new adviser Max Born.

Born's call for a revolution in quantum theory was not new. It was first expressed in a letter to Pauli of 1919, under the following interesting circumstances. The young prodigy had just published an in-depth criticism of Weyl's unified theory of gravitation and electricity. Among other technical reproaches, he wrote:

> There is a physical-conceptual objection that should not be forgotten. In Weyl's theory we constantly operate with the field strength in the interior of the electron. For a physicist the field strength is only defined as a force on a test body;

181. Pauli to Landé, 23 Sept. 1923, *PB*, no. 46.
182. Pauli alluded to this duality of $k$ in his letter to Landé, ibid.: "This conception [dual $k$] also seems to be in better agreement with the facts regarding Röntgen spectra." A thorough discussion of this point is in Pauli 1926a, 204–205; the original X-ray doublet formulae are in Sommerfeld and Wentzel 1921 and Bohr and Coster 1923.

and since there are no smaller test bodies than the electron itself, the concept of the electric field strength in a mathematical point seems to be an empty meaningless fiction. One should stick to introducing only those quantities in physics that are observable in principle.[183]

Pleased by this Machian statement, Born immediately extended it to atomic theory and even dreamed of a completely discrete world, perhaps one based on an analogy with the main object of his past researches, crystal lattices. In his reply to Pauli's letter he wrote:

> I have especially been interested by your remark at the end, that you regard the application of the continuum theory to the interior of the electron as meaningless, because one is there dealing with things which are unobservable in principle. I have pursued exactly this idea for some time.... The way out of all quantum difficulties must be sought by starting from entirely fundamental points of view. One is not allowed to carry over the concept of space-time as a four-dimensional continuum from the macroscopic world of experience into the atomistic world; the latter evidently demands another type of number-manifold to give an adequate picture [Bild]. ... Although I am not yet old, I am already too old and burdened to arrive at the solution. That is your task; according to what I have heard about you, to solve such problems is your calling.[184]

However, when Pauli collaborated with Born in 1921–22, it was not in an attempt to overthrow space-time concepts but rather to improve the mechanics of electronic orbits. In the meantime, Born had learned to appreciate the qualities of the Bohr-Sommerfeld theory, in spite of the persistent difficulties with the radiation problem. For a short while he was even more optimistic than Bohr about the ability of this theory to represent atomic structure and spectra. When, in early 1923, the helium atom resisted such a treatment, Born quickly returned to his original extremist standpoint. In the summer of the same year he declared that "not only new physical assumptions would be needed but *that the entire system of concepts would have to be restructured in its foundations.*" On the contrary Bohr believed, as already mentioned, that the basic principles of the quantum theory could very well accommodate the failure of classical mechanics in the stationary states. And he kindly reproached Born for his extreme attitude.[185]

183. Pauli 1919, 749–750. For the roots of Pauli's operationalism see Hendry 1984.
184. Born to Pauli, 23 Dec. 1919, *PB*, no. 4.
185. Born 1922, 542. See Bohr to Born, 2 May 1923, *BCW* 4:[673]: "However, I do not see this difficulty quite in the same light as you indicate."

The Born-Pauli formalism, or any calculation of energy spectra in the Bohr-Sommerfeld theory, rested on the more or less tacit assumption that the anticlassical character of radiation processes did not contaminate the Coulomb interaction between atomic constituents. Once confronted with the impossibility of proper quantum-theoretical models for helium and anomalous Zeeman patterns, Born adopted the opposite assumption. As a heuristic principle, he assumed that the discrete character of radiation processes implied by the relation $\Delta E = h\nu$ had a counterpart in nonradiative interactions. This took him back to the "number manifold" of 1919, with the new leitmotiv "discretization of atomic physics." In the future theory, the continuous electronic orbits of the Bohr atom would naturally disappear; in general, unobservable quantities would disappear, as he had argued during his early exchange with Pauli.[186]

With the help of Pauli's idea of a dual $j$, in October 1923 Heisenberg gave a first boost to Born's new program. The first assumption of the new theory read: "The model representations have only a symbolic meaning; they are the classical analogue of the 'discrete' quantum theory." In this way Heisenberg could deny a direct physical meaning to previous atomic models, but he nevertheless extracted from them a symbolic content, to be integrated into a new, completely discrete theory. In other words, he was trying to design for Pauli a proper mechanical version of the correspondence principle—that is to say, a formal analogy between classical mechanics (without radiation) and a new discontinuous mechanics.[187]

In the case of the anomalous Zeeman effect, Heisenberg gave the specific correspondence between the Landé-Pauli vector model and its discrete version, in terms of a "new quantum principle" inspired by Born's idea of a structural analogy between mechanical and electrodynamic interactions.[188] Just as radiation energies were expressed in terms of differences $(h\nu = \Delta E)$, the true energies $E$ of the stationary states had to be given by a finite difference of a certain function $F$. More specifically, Heisenberg took

$$E(R, K, J, m) = F(R, K, J + \tfrac{1}{2}, m) - F(R, K, J - \tfrac{1}{2}, m), \qquad (178)$$

with

$$F(R, K, J, m) = \int \delta_B E(J)\, dJ, \qquad (179)$$

186. Born 1924, 379; Heisenberg to Pauli, 9 Oct. 1923, PB, no. 47.
187. Heisenberg to Pauli, 9 Oct. 1923, PB, no. 47.
188. Ibid.; Heisenberg 1924. Presumably discouraged by Bohr's and Pauli's comments, Heisenberg sent his paper for publication only in June 1924.

where $\delta_B E$ is the energy value (172) given by the vector model

$$\delta_B E = m h \bar{v}_L \left( \frac{3}{2} - \frac{1}{2} \frac{K^2 - R^2}{J^2} \right).$$

This choice reproduced Landé's empirical formula (164), in Landé's original half-integral normalization $(R = r + \frac{1}{2}, K = k - \frac{1}{2}, J = j - \frac{1}{2})$. Obviously, Heisenberg had been inspired by Pauli's remark (175), that

$$\frac{1}{j(j-1)} = \Delta \int \frac{dj}{j^2}.$$

By the time Heisenberg finally published these considerations, he and Landé had extracted from empirical data the "branching rule," according to which an ion with angular momentum $J$ gave rise, by addition of an electron with a given (sufficiently high) value of $K$, to two multiplets corresponding to $R = J + \frac{1}{2}$ and $R = J - \frac{1}{2}$. This gave immediate support to the new quantum principle, according to which the spectroscopic value $J$ of the angular momentum of the ion corresponded to *two* values of $J$ in the vector model, $J + \frac{1}{2}$ and $J - \frac{1}{2}$. In return, this principle permitted a formal generalization of Bohr's principle of the permanence of quantum numbers to the angular momentum of the atomic rest (the rest being what is left of an atom after removing one electron): the quantum number $R$ of a given element was equal to the quantum number $J$ of the normal state of the previous element. For instance, in the case of calcium and sodium, the fundamental state of sodium is a singlet state, with $K = \frac{1}{2}$, $R = 1$, and $J = 1$ ($k = 1$, $r = \frac{1}{2}$, $j = \frac{3}{2}$ in the Bohr-Pauli normalization); according to Heisenberg, $J = 1$ should be replaced by the couple of values, $J = (\frac{1}{2}, \frac{3}{2})$, which gives rise to $R = \frac{1}{2}, \frac{3}{2}$ for calcium; these values correspond to the observed singlets and triplets.[189]

Bohr was not as pleased by this accomplishment as Heisenberg expected. Bohr wrote to Heisenberg that "he would welcome his proposed solution if only he could grasp sufficiently the formal as well as the physi-

189. Landé and Heisenberg 1924; Heisenberg 1924, 300–301. In the example of sodium and calcium, Heisenberg's branching rule simply corresponds to the composition of the spins of the last two bound electrons, which determines the multiplicity of the complex structure as long as the coupling is of the $L \cdot S$ type. However, in other cases (multiplets "höherer Stufe" for which the single-ionized atom is not in a S state, or "normal" multiplets the coupling of which is not of the $L \cdot S$ type), Heisenberg's rule did not give the correct multiplet grouping of terms; it just gave the correct values of the total angular momentum. In determining these values, the order of composition of the partial angular momentum is irrelevant, so that one may compose, in the modern terminology, the orbital momentum of one electron with that obtained by composing the spin of this electron with the total angular momentum of the rest of the electrons (this kind of composition generally differs from that prescribed by $L \cdot S$ or $j \cdot j$ coupling). See Pauli 1926a, 260.

cal side." Pauli was even more critical: "I consider [Heisenberg's theory] to be an ugly theory. For, in spite of radical assumptions, it does not provide an explanation of the half-integral quantum numbers and of the failure of Larmor's theorem [the double magnetism of the core]." He nevertheless approved something fundamental in Heisenberg's strategy, as appears in a letter to Bohr of February 1924:

> To me the most important question is: To what extent one can speak of well-defined trajectories of the electrons in the stationary states. I do not think that this can be posited as obvious at all, especially with regard to your considerations about the balance of statistical weights in the coupling [leading to the Zwang]. In my opinion Heisenberg hits the truth precisely when he doubts that it is possible to speak of determinate trajectories.

Indeed, Heisenberg regarded his new quantum principle as something general. Not only the vector model but also all orbital models became purely symbolic. In Landé's terms, they were "Ersatz-models" that had nothing to do with the real, yet unknown nature of motion inside atoms.[190]

### A RELATIVISTIC CONCEPTION

In his antimodel war, Pauli was even more radical than Heisenberg. To him a "symbolic model" was still too much of a model. He believed that even the orbital meaning of the azimuthal quantum number (through the correspondence principle) was lost forever, as he wrote in his letter to Bohr of February 1924: "Against the point of view which you were still holding last fall, I now believe that even for the quantum number $k$ (not only $j$) essential features of the true laws cannot be reproduced by the theory of multiperiodic systems." As indicated above, Pauli had derived this judgment from the appearance of $k(k-1)$ in his expression of the Landé factor, and from the ambiguity of $k$ in the X-ray doublet spectra. There was also, as he explained in the above letter, the persistent lack of any explanation of the Zwang on the outer electrons ($2k \rightarrow 2k - 1$) on the basis of a multiperiodic model.[191]

At this point some remarks should be made about the nature of the disagreement between Bohr and Pauli. I have insisted that from an early period Bohr did not attribute to the electronic orbits of his theory more than a "formal value," since, as he agreed with Campbell in 1921, they

190. Bohr to Heisenberg, 31 Jan. 1924, AHQP; Pauli to Landé, 14 Dec. 1923, *PB*, no. 51; Pauli to Bohr, 21 Feb. 1924, *PB*, no. 56; Landé 1923c.
191. Pauli to Bohr, 21 Feb. 1924, *PB*, no. 56.

could not be considered to be the sources of the radiation emitted by atoms according to ordinary electrodynamics. Moreover, as explained in "On the quantum theory of line spectra" (1918) and repeated in "The fundamental postulates" (1923), this type of model could be operational only in the limit for which the Coulomb interaction can be separated from the radiative interaction.[192]

Nevertheless, when Heisenberg spoke of a "symbolic" character of the electronic orbits, and Pauli of the nonexistence of these orbits, they went further than Bohr: they believed that the idea of definite orbits was not even relevant to an approximate determination of stationary states through ordinary mechanics or any extension of it. In Pauli's opinion, Bohr was wrong to retain classical *concepts* while he gave up classical *laws*. In a letter of 1923 to Eddington he alleged that this was the source of all quantum paradoxes. A proper quantum theory, he believed, had to start from an entirely new set of concepts.[193]

As the encyclopedist of relativity, Pauli found the strongest argument against orbits in the relativistic explanation of optical doublets given by Landé (and, independently, by Millikan and Bowen) in April 1924. In December 1924 he declared to Bohr: "The relativistic doublet formula seems to me to show without any doubt that not only the dynamic concept of force but also the classical theory's kinematic concept of motion will have to be profoundly modified." According to the experiments made by Millikan and Bowen, the doublet spectrum of highly ionized atoms (which lies in the far UV region) appeared to be very similar to that given by the X-ray emission from atoms with an incomplete internal shell. In fact both types of spectra could be described by a formula of the type (177) (up to a sign difference), if only $j$ was identified with $k_2$, and $k$ with $k_1$ (with the proper normalization), in harmony with the selection rules $\Delta k_1 = \pm 1$ and $\Delta k_2 = 0, \pm 1$. This suggested an extrapolation of the relativistic interpretation of X-ray doublets to optical doublets, but again only at the expense of employing the strange ambiguity of the azimuthal quantum number noticed by Pauli in the X-ray case.[194]

There was an important reason, Pauli continued, to favor the relativistic explanation of doublets over the one based on the concept of a magnetic core (*Neigungsgesichtspunkt* in Pauli's terminology), both in the optical and in the X-ray case. This reason had been found by Landé in the

192. See p. 157; p. 122; Bohr 1923b, trans., 10.
193. Pauli to Eddington, 20 Sept. 1923, *PB*, no. 45.
194. Pauli to Bohr, 12 Dec. 1923, *PB*, no. 74; also Pauli to Landé, 24 Nov. 1924, *PB*, no. 71; Landé 1924a, 1924b; Millikan and Bowen 1924a, 1924b, 1924c. See Pauli 1926a, 210–212.

Z-dependence of the width of doublets, where Z is here the effective charge
number perceived by the emitting electron. In the X-ray case, the relativistic correction implied by Sommerfeld's formula is proportional to $Z^4$,
as experimentally verified. Instead, the magnetic-core model gives a $Z^3$
dependence, since the corresponding perturbation is due to the magnetic
field created by the outer electron, which behaves like

$$B_i \propto e \frac{v}{a^2} \propto Z^3 \tag{180}$$

in the simple case of a circular orbit (the velocity $v$ is proportional to Z,
and the radius $a$ is inversely proportional to Z).[195]

In the optical case, a nontrivial Z-dependence occurs only in the case
of dipping orbits. If, for simplicity, we limit ourselves to deeply dipping
orbits, the relativistic correction is most important in the part of the orbit
which is closest to the nucleus. Calling $Z_i$ and $t_i/t_0$ the charge number
perceived and the fraction of time spent by the electron in this inner part,
the corresponding energy-shift is proportional to $Z_i^4(t_i/t_0)$. Since, in a
Kepler orbit, the time of revolution is proportional to $Z^{-2}$, the relativistic
doublet width must be proportional to $Z_i^2 Z_0^2$, where $Z_0$ is the effective
charge number corresponding to the outer part of the orbit. Now, in the
magnetic-core model this width behaves like the magnetic field created
by the inner part of the orbit:

$$B_i \propto Z_i^3(t_i/t_0) \sim Z_i Z_0^2. \tag{181}$$

Here again spectral data confirmed the relativistic conception.


THE EXCLUSION PRINCIPLE

In November 1924 Pauli found a more direct argument to exclude a contribution of the atomic core to alkali doublets.[196] In heavy atoms the
velocity of the electrons in the K-shell of the core must be very high, as
can be appreciated from the expression

$$v = \alpha c Z \tag{182}$$

for the velocity of the electron in the fundamental state of a hydrogenlike
ion with the nuclear charge Z ($\alpha$ is the fine-structure constant). To this

195. Pauli to Bohr, 12 Dec. 1923, PB, no. 74; Landé 1924b, to which Pauli reacted enthusiastically (Pauli to Landé, 30 June 1924, PB, no. 63): "I was highly excited by your new
work, I congratulate you and admire your courage to express such a thing without fear, even
though you do know how crazy it is." See Pauli 1926a, 212.
196. Pauli to Landé, 10 Nov. 1924, PB, no. 68; Pauli 1925a.

value of the electronic velocity corresponds, in a rough estimate, a relativistic modification of the Larmor frequency of the K-shell given by

$$\bar{v}_L = \bar{v}_L^0 \sqrt{1 - v^2/c^2} \tag{183}$$

or, in a first approximation,

$$\bar{v}_L = \bar{v}_L^0 (1 - \tfrac{1}{2}\alpha^2 Z^2). \tag{184}$$

If the K-shell is entirely responsible for the magnetism of the core, as was commonly assumed in the case of alkali atoms (in harmony with Bohr's second atomic theory), this modification leads to a correction $\delta g$ of the Landé factor, according to

$$\delta g/g = -\tfrac{1}{2}\alpha^2 Z^2. \tag{185}$$

Prompted by Pauli, Landé quickly concluded that no such effect existed in the empirical data. Consequently, Pauli affirmed, atomic cores homologous to noble gases could not contribute to the complex structure or the anomalous Zeeman effect. Furthermore, he wrote to Landé, Bohr's assumption that the K-shell (as the normal state of helium) had a unit angular momentum, while the other shells had a zero angular momentum, presented an asymmetry in classification that had no empirical counterpart, since all noble gases, including helium, were known to be diamagnetic.[197]

As Heisenberg later remarked, this argument was not very strict, since neither the *Rumpf* model nor the vector model necessarily attributed the magnetism of the core to a nonvanishing value of the angular momentum of the K-shell. On the contrary, Heisenberg's original *Rumpf* model regarded the angular momentum of the core as borrowed from the outer electron. In Pauli's eyes, however, the objection was minor. His new relativistic argument was just intended to reinforce the more definitive argument about the relativistic theory of doublets.[198]

In a quite remarkable letter to Landé of 24 November 1924, Pauli summarized his previous criticism of the magnetic-core models, putting special emphasis on the relativistic doublet formula, and declared:

> Without attempting to explain this relativistic representation in any manner, I have tried to posit it provisionally as a very deep result, and to bring it into a certain logical relation with other empirical results (in particular the breakdown of Larmor's theorem). . . . As a point of departure I will assume the following: *In alkali atoms the optical electron is itself responsible for both complex structure and anomalous Zeeman effect. There is no question of a coupling*

197. Pauli to Landé, 10 Nov. 1924, *PB*, no. 68, for the question, 14 Nov. 1924, *PB*, no. 69 for the reaction to the answer; Pauli 1925a, 373; Pauli to Landé, 24 Nov. 1924, *PB*, no. 71.
198. Heisenberg to Pauli, 26 Feb. 1925, *PB*, no. 85.

*with the noble-gas-like atomic core (even in other elements). The optical elec-tron is able, in a mysterious unmechanical way, to appear in two states (with the same [spectroscopic] k) of different [angular] momentum.*[199]

This assumption had the great advantage, Pauli continued, of giving a common origin to all known anomalies in the field of complex spectra. It directly provided the quantum numbers $k_1$ and $k_2$ needed to represent screening and relativistic doublets in the relativistic doublet formula. It also took care of the violation of Larmor's theorem, as Pauli could show in the case of the Paschen-Back effect. One just had to interpret the formula (155),

$$\delta E = h\bar{\nu}_L(m \pm \tfrac{1}{2}),$$

as expressing the need for two different values of the projection (along the magnetic field) of the orbital angular momentum: one, $m_1 = m$, gave the "dynamic reaction," or what was left of it in a model-independent approach, the selection rule $\Delta m = 0, \pm 1$; the other, $m_2 = m_1 \pm \tfrac{1}{2}$, gave the magnetic energy. Finally, Pauli's assumption automatically implied, without Bohr's *Zwang*, the conservation of statistical weights during an adiabatic binding of an outer electron, because the weight $2(2k - 1)$ of a given value of $k$ (or $k_1$ in Pauli's new notation) was now entirely attributed to the outer electron, the weight of the core remaining 1 during the binding process.

For his audacious unification of quantum-theoretical troubles, Pauli found a spectacular application: on the length of the periods in Mende-leev's table. During his stay in Copenhagen, he had convinced himself that no satisfactory a priori explanation of these periods had yet been given. After the failure of the helium atom, what was essentially left from Bohr's second atomic theory was the *induced* rule: "Never [will] orbits belonging to two different groups or subgroups of an atom [be found] with the same two quantum numbers k and n."[200]

In other words, the quantum numbers satisfying the principle of per-manence, $n$ and $k$, could serve as labels of the various electronic groups and subgroups. But the number of electrons in a given $n_k$ subgroup re-mained a matter of speculation. In July 1924, a young British physicist, Edmund Stoner, proposed the value $2(2k - 1)$ for this number, which varied from Bohr's but was in harmony with the new phenomenology of X-ray absorption edges (the intensity of each edge being assumed to be a function of the number of electrons in a given $n_k$ subgroup). This choice

199. Pauli to Landé, 24 Nov. 1924, *PB*, no. 71.
200. Bohr [1921d], [152].

also gave the correct length, $2n^2$, of chemical periods, since

$$\sum_{k=1}^{n} 2(2k - 1) = 2n^2. \tag{186}$$

More interestingly, Stoner noted that the number $2(2k - 1)$ also represented the number of Zeeman components of an alkali term with given
$n$ and $k$.[201]

Such a coincidence could hardly be attributed to chance. Bohr, however,
who was soon convinced (although somewhat less than Sommerfeld) of
the superiority of Stoner's scheme for chemical and physical reasons, failed
to perceive something fundamental in the connection between the population of electronic subgroups and the statistical weight of terms. In his
opinion the latter number was subjected to the *Zwang* and therefore could
not be a stable feature of electronic subgroups in the building process of
atoms (*Aufbau*).[202]

On the contrary, in Pauli's new scheme there was no *Zwang;* the multiplicity $2(2k - 1)$ was a characteristic of the electron itself. And the quantum numbers corresponding to the corresponding levels in a magnetic field
could be regarded as satisfying the principle of the permanence of quantum numbers, without infraction of the rule of the adiabatic conservation of statistical weights. Consequently, Pauli labeled the electrons inside
atoms with *four* quantum numbers, $n$, $k_1$, $m_1$, $m_2$ corresponding to the
Paschen-Back components of alkali terms, and deduced Stoner's rule from
the following commandment: "It shall be forbidden that an electron with
the same $n$ belongs to the same values of the three quantum numbers
$k_1$, $m_1$, $m_2$ (equivalence). When an electron corresponds to a definite
$n(k_1, m_1, m_2)$ state, this state is 'occupied.'"[203]

Pauli had no doubt about the importance of the progress brought by
this exclusion principle. He nevertheless realized the shortcomings of his

201. Stoner 1924. See Heilbron 1983.

202. Bohr to Coster, 10 Dec. 1924, BCW 4: [680]–[681]: "I have from the first understood
the formal beauty and simplicity of his classification of the levels; however, ... from the
point of view of quantum theory, it cannot mean a final solution of the problem, since we
do not yet possess any possibility of connecting the classification of levels in a rational
manner with a quantum-theoretical analysis of electron orbits." See Heilbron 1983, 286–287.

203. Pauli to Landé, 24 Nov. 1924, PB, no. 71. In the published version of this principle,
Pauli used the quantum numbers $n$, $k_1$, $k_2$, $m_1$ instead of $n$, $k_1$, $m_1$, $m_2$. In conformity with
the general connection between Zeeman and Paschen-Back levels (established in Pauli 1924),
this gives the same multiplicity $2(2k - 1)$ of a $n_k$-subgroup but has the advantage of retaining
Stoner's fruitful notion of $n(k_1, k_2)$ subgroups. Originally, Pauli preferred to use the Paschen-
Back levels presumably because they were a more direct expression of the *Zweideutigkeit*
of the electron, and also because cases were known (corresponding, for instance, to today's
j · j coupling) for which the attribution of definite values to $k_1$ and $k_2$ (or $k$, $j$) was unnatural.

new "relativistic" approach. To assume "a duality of the quantum properties of electrons that elud[ed] classical description" was also to renounce the explanation of the selection rules ($\Delta j = 0, \pm 1; \Delta m = 0, \pm 1$, etc.) provided by the correspondence principle. The connection of $j$ and $m$ with precessions of the orbital plane of the outer electron was lost, and Pauli had nothing to replace it, since he a priori rejected any interpretation of the new duality of electron properties in terms of ordinary mechanical concepts. Neither could he account for most of the regularities of complex structure and anomalous Zeeman spectra that were explained by the vector model. He summarized the situation to Landé: "I should immediately remark that, for the moment, [my] conception fails wherever the previous conception was particularly useful; however, it seems to serve its aim wherever the previous conception falls short."[204]

In his final paper Pauli pointed to a possible outcome: "It is not excluded that the future will bring us some kind of fusion of these two conceptions." More casually, he wrote to Bohr: "My non-sense is conjugated to the previous non-sense. . . . The physicist who will manage to add these two non-senses shall reach the truth."[205]

Heisenberg, who had been upset by Pauli's cold reception of his new quantum principle, at first responded ironically to Pauli's new "swindle" and welcomed him in the "land of formalism-Philistines," as Pauli must have called the Göttingen theorists. "Swindle times swindle," he continued, "gives nothing right."[206] Nevertheless, he soon recognized that there was more good physics than arbitrary formalism in the dual electron idea, and he tried his best to realize Pauli's prophecy. From such efforts resulted, in April 1925, a new theory of multiplets and their Zeeman effects. The central assumption read:

> Let an atomic core and an electron interact; the energy of this interaction displays a reciprocal duality of the following sort: To given definite stationary states of the core and the outer electron correspond two values of the interaction energy, and, accordingly, two stationary states of the global atom; conversely, to one value of the interaction energy correspond two systems of stationary states of the electron or the core.

This unmechanical duality could be realized in two different symbolic models, either a model with dual core similar to Heisenberg's previous

204. Pauli 1925a, 385; Pauli to Landé, 24 Nov. 1924, *PB*, no. 71.
205. Pauli 1925b, 771; Pauli to Bohr, 12 Dec. 1924, *PB*, no. 74; see also Pauli 1925a, 385.
206. Heisenberg to Pauli, 15 Dec. 1924, *PB*, no. 76. Heisenberg accused Pauli of having introduced electrons with four degrees of freedom. This was not quite true: Pauli only introduced electrons with four *quantum numbers* and avoided any terminology reminiscent of ordinary mechanical concepts, including the notion of degree of freedom.

theory of the anomalous Zeeman effect (based on the vector model and the new quantum principle) or a model with a dual electron akin to Pauli's relativistic conception.[207]

Pauli certainly introduced his property of quantum-theoretical duality as an irreducibly nonmechanical feature of the electron. Yet, this did not imply that his "relativistic" conception was completely independent of any mechanical model. On the contrary, the proof of the relativistic doublet formula relied on a specific mechanical model: that of a relativistic electron in a central field. Heisenberg interpreted the relation between this model and the empirical doublet formula as being of the same nature as the relation between the vector model and the empirical g-factor: it necessitated the introduction of an ambiguity (respectively of $j$ and $k$) which limited the connection of the model with the real atom to one of pure symbolism. Pauli would have added that, even with such restriction, the recourse to pictorial models could be only provisional and heuristic.

Not only were Heisenberg's models limited to *symbolische model-mässige Bilder,* they were not unique, since two different models (at least) were needed to reproduce all empirical data on complex spectra and anomalous Zeeman effect. From Copenhagen Heisenberg commented:

> Clearly the two pictures stand equal to each other; as a consequence of the definiteness of the stationary states of the global atom their consequences cannot contradict each other. Rather, the two pictures will have to complement each other . . . in such a way that the quantities that remain undetermined in one scheme will be determined in the other and vice versa.

By this somewhat enigmatic and very Bohr-like utterance, Heisenberg probably meant that, as a general consequence of the duplicity of the interaction, the uniqueness of the energy of the global system entailed a duplicity and indefiniteness of the states in one part of the system, if the states in the other part were defined; even though a given part of the system received two different pictures in Heisenberg's two models, contradictions between the two models could have arisen only if the respective states of a given part had been unambiguously defined at the same time.[208]

207. Heisenberg 1925b, 841–842.

208. Ibid., 842. This interpretation is confirmed by Fowler's account of conversations with Heisenberg in Copenhagen (in "Recent developments of quantum theory," notes of Fowler's lectures taken by Thomas [1925], AHQP): "The atom as a whole must be supposed to have unique states, and the duplexity only comes in when it is regarded as made up of parts, a core and a series electron; the duplexity comes in because we have carried the analysis too far.—Postulating this we can put the thing in either of two ways which ought to lead to contradictory properties. . . . We now try to build up an atom with unique stationary states to conform to both pictures; this can be done provided the two pictures are as it were complementary and never give unambiguous results in conflict, when we may use either scheme at will."

Heisenberg further emphasized that there was always one picture (even in the case of several outer electrons) for which the orbital motion of the outer electrons was unambiguous, and which allowed an explanation of selection rules through the correspondence principle. In other words, the opposition between Bohr's and Pauli's views seemed to dissolve in some kind of complementarity. Heisenberg pursued this line of thought with considerable empirical success, and also, as he had hoped, with Bohr's benediction and Pauli's tolerance. Nevertheless, he was not completely satisfied. According to the correspondence principle, he argued, one had to expect "a great simplicity of the quantum-theoretical laws governing the interactions inside atoms," one that would reflect the simple regularities of observed spectra. He therefore complained: "It seems that at the moment there exists no way to interpret these laws other than to employ model-dependent pictures of a symbolic nature in which this simplicity is hardly reflected in a satisfactory manner." This remark echoed one already made by Pauli in December 1924: "One now has the strong impression that in all models we speak a language that is not adequate to express the simplicity and beauty of the quantum world."[209]

I now return to Pauli's exclusion principle and to its reception in Copenhagen. Bohr first learned about Pauli's new point of view from a rather aggressive letter, presenting the success of the exclusion principle as implying a major failure of the correspondence principle:

I have often told you that, in my opinion, the correspondence principle has nothing to do with the problem of the completion of electron groups in atoms. At that time you always answered that in this case I was *too* critical—but now I believe my point to be rather well established. The exclusion of certain stationary states (*not* transitions [as would have been the case for exclusions derived from the correspondence principle]) which I propose displays a fundamental similarity with the exclusion of the states $m = 0$ and $k = 0$ in the hydrogen atom, much more than it does with, for instance, the selection rule $\Delta k = \pm 1$.[210]

In his reply, Bohr admired the "numerous beautiful novelties" in Pauli's considerations, but he disparaged the attack on the correspondence principle:

I am not quite sure whether you do not [read: I am positively convinced that you do] cross a dangerous border when you pronounce—in the spirit of your

209. Heisenberg 1925b, 860; Pauli to Sommerfeld, 6 Dec. 1924, *PB*, no. 72.
210. Pauli to Bohr, 12 Dec. 1924, *PB*, no. 74.

old "Carthaginem esse delendam"—a final death sentence on explanations of the completion of groups based on the correspondence principle.[211]

In his next letter to Bohr, Pauli maintained his conviction, and even declared himself "undisturbed" by his crossing of the "dangerous border." His "physical intuition" excluded a further justification of the exclusion principle on the basis of the correspondence principle. More generally, he believed that the first step to be taken toward a fundamental elucidation of the quantum enigma was not an extension of the correspondence principle but a development of purely quantum-theoretical concepts. In this respect he approved Sommerfeld's opinion that "the greatest hope was to be placed in the magic power of quanta, not on considerations of correspondence or stability." As he further explained to Bohr, a later relevance of the correspondence principle was almost certain, but only after a proper reformulation of all quantum laws in terms of true quantum concepts, that is, only after passing through a rejection of the ordinary concept of motion in favor of a "new kinematics":

> There is no doubt that the correspondence principle is not limited to multiperiodic systems but is also valid for all atoms in some form. But we should not abuse ourselves: no exact formulation of this principle in the case of non-multiperiodic systems is yet available. Instead we must *first* search for such a formulation. . . . Not only the dynamic concept of force but also the kinematic concept of motion of the classical theory will have to undergo profound modifications. . . . Since the concept of motion lies at the foundation of the correspondence principle, the efforts of theoreticians must focus on its clarification. I think that the energy and [angular] momentum values of the stationary states are something much more real than the "orbits." The (not yet reached) aim must be to deduce these and all other physical and real observable properties of the stationary states from the (integral) quantum numbers and to deduce quantum-theoretical laws. But we should not tie the atoms in the chains of our prejudices (to which in my opinion belongs the existence of electron orbits in the sense of ordinary kinematics); on the contrary, we must adapt our concepts to experience.[212]

Hence, at the climax of the crisis of quantum theory, Pauli gave priority to extricating quantum laws from the morass of classical concepts, an operation that he hoped would eliminate current paradoxes. He therefore

211. Bohr to Pauli, 22 Dec. 1924, *PB*, no. 77.
212. Pauli to Bohr, 31 Dec. 1924, *PB*, no. 79; Sommerfeld's opinion is in Pauli to Sommerfeld, 6 Dec. 1924, *PB*, no. 72; Pauli to Bohr, 12 Dec. 1924, *PB*, no. 74. See also Pauli 1925b, 771: "Here we encounter a difficult problem: *how to interpret the coming into play* [Auftreten] *of the type of motion of the series electron required by the correspondence principle independently of its* [the motion's] *previous special dynamic interpretation* [Deutung], *which can hardly be maintained*" (Pauli's emphasis).

recommended "a radical sharpening of the opposition between classical and quantum theory." At the same time, as will presently be seen, Bohr and Heisenberg recommended a "sharpening of the correspondence principle." Quantum mechanics would emerge from a combination of the two attitudes.[213]

SUMMARY

In late 1922 Kramers managed to calculate the ionization energy of his and Bohr's model of helium with enough precision to assure a clash with known experimental results. Bohr immediately found a reason for this failure: the normal state of the helium model was mechanically unstable, and this excluded the kind of continuous deformation necessary for defining the energy of stationary states. But he continued to believe that electronic orbits were useful, if only they were sufficiently emancipated from the laws of ordinary mechanics. He even hoped that the classical helium calculations would still provide a correct *qualitative* description of the types of orbits, including the selection and exclusion rules.

The latter hope quickly vanished when, in early 1923, Born and Heisenberg, thanks to their systematic perturbation theory, could prove that the two classes of multiperiodic motions privileged by Bohr and Kramers were *not* disconnected, which ruined Bohr's correspondence argument about the noncombination of o-helium and p-helium spectra. Being aware of the paradigmatic value of helium, Born called this a catastrophe and questioned the whole quantum theory. Bohr reacted more moderately. He admitted a momentary failure of his deductive use of the correspondence principle in atom-building but remained confident in his inductive results, especially the orbital significance of the quantum numbers $n$ and $k$. More generally, he emphasized the compatibility of his general postulates and principles with the new situation. His formulation of the postulates, from 1918 on, had been independent of the applicability of ordinary mechanics to the motion in stationary states. So was the correspondence principle, which related spectral characteristics to the electronic motion, whatever this motion could be. Finally, the adiabatic principle could be purged from any reference to ordinary mechanics, if it were formulated as a general "principle of the permanence of quantum numbers" during all (slow or fast) small perturbations. This was enough to save most of the reasonings

213. Pauli to Sommerfeld, 6 Dec. 1924, *PB*, no. 212. For the sharpening of the correspondence principle, see the next chapter.

leading to the classification of elements, except for the deductive use of the correspondence principle.

Although not for an atom as simple as helium, Bohr had been expecting violations of ordinary mechanics in stationary states for a long time. Already in 1918, on the basis of the correspondence principle he pointed to a necessary violation of Larmor's classical theorem (concerning the effect of a magnetic field on a system of charges) by the assumed orbits in the case of the anomalous Zeeman effect (magnetic splitting differing from the classical triplet). He also expected a violation of ordinary mechanics to occur in the interaction between the core and the outer electron of alkali atoms, because in general the core was mechanically unstable (helium seemed to escape these conclusions because its observed Zeeman patterns were normal, and its hydrogen "core" was obviously stable). Bohr even suspected some unity in these violations, since the Zeeman anomaly appeared only when the unperturbed line displayed a "complex structure" which Bohr attributed to a special interaction between core and outer electron.

In the period 1920–1922 Sommerfeld's school attacked the complex structure and the related anomalous Zeeman effects with a characteristic mixture of multiperiodic models and quantum-number phenomenology, which Heisenberg called "atomystic." These methods led to surprisingly simple regularities, thus creating a gap between what could be simulated through *ad hoc* models and what could be derived from Bohr's general principles. Even though in Bohr's opinion this evolution was not the most important symptom of the crisis (the helium failure was), it certainly played a role in clarifying the nature and extent of the breakdown of classical mechanics in the description of atomic motion. In turn this information would spur the evolution of the correspondence principle.

Sommerfeld and Landé first found a system of atomic levels, quantum numbers, and selection rules which, through Bohr's frequency rule, conveniently summarized empirical data. Heisenberg further imagined a quantized pseudomechanical model, a spinning core (*Rumpf*) coupled with outer electrons, which matched his colleagues' formulae, but at an exceedingly high price. He violated no fewer than three fundamental principles and used a half-integral $k$, thus contradicting Bohr's atomic theory in fundamental ways.

In late 1922 Pauli collaborated with Bohr in trying to find what they considered a decent multiperiodic model of the anomalous Zeeman effect, that is, one with the mildest possible violation of mechanical laws, and with integral quantum numbers. It was a failure, but an instructive one.

It led Bohr, in the spring of 1923, to the notion of *unmechanischer Zwang,* a sort of nonmechanical stress between atomic core and outer electron, or, less poetically, a precise rule for deriving the Zeeman multiplicity of alkali spectra. A little later, focusing on the Paschen-Back effect (the large-field limit of the Zeeman effect) and on Landé's new model and formula for the anomalous Zeeman effect (the famous g-factor), Pauli managed to improve greatly on Heisenberg's *Rumpf* model. In the new "vector model" the coupling between core and outer electron respected mechanical laws; but the magnetism of the core had twice its classical value, and, even more strangely, the square of the total angular momentum, $j^2$, had to be replaced with $j(j-1)$, suggesting that $j$ was ambiguous.

While he was still in Copenhagen, Pauli refrained from extending this ambiguity to the quantum number $k$, which would have upset Bohr's correspondence principle. But at the end of 1923 he took this heretical step because it made Landé's g-factor more symmetric (though not empirically better) and because it seemed to be needed for the relativistic explanation of the doublet structure of X-ray spectra: in a formula that normally should have contained *one* value of $k$, different values had to be injected in the parts of the formula to obtain agreement with the empirical results. In 1924 Millikan, Bowen, and Landé found this type of relativistic formula to apply also to the doublet structure in the optical domain, which challenged the magnetic core models and reinforced Pauli's conviction about the ambiguity of $k$. At the end of the year he rejected the idea that the cores in alkali atoms had anything to do with the complex structure and Zeeman effects. As a unique explanation of all encountered anomalies he proposed a "mysterious" *intrinsic* ambiguity of electrons. In this conception the number of different quantum states accessible to a $k$ electron became $2(2k-1)$, an expression in which Pauli immediately recognized Stoner's semiempirical value for the maximal population of electronic subgroups. Consequently, he pronounced the exclusion principle, according to which two electrons could never occupy the same quantum state in a given atom.

Meanwhile in Göttingen, since the summer of 1923 Max Born had been dreaming of a revolution in which mechanical laws, which had failed so dramatically in the helium problem, would be replaced with discrete laws, just as electrodynamic laws had been replaced with Bohr's postulates. Inspired by this idea and by Pauli's ambiguous $j$, Heisenberg proposed a "new quantum principle" that made the energy levels of the new theory equal to some sort of average between successive energy levels of Bohr's "old" theory. This move put so much more distance between the classical model and the final spectrum that Heisenberg declared the model representations to have only a *symbolic* meaning: they were only the "classical analogues of the discrete quantum theory." Bohr judged Heisenberg's theory too formal. Pauli appreciated the rejection of definite orbits but condemned the overall ugliness and arbitrariness of the new quantum principle.

Heisenberg returned the compliment when he first heard about Pauli's intrinsically ambiguous electrons. Yet he soon explored a compromise: the *interaction* between atomic core and outer electron was ambiguous, which implied either an ambiguity of the core's state or an ambiguity of the outer electron's state (in the alkali case). The first alternative corresponded to Heisenberg's previous use of the "new quantum principle," while the second corresponded to Pauli's new point of view. The two conceptions, Heisenberg commented from Copenhagen, were not contradictory; rather, they "complemented each other" since they answered different empirical questions. By the spring of 1925 this symbolic multimodel approach led to the best theoretical coverage of "term zoology and Zeeman botany" ever reached.

In this theory Heisenberg paid due respect to the correspondence principle. Among the complementary pictures he used, there was always one in which the orbital motion of the outer electrons was unambiguous, so that the traditional derivation of selection rules on the basis of the correspondence principle could be saved. He also suggested that a sharper use of this principle would lead to a much simpler theory, one directly reflecting the simple empirical regularities. Pauli thought differently. He interpreted his exclusion principle as pointing to a major failure of the correspondence principle and believed that this principle would be of no avail until quantum theory should be purged of all classical prejudices and grounded on purely quantum-theoretical notions. Naturally, Bohr warned Pauli that he had crossed a dangerous line. He still believed that the correspondence principle was the best guide for progress in the quantum theory, the more so because of the new radiation theory that will now be discussed.

# CHAPTER IX

# The Virtual Orchestra

Until the crisis of 1922–23 Bohr refrained from any specific suggestions about the mechanism of radiation (meaning the mutual interaction between atoms and radiation), because, thanks to the correspondence principle, much could be said about the radiation emitted by atoms without such knowledge, as long as the electronic motion in stationary states could be calculated on the basis of ordinary mechanics. With the failure of the helium atom, this strategy came to a dead end. No general a priori procedure was left to the calculating physicist for determining atomic spectra. Quantum discontinuity now seemed to contaminate the whole of atomic theory.

Moreover, the need for a solution to the empirical paradoxes of radiation phenomena was becoming urgent. Studies of the properties of high-frequency radiation, mainly Maurice de Broglie's and Compton's, gave more direct support to Einstein's light quantum in the years 1921–1923—and a lot to authorities in the quantum theory to think about. For instance, in the third edition of his *Atombau* (1922), Sommerfeld denounced the "dilemma" that made light quanta or waves only "one half of the truth"; Einstein tried to imagine a "ghost field" that would guide light quanta so as to produce interference patterns—and, in a letter to Ehrenfest, declared himself "ready for the mad-house"; in 1923 Pauli delighted in a new derivation of Planck's law based only on light quanta and Compton processes. Bohr himself began to believe that the time was ripe to search for a new radiation theory.[214]

## THE THEORY OF BOHR, KRAMERS, AND SLATER (BKS)

I have already described how, in spite of his acute awareness of fundamental difficulties, Bohr publicly rejected Einstein's and Rubinowicz's conceptions of radiation. He saw them as self-contradictory or strategically impotent.[215] However, from contemplation of his opponents' arguments he drew some essential characteristics of a future theory of radiation.

### CONSERVATION LAWS AND SPACE-TIME DESCRIPTION

First of all, from Einstein's proof that the conservation of energy-momentum in radiation processes implied their directed character, Bohr concluded that conservation laws were violated during a quantum jump. To him there was no milder escape from absurdity, for as Lorentz had argued in 1910, light quanta seemed incompatible with interference phenomena. This private opinion of Bohr was expressed, for instance, in an early draft of "On the quantum theory of line spectra" (but withdrawn from the final version):

> Reversing the line of argument in Einstein's paper, it might be said that Einstein's result combined with interference phenomena would seem to prove that conservation of momentum cannot hold for a single process of radiation. . . . It would seem that any theory capable of an explanation of the photoelectric effect as well as the interference phenomena must involve a departure from the ordinary theorem of conservation of energy as regards the interaction between radiation and matter.[216]

Bohr publicized this opinion in the context of a broader discussion of radiation phenomena published in 1923 in "The fundamental postulates," under the heading "On the formal nature of the quantum theory." In his

214. Sommerfeld 1919, 3d ed. (1922), 311–312; Einstein to Ehrenfest, 15 March [1921], AHQP; Pauli 1923b. On the empirical studies of high-frequency radiation, see Wheaton 1983; on Einstein's "Gespensterfeld" see Lorentz 1927, pars. 50–53; Hendry 1984, 16–19; Klein 1970a.

215. See above, pp. 140–142. Bohr expressed a similar opinion in a never-sent letter to Darwin of 1919, BCW 5:[15]–[16].

216. BMSS, partly quoted in BCW 5:[15].

quantum theory there were already principles, but, he admitted, there was "no consistent picture of radiation phenomena with which these principles [could] be brought into conformity." This is why he spoke of the "formal nature" of the quantum theory (in the same sense as he earlier spoke of the formal nature of electronic orbits). With his usual optimism, Bohr hoped that the future and the correspondence principle would bring the lacking "picture," in the broadest sense of the word, that is, a "description," as he would soon prefer to say. However, he very much doubted that a picture in the narrow spatiotemporal meaning of the word would ever be attainable:

> The satisfactory manner in which the [light-quantum] hypothesis reproduces certain aspects of the phenomena is rather suited for supporting the view, which has been advocated from various sides, that, in contrast to the description of natural phenomena in classical physics in which it is always a question only of statistical results of a great number of individual processes, a description of atomic processes in space and time cannot be carried through in a manner free from contradiction by the use of conceptions borrowed from classical electrodynamics, which up to this time have been our only means of formulating the principles which form the basis of the actual applications of the quantum theory.[217]

Here, Bohr was still careful: he did not quite exclude the possibility of a space-time description that was based on a theory deviating from classical electrodynamics. However, he was more radical in private, for instance in a letter to Harald Høffding written in September 1922:

> It is my personal opinion that these difficulties [in the atomic theory] are of such a nature that they hardly allow us to hope that we shall be able, within the world of the atom, to carry through a description in space and time that corresponds to our customary sensory images.[218]

A few months later, after the light-quantum interpretation of the Compton effect was established, and after conversations with Pauli, he also wrote:

> It is . . . probable that the chasm appearing between these two different conceptions of the nature of light [corpuscular and wavelike] is an evidence of unavoidable difficulties of giving a detailed description of atomic processes

217. Bohr 1923b, trans., 34, 35. As noted in Bohr, Kramers, and Slater 1924, 190n, in 1916 O. W. Richardson had written: "At the present there seems no obvious escape from the conclusion that the ordinary formulation of the geometrical propagation involves a logical contradiction and it may be that it is impossible consistently to describe the spatial distribution of radiation in terms of three-dimensional geometry" (Richardson 1916, 507–508).
218. Bohr to Høffding, 22 Sept. 1922, AHQP.

without departing essentially from the causal description in space-time that is characteristic of the classical mechanical description of nature.[219]

To avoid a common misinterpretation of the latter statement, I will recall that Bohr's rejection of any space-time description of radiation processes in late 1922 did not concern his previous use of space-time pictures for electronic motion and freely waving electromagnetic fields. In his opinion these pictures remained the necessary basis for the application of the correspondence principle; but their validity was to be limited to the approximation where the interaction between the two entities in question, the radiation field and the electronic orbits, could be neglected; and they were of a purely formal nature, since the "correspondence" between the electronic orbits and the emitted radiation could not be deduced from a causal mechanism occurring in space and time.

In fact the correspondence principle, with its limited recourse to classical pictures, and the adiabatic principle, in its updated version as a principle of permanence, were the only general principles that could guide the future elaboration of the quantum theory, as Bohr concluded in his "Fundamental postulates": they were true principles of the quantum theory, whereas the energy principle and the implicit principle of visualizability were only principles of the classical theory—which were likely to collapse along with the classical theory. As we shall presently see, the BKS theory allowed for violation of the energy principle, while maintaining the space-time picture of radiation to a degree somewhat higher than originally expected by Bohr.[220]

LATENT FORCES

Bohr derived a positive characteristic of his developing conception of radiation from Rubinowicz's coupling theory. In "The fundamental postulates" this theory received a more sympathetic review than it had earlier, for it addressed the problem of the coupling between atoms and radiation in terms analogous to those of the new "quantum kinetics" with which Bohr hoped to describe atomic collisions. The object of the coupling theory was to give a measure of the statistical connection between in- and outgoing stationary states of a system consisting of a quantized atom and

---

219. Bohr [1924a], [571]. This manuscript must have been written in early 1924, for it refers to the basic ideas of the BKS paper (without naming Slater): "The interaction between radiation and atom is uniquely determined, as far as the continuous change of the radiation field is concerned, by the state of this field and the instantaneous state of the atom."
220. Bohr 1924b, trans., 42.

quantized radiation in an enclosure, while the object of the quantum kinetics was to give a similar type of connection between the stationary states of two colliding atoms. Moreover, Bohr emphasized, the coupling theory could be properly connected to the correspondence principle, thanks to the following consideration:

> Just as in classical electrodynamics the so-called force of reaction of the radiation conditions the immediate coupling between the field of radiation and the various harmonics of motion of the atom, so we shall assume that the probability of the occurrence of the various processes of exchange between the atom and the enclosure [the wave modes of which are quantized] is controlled by "latent" reactions of radiation, which answer to the harmonic components [rather: to the frequencies] corresponding to the respective processes of transition.[221]

The idea of a "latent" force may be regarded as an obscure prefiguration of the later notion of "virtual fields." Indeed, it served the same function of controlling transition probabilities and was also answering oscillations at the atomic frequencies $(E_n - E_m)/h$.[222]

In spite of this vague formal conciliation with the correspondence principle, Bohr could not accept Rubinowicz's viewpoint on a more fundamental level, for it discarded nonstationary radiation fields and presupposed a full parallelism between radiation and matter, which was at variance with "the pronounced dualism already present in the classical theory between the description of the motion of systems constituted of electrified particles, on the one hand, and the spreading propagation of radiant energy in free space, on the other hand." Until further proof of the contrary (the failure of the BKS theory), Bohr believed that the latter dualism had to remain a feature of the quantum theory.

After his consideration of the coupling principle, Bohr reviewed dispersion phenomena. He again discarded Debye's old theory, with its overly literal application of the orbital model (discussion with Oseen mentioned in chapter V), and praised Ladenburg's "very interesting and promising theory," which assimilated the dispersing atoms to a set of "Ersatz-oscillators" at the observed atomic frequencies. Ladenburg had arrived at this conception in 1921 by transposing the classical connection between selective absorption (resonance) and dispersion, while replacing the electronic motion with formal classical oscillators at the absorption frequencies. The empirical pertinence of Ladenburg's formula, as Bohr wrote to its inventor in May 1923, suggested that the action of an incoming light

221. Ibid., 36.
222. Ibid., 37.

wave on an atom was not to directly induce quantum jumps; rather, it exerted a continuous action according to an unknown mechanism. In this way dispersion remained a largely classical phenomenon, provided that the electronic motion in a stationary state was replaced by a "virtual orchestra" (as Landé later put it). To Bohr and Kramers this must have been a sign of a possible extension of the correspondence principle.[223]

## SLATER'S IDEA

In December 1923 a young American physicist, John Slater, sent a letter to Kramers summarizing new ideas on radiation and announcing his imminent arrival in Copenhagen. During a short stay in Cambridge (England) he had developed a new conception of light with "both the waves and the particles." As far as the propagation of light in free space is concerned, Slater's scheme was very similar to Einstein's "ghost field" or to de Broglie's "atomes couplés en onde" (but it is not known to what extent Slater was aware of these anterior considerations). Energy-carrying light quanta were assumed to follow a path always tangent to the Poynting vector of an energyless electromagnetic field, so that regions of high field-intensity had to correspond to a high concentration of quanta.[224]

Slater's real originality, as perceived by Bohr and Kramers, was a new assumption about the interaction between field and atoms. In order to save the connection between line width and emission time, the field had to be emitted by atoms *during their sojourn in stationary states* and not— contrary to Bohr's original assumption—during the quantum jumps from state to state. The frequencies and intensities of this field were those implied by "motions with the frequency of possible emission lines," the amplitudes of which were (approximately) given by the correspondence principle. Light-quantum emission occurred at random, with a probability proportional to the total flux of the Poynting vector across a sphere surrounding the source atom; at the same time a quantum jump took place

223. Ibid., 39; Bohr to Ladenburg, 17 May 1923, BCW 5:[399]–[400]; "Virtual orchestra" is in Landé 1926, 456.
224. Slater to Kramers, 8 Dec. 1923, BCW 5:[492]–[493]; Slater to his mother, 8 Nov. 1923, quoted in Stolzenburg 1985, [7]; J. C. Slater, manuscripts of 1 Nov. 1923 and 4 Nov. 1923, quoted ibid., [7]–[8]. As a consequence of C. D. Ellis's visit to Maurice de Broglie's laboratory in 1923, an English summary of Louis de Broglie's ideas dated 1 Oct. 1923 was published in the *Philosophical Magazine* in February 1924 (de Broglie 1924), and a shorter one in a letter to *Nature* published on 13 Oct. 1923 (de Broglie 1923). Einstein's idea of the *Gespensterfeld* had been expounded by Lorentz at Caltech in 1922 (Lorentz 1927). For the history of BKS, see Stolzenburg 1985; Klein 1970a; Konno 1983; Dresden 1987. Slater's guiding mechanism was not relativistic-invariant: the Poynting vector is part of a tensor of the second rank, whereas the momentum (velocity) of a light quantum is part of a 4-vector.

in the source atom so that energy was conserved (according to $\Delta E = h\nu$). Similarly, the absorption of a light quantum necessitated a quantum jump in the absorbing atom toward a higher stationary state. Slater further commented that the only place where "chance" entered his theory was in the emission process, and he hoped that "when the dynamics inside atoms [would be] better known, chance might be eliminated there also."[225]

VIRTUALITY

Slater's idea landed in Copenhagen just at the right time: Bohr and Kramers were starting to speculate on a new theory of radiation. Kramers and Bohr immediately criticized Slater's recourse to light quanta, for it contradicted the classical character of (free) electromagnetic radiation assumed (at an a priori level) in the correspondence principle. But they noticed that the rest of his scheme was in fact independent of the assumption of light quanta. To "correct" this theory, they thought, for Slater's picture of individual light quanta being absorbed one just had to substitute a statistical action of the field, inducing quantum jumps according to Einstein's probability laws.[226]

The essential characteristic of the resulting conception was "the connection of the spontaneous radiation with the stationary states themselves and not with the transitions," as Bohr noted in a letter to Slater, with the comment: "Especially I felt it was more harmonious from the point of view of the correspondence principle."[227] This radiative activity of stationary states saved indeed a good part of the continuity found in classical electrodynamics. With the outstanding exception of sudden switches of the field's sources (corresponding to the quantum jumps), a space-time description of radiation processes seemed to be possible, notwithstanding Bohr's earlier intuition to the contrary. In a summary of what was left of his ideas sent to *Nature* in late January 1924, Slater faithfully reproduced Bohr's judgment (probably dictated, considering the style):

> On the basis of Bohr's correspondence principle it seems possible to build up a more adequate picture of optical phenomena than has previously existed, by

225. Slater to Kramers, 8 Dec. 1923, *BCW* 5:[492]–[493]; Slater, manuscript of 4 Nov. 1923, quoted in Stolzenburg 1985, [7]–[8].

226. Slater to his parents, 18 Jan. 1924, quoted in Stolzenburg 1985, [11n]: "I have finally become convinced that the way they [Bohr and Kramers] want things, without the little lumps carried along the waves, but merely the waves which carry them, is better."

227. Bohr to Slater, 10 Jan. 1925, *BCW* 5:[66]–[68]. See also Bohr, Kramers, and Slater 1924, 786: "The essentially new assumption . . . that the atom, even before a process of transition between two stationary states takes place, is capable of communication with distant atoms through a virtual radiation field, is due to Slater."

associating the essentially continuous radiation field with the continuity of existence in stationary states, and the discontinuous changes of energy and momentum with the discontinuous transitions from one state to another.[228]

In the BKS paper Bohr further explained how the new theory proceeded from a refinement of the "natural generalization" of classical electrodynamics expected on the basis of the correspondence principle. According to the narrowest form of this principle, to the intensity of a harmonic component of the classical electric moment in a stationary state "corresponded" the probability of a transition from this state to another. In the new refinement, to the harmonic component $\tau$ of the motion in a given stationary state $n$ corresponded a "virtual oscillator" at the frequency $(E_n - E_{n-\tau})/h$, the function of which was to emit or absorb, during the lifetime of the stationary state, a "virtual field" obeying Maxwell's equations in free space. Moreover, the connection between the virtual field and transition probabilities was also determined by analogy with classical electrodynamics, that is to say, in a manner similar to that of Einstein's paper of 1917 (absorption, stimulated and spontaneous emission).[229]

The introduction of the word "virtual" was probably Bohr's. It corresponded to Ladenburg's *Ersatz* in the oscillator case, and to Einstein's *Gespensterfeld* in the case of radiation. The oscillators were virtual in reference to the more real electronic orbits, and also, as will later appear, because they did not interact with surrounding fields in a classical manner. The fields were virtual not only because they carried no energy, as was the case with Einstein's ghost field, but also because they emanated from the stationary states, in contrast with Bohr's earlier conception, according to which field emission could occur only during the quantum jumps.

In such circumstances one might wonder what was "real" or at least observable in the BKS theory. Although Bohr did not explicitly address this question, the answer can be inferred from his first postulate, which gave the most central position to the concept of stationary state, and from the adiabatic principle (rather: the postulate of the permanence of quantum numbers), which provided the kind of continuity needed for the definition of the energy of these states, without reference to radiation. In short, the stationary states were real, because they were stable (by definition), and could be compared with one another in a continuous way. Instead the virtual fields were accessible only statistically, through their effect on the distribution of stationary states of a large number of atoms. This sta-

228. Slater 1924, 307.
229. Bohr, Kramers, and Slater 1924, 789–790.

tistical feature, since it was integrated in a coherent picture of radiation phenomena (the BKS picture), now seemed an ineluctable consequence of any attempt to employ the quantum postulate together with the correspondence principle.[230]

The first consequence of the new picture was, Kramers pointed out to Slater, "a much greater independence between transition processes in distant atoms" than Einstein had deduced from the light-quantum hypothesis. Indeed, according to the latter hypothesis, a transition in a given atom could occur only if a previous transition had occurred in another atom, in order to provide the energy of a connecting light quantum. According to the BKS theory such a correlation did not exist; only the *probabilities* of transitions in distant atoms could influence one another. Accordingly, energy-momentum was conserved only statistically, not for individual processes. Precisely this point would allow a later discrimination between Bohr's and Einstein's conceptions of radiation.[231]

## THE BKS PAPER

The BKS paper was written in an unusually short amount of time (for Bohr), and in Bohr's characteristic style. The entire text was almost bereft of mathematical formulae, quantitative applications of the qualitative scheme being left to further studies. Bohr first related the new conception to his favorite themes. The "formal character" of the quantum theory, he wrote, was not removed in any manner: the discontinuous processes, "at the present state of science," still eluded a detailed mechanism and had to be described in a statistical manner. In fact, Bohr now believed the chances of a causal space-time description of the interaction between matter and radiation to be very scant.[232]

The new theory nevertheless allowed for "a consistent description of optical phenomena," the harmony of which was warranted by the close analogy with classical electrodynamics. The rest of the paper was dedicated to a discussion of known optical phenomena, in the course of which

230. According to some commentators on the BKS paper, this theory could not be understood without an implicit introduction of *real* electromagnetic fields on top of the virtual ones. This interpretation is sharply contradicted by Bohr's and Heisenberg's later treatment of the fluorescence light (see pp. 239–242). On the necessarily statistical nature of the "correspondence" see BKS 1924, 790; cf. p. 126 above.

231. Kramers's reaction is reported in Slater 1924, 308.

232. BKS 1924, 790, 788. On the context of the BKS paper and on the early role of the correspondence principle, see Wassermann 1981.

more specific assumptions were made about the interaction between atoms and virtual fields.[233]

As already assumed by Slater, Einstein's coefficient of spontaneous emission, $A_m^n$, had to give the total flux (of the Poynting vector) of the virtual field emitted in the stationary state $n$ by the virtual oscillator associated with the (virtual) transition $n \to m$; the virtual field at the place of a given atom (in a stationary state $n$) had to induce positive (toward an upper state $n'$) and negative (toward a lower state $n''$) transitions in this atom with a probability proportional, in Einstein's manner, to the spectral density of the field at the corresponding resonance frequencies $((E_{n'} - E_n)/h$ and $(E_n - E_{n''})/h)$. Further, in order to explain ordinary (non-resonant) dispersion, the virtual field had to be able to interact directly with the virtual oscillators of the encountered atoms. To serve this end, and by analogy with Einstein's introduction of two types of resonant absorption (true absorption and stimulated emission), BKS associated two types of virtual oscillators with a given stationary state, "positive" ones corresponding to transitions from a given state to lower ones, and "negative" ones corresponding to transitions from this state to higher ones; near resonance the former reinforced the incoming virtual field, the latter attenuated it.[234]

This distinction between positive and negative oscillators, even though it had a rough (only near the resonance) classical counterpart in the notion of being in and out of phase, prevented Bohr and Kramers from regarding the virtual oscillators as entities obeying the equations of classical electrodynamics. Indeed, in the classical theory the phase relation between an electric oscillator and the incoming radiation has nothing to do with the frequency of this oscillator; any classical oscillator can both reinforce and attenuate the incoming radiation. Concerning this, Bohr commented:

> It must be remembered that the analogy between the classical theory and the quantum theory as formulated through the correspondence principle is of an essentially formal character, which is especially illustrated by the fact that on the quantum theory the absorption and emission of radiation are coupled to different processes of transition, and thereby to different virtual oscillators.[235]

Just this point, Bohr added, permitted a new quantitative theory of dispersion, the one Kramers would publish in April 1924.

233. BKS 1924, 785–786.
234. Ibid., 793.
235. Ibid., 797.

More urgently, BKS had to integrate into their theory what was considered by most to be conspicuous support for the light quantum, that is to say, the Compton effect. As Compton had noted himself, the Compton radiation can be interpreted as the radiation emitted by the forced oscillations of an electron moving with the velocity $\beta c = ch\nu/(h\nu + mc^2)$ (a hardly natural choice) away from the source of the incoming radiation (with frequency $\nu$). Indeed, the frequency of the forced oscillations is

$$\bar{\nu} = \nu \frac{\sqrt{1 - \beta^2}}{1 + \beta}, \tag{187}$$

while the frequency of the radiation emitted at the azimuth $\theta$ (for an observer at rest with respect to the source of the primary radiation) is

$$\nu' = \bar{\nu} \frac{\sqrt{1 - \beta^2}}{1 - \beta \cos \theta} = \nu \frac{1 - \beta}{1 - \beta \cos \theta}. \tag{188}$$

This gives, for the wave-length shift,

$$\lambda' - \lambda = \lambda \left( \frac{\nu}{\nu'} - 1 \right) = \lambda \frac{\beta}{1 - \beta} (1 - \cos \theta) \tag{189}$$

or, with the above-given choice of $\beta$,

$$\Delta\lambda = \frac{h}{mc} (1 - \cos \theta), \tag{189'}$$

which is Compton's result.[236]

The velocity $\beta c$ of the scattering electron, absurd from the classical point of view, could very well fit in the BKS theory, as just one more formal virtue of the corresponding virtual oscillator.[237]

The Compton experiment was only one among other phenomena discussed in the BKS paper. With great satisfaction Bohr verified in every instance that, thanks to the relaxation of the energy principle, the continuous and discontinuous aspects of radiation were no longer conflicting. They appeared to coexist harmoniously in what Bohr now judged to be the best possible "translation" of classical electrodynamics in the terms of the two quantum postulates:

> Using a metaphor, we may say that we are dealing with a translation of the electromagnetic theory into a language alien to the usual description of nature, a language in which continuities are replaced by discontinuities and gradual changes by immutability, except for sudden jumps, but a translation in which

236. Compton 1923, 487. See Stuewer 1975.
237. BKS 1924, 793, 799.

nevertheless every feature of the electromagnetic theory, however small, is duly recognized and receives its counterpart in the new conceptions.[238]

## DISPERSION THEORY

The problem of dispersion was part of the general puzzle of the interaction between light and atoms and for this reason had come to the fore of quantum theory even before the BKS paper. In 1921 Ladenburg had derived an empirically successful formula for the polarizability of an unexcited Bohr atom:

$$\alpha = \sum_n f_n \frac{e^2}{\mu} \frac{1}{4\pi^2(v_{n0}^2 - v^2)}, \tag{190}$$

where $v_{n0}$ is the frequency of the transition $0 \to n$, $v$ is the frequency of the dispersed light, and the $f_n$'s are positive coefficients related to Einstein's $A_0^n$ through

$$f_n = \frac{3mc^3}{8\pi^2 e^2} \frac{A_0^n}{v_{n0}^2}. \tag{191}$$

Ladenburg reasoned in the following way.

Far from resonance, the equation of motion of an elastically bound electron in the presence of an electric field E has, for the coordinate $x$ along E, the classical form:

$$\ddot{x} + (2\pi v_0)^2 x = -\frac{e}{\mu} E. \tag{192}$$

The resulting expression for the polarizability $\alpha = -ex/E$ is

$$\alpha = \frac{e^2}{\mu} \frac{1}{4\pi^2(v_0^2 - v^2)}. \tag{193}$$

In order to reach a quantum-theoretical generalization, Ladenburg associated with every transition $n \to 0$ of a Bohr atom a number $f_n$ of classical "Ersatz-oscillators" of the above type with the frequency $v_{n0}$.[239]

According to Einstein, the energy *spontaneously* emitted in a unit of time through such transitions is

$$\dot{\varepsilon} = A_0^n h v_{n0}. \tag{194}$$

---

238. Bohr 1925b, 15; also in *BCW* 5:[140].
239. Ladenburg 1921; Ladenburg and Reiche 1923 for the empirical verification of Ladenburg's formula.

At the same time, the Ersatz-oscillators emit the energy

$$\acute{\varepsilon}' = f_n \frac{2}{3c^3} (e\ddot{x})^2 \tag{195}$$

or, by averaging over a great number of periods,

$$\overline{\varepsilon}' = f_n \frac{e^2}{3c^3} (2\pi\nu_{n0})^4 x_0^2, \tag{196}$$

where $x_0$ is the amplitude of the oscillation. In order to connect $f_n$ to $A_0^n$, Ladenburg identified $\acute{\varepsilon}$ with $\overline{\varepsilon}'$ for the value of $x_0$ corresponding to an elastic energy $h\nu_{n0}$, which is given by

$$h\nu_{n0} = \tfrac{1}{2}\mu(2\pi\nu_{n0})^2 x_0^2. \tag{197}$$

Equations (194), (196), and (197) then give the relation (191).

KRAMERS'S FORMULA

The above-described (formal) recourse to classical dispersion theory suggested a more direct approach through the correspondence principle. Before the end of 1923 (therefore before Slater's arrival in Copenhagen), Kramers reached in this manner a new dispersion formula that would play an essential role in the subsequent developments of quantum theory. Since the details of his original reasoning are not known, I will try to reconstruct them by extracting the part of his later published reasonings which is independent of the BKS theory.[240]

According to the strategy recommended by the correspondence principle, Kramers first had to derive the classical dispersion formula for a nondegenerate multiperiodic system. For an expert in canonical perturbation theory, as he was, this was an easy matter.[241] Here I will start from the time-independent perturbation theory introduced in chapter 6. The generalization to a time-dependent perturbation $\varepsilon\Omega(q, t)$ is straightforward: one just has to use a time-dependent generating function $\varepsilon f(w^0, J^0, t)$ and add $\partial\varepsilon f/\partial t$ to the transformed Hamiltonian $H$. In the present case of a harmonic perturbation

$$\varepsilon\Omega = -\mathbf{P} \cdot \mathbf{E}_0 e^{i2\pi\nu t} \tag{198}$$

240. Chronology from Slater to van Vleck, 27 July 1924, BCW 5:43: "You perhaps noticed his [Kramers's] letter to Nature on dispersion; the formulas in that [letter] he had before I came, although he did not see the exact application."

241. Indeed, this calculation had already been made by Epstein 1922, although with different notations.

(P stands for the electric polarization of the system, and $E_0 e^{i2\pi vt}$ for the external electric field), this modification amounts to the substitution

$$w \cdot \tau \rightarrow w \cdot \tau + vt \tag{199}$$

in the equations (131), which gives, for the generating function corresponding to (134),

$$\varepsilon f(w, J, t) = -\frac{1}{2\pi i} \sum_\tau \frac{\varepsilon \Omega_\tau}{\bar{v} \cdot \tau + v} e^{2\pi i(w \cdot \tau + vt)} \tag{200}$$

in the nonresonant case for which the denominators never vanish.

The polarization P admits a Fourier decomposition in terms of the original action-angle variables $(w^0, J^0)$,

$$P = \sum_\tau C_\tau(J^0) e^{2\pi i \tau \cdot w^0} \qquad (C_{-\tau} = C_\tau^*), \tag{201}$$

the functions $C_\tau(J^0)$ being the ones used in the derivation of intensities through the correspondence principle.

Under the effect of the perturbation, $w^0$ and $J^0$ are no longer action-angle variables, and their variation in time is obtained via their relation (the inverse of (12g)) to the new action-angle variables $(w, J)$:

$$w^0 = w - \varepsilon \frac{\partial f}{\partial J}, \qquad J^0 = J + \varepsilon \frac{\partial f}{\partial w}.$$

This adds to the unperturbed polarization $P^0$ a first-order correction:

$$P^1 = \varepsilon \left( \frac{\partial f}{\partial w} \cdot \frac{\partial P^0}{\partial J} - \frac{\partial f}{\partial J} \cdot \frac{\partial P^0}{\partial w} \right) \tag{202}$$

or, according to (200) and (201),

$$P^1 = \sum_{\tau\tau'} \left[ \tau' \cdot \frac{\partial C_\tau}{\partial J} \frac{C_{\tau'} \cdot E_0}{\tau' \cdot \bar{v} + v} - C_\tau \tau \cdot \frac{\partial}{\partial J} \left( \frac{C_{\tau'} \cdot E_0}{\tau' \cdot \bar{v} + v} \right) \right] e^{2\pi i[(\tau + \tau') \cdot w + vt]} \tag{203}$$

This expression contains terms with the frequency $vt$ and also terms with the frequencies $\tau'' \cdot \bar{v} + v$, where $\tau''$ is a nonzero sequence of integers (with both signs) and $\bar{v}$ is the sequence of fundamental frequencies of the multiperiodic system (which can easily be seen to remain unchanged to the first order of perturbation, as a consequence of the vanishing of the time average of the perturbation). In the case of ordinary dispersion the only terms of interest are the ones of the first type. The corresponding part of $P^1$ is:

$$P_v^1 = \sum_\tau \left[ \tau \cdot \frac{\partial C_\tau}{\partial J} \frac{C_\tau^* \cdot E_0}{\tau \cdot \bar{v} - v} + C_\tau \tau \cdot \frac{\partial}{\partial J} \left( \frac{C_\tau^* \cdot E_0}{\tau \cdot \bar{v} - v} \right) \right] e^{2\pi i vt}, \tag{204}$$

or

$$P_\nu^1 = \sum_\tau \tau \cdot \frac{\partial}{\partial J} \left( C_\tau \frac{C_\tau^* \cdot E_0}{\tau \cdot \bar{\nu} - \nu} \right) e^{2\pi i \nu t}. \tag{205}$$

For the sake of simplicity, Kramers first limited the rest of the discussion to the case of parallel directions for $E_0$ and $C_\tau$. Then (205) gives the polarizability

$$\alpha = \sum_\tau \tau \cdot \frac{\partial}{\partial J} \frac{|C_\tau|^2}{\tau \cdot \bar{\nu} - \nu}. \tag{206}$$

Separating the terms with $\tau \cdot \bar{\nu} > 0$ from those with $\tau \cdot \bar{\nu} < 0$ and substituting $-\tau$ for $\tau$ in the second type of terms gives

$$\alpha = \sum_\tau' \tau \cdot \frac{\partial}{\partial J} \frac{2\tau \cdot \bar{\nu} |C_\tau|^2}{(\tau \cdot \bar{\nu})^2 - \nu^2}, \tag{207}$$

where the sum $\sum'$ is limited to the values of $\tau$ for which $\tau \cdot \bar{\nu} > 0$.

In order to find the quantum-theoretical counterpart of this formula Kramers applied the usual "correspondence"

$$\tau \cdot \bar{\nu} \to \nu_{n, n-\tau} = (E_n - E_{n-\tau})/h, \tag{208}$$

$$|C_\tau|^2 \to \frac{3c^3 h \nu_{n, n-\tau}}{4(2\pi \nu_{n, n-\tau})^4} A_{n-\tau}^n. \tag{209}$$

The coefficient in the latter expression is the one giving asymptotic agreement between Einstein's expression ($A_{n-\tau}^n h \nu_{n, n-\tau}$) for the energy emitted in a unit of time and the corresponding classical expression $((2\pi\tau \cdot \bar{\nu})^4 4|C_\tau|^2/3c^3)$.[242]

In $\tau \cdot \partial/\partial J$ Kramers recognized the operator that gives the frequencies $\bar{\nu}$ when applied to $H$. To Bohr's favorite correspondence

$$\tau \cdot \frac{\partial}{\partial J} H \to \frac{1}{h} (E_{n+\tau} - E_n) \tag{210}$$

Kramers admitted the natural generalization

$$\tau \cdot \frac{\partial}{\partial J} \to \frac{1}{h} \Delta_\tau \tag{211}$$

where $\Delta_\tau$ is the finite difference by the increment $\tau$.

242. The total energy radiated in a time unit by the classical dipole of the system is $(2/3c^3)\ddot{P}^2$, with $P = \sum_\tau C_\tau e^{2\pi i \tau \cdot \bar{\nu} t}$. Since $P = P^*$, we have

$$\ddot{P}^2 = \ddot{P}\ddot{P}^* = \sum_{\tau\tau'} C_\tau C_{\tau'}^* (2\pi\tau \cdot \bar{\nu})^2 (2\pi\tau' \cdot \bar{\nu})^2 e^{2\pi i(\tau - \tau') \cdot \bar{\nu} t}$$

Time-averaging gives $(2/3c^3)\overline{\ddot{P}^2} = \sum_\tau' (4/3c^3)(2\pi\tau \cdot \bar{\nu})^2 |C_\tau|^2$, as assumed in the parenthesis.

When applied to (207), the substitutions (208), (209), and (211) give

$$\alpha = \sum_{\tau}' \frac{1}{h} \Delta_{\tau} \left( \frac{3c^3 h}{2(2\pi)^4} \frac{A_{n-\tau}^n}{v_{n,n-\tau}^2} \frac{1}{v_{n,n-\tau}^2 - v^2} \right), \tag{212}$$

or

$$\alpha = \frac{3c^3}{2(2\pi)^4} \left( \sum_{\tau}'' \frac{A_n^{n+\tau}}{v_{n+\tau,n}^2} \frac{1}{v_{n+\tau,n}^2 - v^2} - \sum_{\tau}'' \frac{A_{n-\tau}^n}{v_{n,n-\tau}^2} \frac{1}{v_{n,n-\tau}^2 - v^2} \right), \tag{213}$$

in which the sums $\sum''$ must naturally be restricted to values of $\tau$ for which the quantum-theoretical frequencies $v_{n+\tau,n}$ and $v_{n,n-\tau}$ are defined. This is the so-called Kramers formula. In the large-$n$ limit it gives back the classical dispersion formula, since it was derived from that formula by a "correspondence" translation. For $n = 0$ (fundamental state), it gives back Ladenburg's formula, because the terms with the negative sign disappear.[243]

Originally, the occurrence of the latter type of terms must have puzzled Kramers: they were absurd from the point of view of a classical oscillator model. Fortunately, the BKS theory soon brought some kind of physical explanation. "The main characteristic of this theory," as Kramers and Heisenberg put it, was "the assumption that the reaction of an atom to the radiation field is essentially a reaction in a definite stationary state." For an atom in the stationary state, say $n$, this conception suggested two types of virtual oscillators with, correspondingly, two types of reactions to the virtual field: one corresponding to the frequencies $v_{n+\tau,n}$, the other to the frequencies $v_{n,n-\tau}$. Kramers interpreted his formula as the precise expression of this anticlassical dichotomy.[244]

Kramers first announced his results in a letter to *Nature* of March 1924. In another letter of July 1924 he gave a sketch of the proof, and the following warning:

> It may be emphasized that the notation "virtual oscillator" used in my former letter does not mean the introduction of any additional hypothetical mechanism, but is meant only as a terminology suitable to characterise certain main features of the connexion between the description of optical phenomena and the theoretical interpretation of spectra.

In other words, the virtual oscillators should be considered not a new classical model but rather (as Bohr had already asserted in the BKS paper)

---

243. The above reconstruction is based on the explicit calculation published in Kramers and Heisenberg 1925.

244. Kramers and Heisenberg, 1925, 682.

another formal product of the correspondence principle applied to the ordinary orbital model.[245]

## THE PAPER BY KRAMERS AND HEISENBERG

In a third paper written in early 1925 in collaboration with Heisenberg, Kramers gave the full demonstration of his formula and a generalization retaining the terms of the induced moment with frequencies differing from those of the incoming radiation. To the frequencies $\tau'' \cdot \bar{\nu} + \nu$ in the classical formula (203) corresponded quantum-theoretical frequencies $\nu + (E_n - E_m)/h$ with $n - m = \tau''$ for the dispersed light. Such an "incoherent" secondary radiation had already been predicted by A. Smekal in 1923, on the basis of the light-quantum hypothesis. In his argument, the frequency shift in the scattered light simply corresponded to an energy difference between in- and outgoing light quanta resulting from a quantum jump in the scattering atom. Smekal also believed that an eventual observation of this effect (as made by Raman in 1928) would support the light-quantum hypothesis.[246]

To Bohr's pleasure, the new argument by Kramers and Heisenberg got rid of the light quanta and made the incoherent radiation a consequence of the correspondence principle: the quantum-theoretical translation of a classical frequency modulation (now called the Brillouin effect).

The way Heisenberg and Kramers reached the expression for the intensity of this radiation deserves special attention, for it anticipates some essential features of the new mechanics invented by Heisenberg a few months later. According to (203), a component of the classical polarization of the perturbed atom with the frequency $\tau_0 \cdot \bar{\nu} + \nu$ has the general form

$$\xi_{\tau_0} = \sum_{\tau + \tau' = \tau_0} \left( \tau' \cdot \frac{\partial \alpha_\tau}{\partial J} \beta_{\tau'} - \alpha_\tau \tau \cdot \frac{\partial \beta_{\tau'}}{\partial J} \right), \qquad (214)$$

where $\alpha_\tau$ is a component of $\mathbf{C}_\tau$, and $\beta_\tau$ is given by

$$\beta_\tau = \frac{\mathbf{C}_\tau \cdot \mathbf{E}_0}{\tau \cdot \bar{\nu} + \nu}. \qquad (215)$$

The rules for the quantum-theoretical translation of $\tau \cdot \bar{\nu}$ and $\tau \cdot \partial/\partial J$ were already known as (208) and (211). For the amplitude $\mathbf{C}_\tau$ and the derived

245. Kramers 1924a, 1924b, 311. Like Bohr, Kramers emphasized the absurdity of the distinction between negative and positive oscillators from the point of view of classical electrodynamics.

246. Kramers and Heisenberg 1925. According to Dresden 1987, Heisenberg's contribution to this paper was minor. See Smekal 1923.

$\alpha_\tau$ and $\beta_\tau$, however, no quantum-theoretical counterpart had yet been defined. Only the corresponding intensity was known to translate into Einstein's $A_{n-\tau}^n$. Since, at the classical level, the phase appeared to play a role in formula (214), Heisenberg and Kramers introduced new "characteristic amplitudes" for the virtual oscillators through the following substitutions:

$$\alpha_\tau \to a_{n+\tau, n} \quad \text{or} \quad a_{n+\tau+\tau', n+\tau'} \tag{216}$$

$$\beta_{\tau'} \to b_{n+\tau', n} \quad \text{or} \quad b_{n+\tau+\tau', n+\tau} \tag{217}$$

With this rule, and with the requirement that the final formula be the simplest possible, the expression (214) for a Fourier component of the electric moment was translated "almost unambiguously" into

$$X_{n+\tau_0, n} = \frac{1}{h} \sum_{\tau+\tau'=\tau_0} \left[ (a_{n+\tau+\tau', n+\tau'} - a_{n+\tau, n}) \frac{1}{2}(b_{n+\tau+\tau', n+\tau} + b_{n+\tau', n}) \right.$$
$$\left. - \frac{1}{2}(a_{n+\tau+\tau', n+\tau'} + a_{n+\tau, n})(b_{n+\tau+\tau', n+\tau} - b_{n+\tau', n}) \right], \tag{218}$$

or

$$X_{n+\tau_0, n} = \frac{1}{h} \sum_{\tau+\tau'=\tau_0} a_{n+\tau_0, n+\tau'} b_{n+\tau', n} - b_{n+\tau_0, n+\tau} a_{n+\tau, n}. \tag{219}$$

With Kramers and Heisenberg's relabeling of stationary states this reads[247]

$$X_{QP} = \frac{1}{h} \sum_R (a_{QR} b_{RP} - b_{QR} a_{RP}). \tag{220}$$

The modern reader recognizes here matrix products and even a commutator.[248] Of course, Heisenberg did not analyze the results in such terms; but he would keep in mind two essential characteristics of the extended theory of dispersion: quantum-theoretical *amplitudes* appeared to play a fundamental role, and they combined only through products of the type $a_{PQ} b_{QR}$, where the stationary states corresponding to the middle indices are identical.

BORN'S "QUANTUM MECHANICS"

In his second letter to *Nature* (July 1924), Kramers wrote:

> [My] dispersion formula . . . possesses the advantage over a formula such as is proposed by Mr. Breit in that it contains only such quantities as allow of a direct interpretation on the basis of the fundamental postulates of the quantum

247. Kramers and Heisenberg 1925, 697, 699.
248. A retrospective comment: this commutator "corresponds" to the Poisson bracket (202) of the unperturbed polarization $P_0$ with the generating function $\varepsilon\varphi$, which gives the first-order polarization $P_1$. See Rüdinger 1985.

theory of spectra and atomic constitution, and exhibits no further reminiscence of the mathematical theory of multiple periodic systems.[249]

This remark should not be interpreted as implying that Kramers had deliberately oriented his calculations so as to eliminate non-quantum-theoretical quantities. He had simply benefited from a happy coincidence: *all* symbols appearing in the classical dispersion formula $(\tau \cdot \bar{\nu}, |C_\tau|^2, \tau \cdot \partial/\partial J)$ had quantum-theoretical counterparts that were already known from previous applications of the correspondence principle. However, the comment was likely to encourage Born's and Pauli's endeavors to eliminate classical concepts from quantum physics.

During a short visit of Kramers to Göttingen,[250] Born came to know the proof of the new dispersion formula and immediately conceived of a generalization that would allow him progress in his program for the "discretization of physics." As already mentioned, he believed that the difficulties experienced in the description of the interaction between atoms and radiation were of the same nature as those concerning the interaction between electrons in an atomic system. After all, he argued, the internal electric fields in an atom with several electrons varied just as fast as radiation fields.[251] He therefore believed that the discontinuity expressed in Bohr's second postulate had to affect the electron-electron interaction. In 1923 he supported Heisenberg's "new quantum principle," which introduced a difference equation analogous to $\Delta E = h\nu$ for the latter interaction.

While not ready to endorse a premature project of discretization, Pauli nevertheless found the analogy between intra-atomic coupling and radiation coupling inspiring. For instance, in June 1923 he wrote to Sommerfeld:

> I often think that not only in [the theory of] dispersion, which deals with a purely external force, but also in the interaction of the electrons in an atom, the individual electronic orbits rather behave like a system of oscillators, the frequencies of which are given not by the motion but by the transitions.[252]

A year later Born found a precise expression of this idea, a formal analogy between Kramers's dispersion theory and the electronic interaction inside atoms. Heisenberg soon joined him in his efforts and revealed

---

249. Kramers 1924b, 311; Breit's formula appeared in Breit 1924. A similar comment was later made by Kramers and Heisenberg 1925, 691: "In particular, we shall obtain, quite naturally, formulae which contain only the frequencies and amplitudes which are characteristic for the transitions, while all those symbols which refer to the mathematical theory of periodic systems will have disappeared."
250. See Born 1924, 380 n. 1.
251. Ibid., 379.
252. Pauli to Sommerfeld, 6 June 1923, *PB*, no. 37.

to Pauli the secret of Born's progress toward a new "quantum mechanics," namely, a generalization of Kramers's substitution rule (211)

$$\tau \cdot \frac{\partial}{\partial J} \rightarrow \frac{1}{h} \Delta_\tau$$

applicable in the canonical perturbation theory. "Born does that and may be right to regard it as the beginning of a reasonable quantum mechanics of [interelectronic] coupling." A month later, in July 1924, Heisenberg expressed a more open optimism: "The nicest thing about the new radiation theory by Bohr and Kramers is simply that one now knows (or surmises), on the basis of Born's calculations, how the quantum mechanics possibly appears."[253]

What Heisenberg meant by "Born's calculations" was a quantum-theoretical version of the classical energy formula for the second-order perturbation of a nondegenerate multiperiodic system. Born and Pauli had already established the classical formula.[254] I give here a simplified derivation, in which the relation to the classical dispersion formula is more transparent.

The following lemma will be needed:

The second-order energy perturbation (for a given numerical value of the action variables) of a nondegenerate multiperiodic system is identical with one-half of the time average of the first-order variation of the perturbing potential. The proof is as follows. According to the adiabatic theorem, the perturbed motion for a given constant value of the action variables may be obtained by adiabatically turning on the perturbation. During this operation the Hamiltonian function is

$$H = H_0(q, p) + \lambda(t)\, \varepsilon \Omega(q), \tag{221}$$

where, as usual, $\lambda(t)$ is a function of time slowly and smoothly varying from zero to one. The corresponding canonical equations imply

$$\frac{dH}{dt} = \frac{\partial H}{\partial \lambda} \dot{\lambda}. \tag{222}$$

---

253. Heisenberg to Pauli, 8 June 1924, *PB*, no. 62 (this letter also documents that the substitution $\tau \cdot \partial/\partial J \rightarrow (1/h)\Delta_\tau$ and the corresponding derivation of the dispersion formula had been privately communicated by Kramers *before* Born started to work on his "quantum mechanics"); Heisenberg to Landé, 6 July 1924, AHQP.

254. At the first order of perturbation the classical energy correction is the zero-frequency component (i.e., the time average) of the perturbation regarded as a function of the unperturbed motion; therefore, no virtual oscillator is involved, and the classical energy formula can be maintained in the quantum theory.

Granted that the variation of $\lambda$ is much slower than the variation of $\partial H/\partial \lambda$, the above relation can be integrated as

$$\Delta H = \int_0^1 \overline{\frac{\partial H}{\partial \lambda}}\, d\lambda \tag{223}$$

where $\overline{\partial H/\partial \lambda}$ represents the secular average of $\partial H/\partial \lambda$. For a given value of $\lambda$, the value of $\partial H/\partial \lambda$ to the second order of perturbation is given by

$$\frac{\partial H^{(2)}}{\partial \lambda} = \varepsilon\Omega(q^0 + \lambda\varepsilon q^1), \tag{224}$$

where $q^0 + \lambda\varepsilon q^1$ is the first-order perturbed motion corresponding to the perturbation $\lambda\varepsilon\Omega$. Calling $\Omega^0$ the value of $\Omega$ for the unperturbed motion, and $\varepsilon\Omega^1$ the difference $\Omega - \Omega_0$ for $\lambda = 1$, one has

$$\frac{\partial H^{(2)}}{\partial \lambda} = \varepsilon\Omega^0 + \lambda\varepsilon^2\Omega^1. \tag{225}$$

According to (223), this implies, for the energy perturbation up to second order,

$$\Delta H^{(2)} = \varepsilon\overline{\Omega^0} + \tfrac{1}{2}\varepsilon^2\overline{\Omega^1}, \tag{226}$$

where the second term has the form that was to be proved.

Consider now the Fourier development of $\varepsilon\Omega$:

$$\varepsilon\Omega = \sum_\tau C_\tau(J^0)e^{2\pi i u^w \cdot \tau}. \tag{227}$$

The average $\overline{\Omega^1}$ is the zero-frequency component of the first-order term of $\Omega$, which is identical with the electric moment $\mathbf{P}^1$ defined on p. 226, up to the formal substitutions

$$C_\tau \to C_\tau, \qquad E_0 \to -1, \qquad \nu \to 0.$$

Transposing in this way the expression resulting from (206) for $\mathbf{P}^1 = \alpha\mathbf{E}_0$, and combining with the formula (226) of the lemma gives, for the second-order energy perturbation:

$$E^{(2)} = -\sum_\tau{}' \, \tau \cdot \frac{\partial}{\partial J}\frac{|C_\tau|^2}{\tau \cdot \nu}. \tag{228}$$

To reach the quantum-mechanical version of this formula, Born used Kramers's recipes (210) and (211):

$$\tau \cdot \bar{\nu} \to (E_n - E_{n-\tau})/h$$

$$\tau \cdot \frac{\partial}{\partial J} \to \frac{1}{h}\Delta_\tau$$

and introduced, as the counterpart of $|C_\tau|^2$, coefficients $\Gamma_{n,n-\tau}$ playing here the same role as the intensities in Kramers's formula. The resulting quantum-mechanical perturbation formula is

$$E_n^{(2)} = -\sum_\tau{}' \left( \frac{\Gamma_{n+\tau,n}}{E_{n+\tau} - E_n} - \frac{\Gamma_{n,n-\tau}}{E_n - E_{n-\tau}} \right) \tag{229}$$

or, assuming a symmetric $\Gamma$,

$$E_n^{(2)} = \sum_m \frac{\Gamma_{nm}}{E_n - E_m}. \tag{230}$$

To make his formula more plausible, Born described it in terms of the virtual oscillators of the BKS theory, of which he was an enthusiastic supporter. He regarded the $\Gamma$ coefficients as some characteristic of the virtual oscillators that were to be responsible for the interplay between electrons in a given atom. Electronic orbits had to disappear from the new quantum-mechanical description, and the virtual oscillators had to be "das Reale, das Primäre."[255]

Furthermore, Heisenberg's new quantum principle could be seen as resulting from the new methods formula. In this case the only relevant quantum number is the total angular momentum $J$. The operator $\tau \cdot \partial/\partial J$ degenerates into $\pm \partial/\partial J$, since the only nonvanishing harmonics of the motion correspond to $\tau = 0, \pm 1$. Therefore, Born's perturbation involves a unit-difference operator acting on a function of $J$, exactly as in Heisenberg's new quantum principle applied to the anomalous Zeeman effect.[256]

Unfortunately, as far as practical calculations were involved, Born's tentative quantum mechanics was completely impotent: his procedure gave no hint about how to determine, either empirically or theoretically, the $\Gamma$-coefficients appearing in the perturbation formula. Only in the special case of perturbation by an external electric field could these coefficients be related to the intensities of spectral lines; even so, they could not be calculated beyond the approximation given by the correspondence principle. Quantum-mechanical equations still had to be found to derive exact intensity formulas and, more generally, to derive the $\Gamma$'s corresponding to any potential. Born was quite aware of this situation: "The problem of the determination of the $\Gamma$'s is closely related to the question of the ratios of the intensities of spectral lines, and it is of the highest importance for the further development of the quantum theory."[257]

255. Born 1924, 387.
256. Ibid., 394–395.
257. Ibid., 388.

## INTENSITIES

### THE UTRECHT SUM RULES

There were, in 1924, other reasons to focus on the problem of intensities. At Utrecht skillful spectroscopists like Ornstein and Burger managed to measure intensity ratios within multiplets and Zeeman components, and, on semiempirical grounds, they even guessed "sum rules" that could completely determine these ratios in most relevant cases. For Zeeman multiplets—to which I will confine my account—the rule reads:

> The sum of intensities of the rectilinear ($\Delta m = 0$) and circular ($\Delta m = \pm 1$) components of the light emitted (or absorbed) by an atom with a given value of the magnetic number $m$ is independent of the value of this quantum number.[258]

For Einstein's emission coefficients (which also give the absorption probability, up to a factor that does not depend on the magnetic numbers in a first approximation), this gives

$$\left.\begin{array}{l} A_m^m + A_{m+1}^m + A_{m-1}^m = A \\ A_m^m + A_m^{m-1} + A_m^{m+1} = A' \end{array}\right\} \tag{231}$$

where $A$ and $A'$ are only functions of the other quantum numbers of the initial and final stationary states, and $A_{m'}^{m''}$ is understood to be zero whenever the corresponding value of $m'$ or $m''$ is forbidden.

Other relations between the $A$ coefficients can be obtained from Bohr's principle of "spectroscopic stability" (1918), which may be paraphrased as:

> The intensities of the polarized components into which an unpolarized spectral line splits under the influence of small external forces will be such, that the ensemble of all components together will show no characteristic polarization in any direction, if small quantities proportional to the intensity of the external forces are neglected.[259]

(Bohr should have specified: under "natural" conditions of excitation for which the number of atoms in the upper quantum states remains proportional to their degree of degeneracy.) With the above convention, this gives

$$\sum_m A_m^m = \sum_m A_{m+1}^m = \sum_m A_{m-1}^m . \tag{232}$$

258. Ornstein and Burger 1924a, 1924b. For a clear account, see Pauli 1926a, 61–68.
259. Bohr 1918b, 85; Kramers 1919, 327–328.

Bohr and Kramers regarded this principle as an important one, for it provided exact relations between intensities where the correspondence principle (in Kramers's hands) provided only approximate values.

As Ornstein and Burger noticed, relations (231) and (232) were sufficient to determine the relative intensities of Zeeman components, if only at least one of the inner quantum numbers of the relevant stationary states did not exceed one unit.[260] Take for instance the case of transitions for which $j$ has the same value, $\frac{1}{2}$, in the initial and final states. The sum rule (231) becomes

$$\left.\begin{array}{l} A_{1/2}^{1/2} + A_{-1/2}^{1/2} = A_{-1/2}^{-1/2} + A_{1/2}^{-1/2} \\ A_{-1/2}^{-1/2} + A_{-1/2}^{1/2} = A_{1/2}^{1/2} + A_{1/2}^{-1/2} \end{array}\right\} \tag{233}$$

which implies

$$A_{-1/2}^{1/2} = A_{1/2}^{-1/2} \quad \text{and} \quad A_{1/2}^{1/2} = A_{-1/2}^{-1/2}. \tag{234}$$

The rule (232) of spectroscopic stability becomes

$$A_{1/2}^{1/2} + A_{-1/2}^{-1/2} = A_{1/2}^{-1/2} = A_{-1/2}^{1/2}. \tag{235}$$

The resulting intensity diagram is (in arbitrary units)

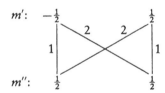

This type of reasoning seemed to provide a powerful alternative to the correspondence principle in the determination of intensities. In fact, the danger for Bohr was great that his critics would interpret the Utrecht sum rules as a new symptom of the impotence of the correspondence principle. Sommerfeld asserted as much in a letter to Kramers of July 1924, with the comment: "The final view should be that the correspondence principle is a (highly valuable) limiting *theorem* of the quantum theory, but not its *foundation*." To which Kramers replied:

> Bohr is far from considering the correspondence principle as a foundation for an axiomatic formulation of quantum theory. Bohr's formulation of the principle is of course everywhere tentative and cautious, and it would, to say

260. Kronig 1925a, 1925b removed the latter restriction and reached general intensity formulae by assuming a polynomial form of the second degree in $m$ (which was known to be valid in the limit of high quantum numbers given in Sommerfeld and Heisenberg 1922). See Pauli 1926a, 67.

the least, be too early to conclude to a "failure" or "inadequacy" of the correspondence principle from the beautiful intensity measurements at Utrecht.[261]


HEISENBERG'S SHARPENING OF THE CORRESPONDENCE PRINCIPLE

In September 1924 Heisenberg moved to Copenhagen for an eight-month stay and started to collaborate with Bohr and Kramers on dispersion theory. Already during his previous collaboration with Born he had observed the important role that the correspondence principle played in the search for a new quantum mechanics.[262] Now working at the main source of quantum-theoretical truths, he soon discovered, to Bohr's and Kramers's great pleasure, that the "Utrecht sum rules" could be derived from the correspondence principle. As he put it in his paper of November 1924, "Various empirical rules about intensities and polarizations can be conceived as a natural sharpening [sinngemässe Verschärfungen] of the correspondence principle."[263]

His reasoning was the following. Consider an alkali atom with a total angular momentum $j$ making an angle $\theta$ with the direction Oz of observation. In Bohr's orbital representation, this momentum is connected with a precession of the orbital plane of the outer electron around it. The Fourier component $C_0 e^{2\pi i \bar{v}_0 t}$ of the electric moment at the original orbital frequency $\bar{v}_0$ is a rectilinear vibration along $j$, while the components $C_{\pm} e^{2\pi i (\bar{v}_0 \pm \bar{v}_j) t}$ with a frequency shifted by plus or minus the precession frequency $\bar{v}_j$ are circular vibrations in a plane normal to $j$. For each value of $\tau$ the monochromatic spatial vibration $C_\tau e^{2\pi i (\bar{v}_0 + \tau \bar{v}_j) t}$ can now be decomposed as a sum of two circular vibrations around Oz ($\sigma_\pm$ components) and a rectilinear one along Oz ($\pi$ component), with the respective amplitudes $C_\tau^\pm(\theta)$ and $C_\tau^0(\theta)$.[264] The expression

$$|C_\tau|^2 = |C_\tau^+(\theta)|^2 + |C_\tau^-(\theta)|^2 + |C_\tau^0(\theta)|^2 \tag{236}$$

represents, on this basis, a quantity proportional to the total energy emitted by the atomic dipole at the frequency $\bar{v}_0 + \tau \bar{v}_j$, and is therefore independent of $\theta$.


261. Sommerfeld to Kramers, 5 July 1924, AHQP; Kramers to Sommerfeld, 6 Sept. 1924, AHQP.
262. Also Heisenberg and Sommerfeld 1922 had already used the correspondence principle to derive *approximate* formulae for the relative intensities in Zeeman multiplets.
263. Heisenberg 1925a, 617.
264. This would in fact be the case of any vector oscillating at a given frequency, as can easily be proved.

Consider now a large collection of such atoms for which j is distributed isotropically. The light emitted in a given direction of space is completely unpolarized, as results both from general symmetry reasons and indirectly from the integration of $|C_\tau^0(\theta)|^2$ and $|C_\tau^\pm(\theta)|^2$ over a solid angle of $4\pi$.[265] It can easily be seen that the absence of polarization in any direction of space is equivalent to the condition

$$\overline{|C_\tau^0|^2} = \overline{|C_\tau^+|^2} = \overline{|C_\tau^-|^2}, \tag{237}$$

where the horizontal bars denote the average over the atomic assembly, or, equivalently, over the angle $\theta$.

A small homogeneous magnetic field is now turned on along Oz. In a first approximation, the only effect of this field is to superpose a global rotation around Oz upon the unperturbed motion of every atom. This operation leaves $C_\tau^0$ unchanged and alters $C_\tau^\pm$ only by a phase factor $e^{\pm 2\pi i \bar{v}_L t}$, where $\bar{v}_L$ is the Larmor frequency. Consequently, relations (236) and (237) remain valid.

According to the correspondence principle, the quantum-theoretical counterpart of these relations is readily obtained by quantizing the angle according to $\cos\theta = m/j$, replacing averages over $\theta$ with sums over $m$, and performing the substitutions

$$|C_\tau^0|^2 \to A_{j-\tau, m}^{j, m} \qquad |C_\tau^\pm|^2 \to A_{j-\tau, m\pm 1}^{j, m} \tag{238}$$

(the proportionality coefficients can be omitted, for they do not depend on $m$ in a first approximation). The result is exactly the sum rules (231) and (232) used by Ornstein and Burger. More generally, all classical relations involving only the Fourier coefficients of the electric moment seemed to have an exact counterpart in the quantum theory, the form of which could be suggested by the correspondence principle.[266]

In the BKS spirit, and in conformity with Born's analogy between dispersion and intra-atomic interactions, Heisenberg attributed this sharpening of the correspondence principle to unknown virtues of the virtual oscillators: "The virtual oscillators of the quantum theory which are responsible for radiation [processes] obey laws such that the closest analogy between the classical theory and the quantum theory is kept valid."[267]

---

265. An elementary calculation gives

—for $\Delta j = 0$: $\quad C_0^\pm = (1/\sqrt{2}) \, C_0 \sin\theta, \qquad C_0^0 = C_0 \cos\theta$

—for $\Delta j = \pm 1$: $\quad \begin{cases} C_{+1}^\pm = (1/2) \, C_{+1}(1 \pm \cos\theta), & C_{+1}^0 = (1/\sqrt{2}) \, C_{+1} \sin\theta \\ C_{-1}^\pm = (1/2) \, C_{-1}(1 \mp \cos\theta), & C_{-1}^0 = (1/\sqrt{2}) \, C_{-1} \sin\theta \end{cases}$

266. Heisenberg 1925a, 618–621.
267. Ibid., 617–618.

However, he did not venture to investigate what these laws could have been.

The same paper by Heisenberg contained another quantitative application of the correspondence principle which justified the title: "An application of the correspondence principle to the polarization of the fluorescence light." According to observations made by Wood and Ellet in 1923, the light scattered by mercury vapor under excitation by polarized light at the resonance frequency (the so-called fluorescence light) was almost completely polarized; but a small magnetic field (not parallel to the polarization of the incident light) substantially reduced this polarization. For theoreticians who believed, like Sommerfeld, that degenerate states (in absence of external fields) were just an isotropic statistical mixture of space-quantized states, these results were quite surprising: Indeed, as they conceived the situation there should have been no difference between the behavior of the vapor without magnetic field and the average behavior of atoms individually subjected to small magnetic fields with random direction (representing the direction of quantization of individual atoms).[268]

To Bohr, instead, the polarization of the fluorescence light was not surprising: he had never believed in a sharp quantization of the electronic motion in degenerate states. After several experimental and theoretical developments of the problem, which I omit, he showed in November 1924 how harmoniously Wood's observations fitted the general point of view expressed in the BKS paper. Only in the nondegenerate case, he argued, were the virtual oscillators and their characteristic polarizations uniquely connected to the harmonic components of the electronic motion in a given stationary state; "we must therefore be prepared to find that the behavior of a degenerate atom, as far as radiation is concerned, is not fixed by the motion in the stationary states in question but requires a further specification of the virtual oscillators."[269]

This further specification was of course to be sought in the correspondence principle, more specifically in an analogy with the classical behavior of a degenerate multiperiodic system submitted to a harmonic perturbation, which Kramers proceeded to examine. In general the mathematics proved to be too complicated.[270] However, Bohr remarked, there was a

268. Wood and Ellet 1923.
269. Bohr 1924b, 1115–1116.
270. Ibid., 1116 n. 6; see also Kramers 1925a, 153; also in *Scientific papers*, [331].

simple example of a classical degenerate system for which the result was obvious: a three-dimensional isotropic oscillator carrying an electric charge. In this case the vibration forced by a polarized electric wave is obviously parallel to the exciting electric vector, and the scattered light is completely polarized. Bohr suggested a natural generalization: the virtual oscillators responsible for the fluorescence in Wood's experiments would be analogous to three-dimensional isotropic oscillators, in such a way that "the vibrational state of these [virtual] oscillators in the activated atoms may depend in our case upon the type of excitation of the atoms, especially upon the direction of the light vector of the exciting radiation."[271]

To the profit of the general idea of a sharpening of the correspondence principle, Heisenberg soon replaced Bohr's exceedingly subtle handling of different levels of analogy with a clear, systematic, and quantitative method of reasoning. The difficulty encountered by Kramers in the determination of the perturbation of a classical degenerate system by a polarized electromagnetic wave, he cleverly noticed, could be circumvented by introducing a small magnetic field parallel to the electric vector of the incident wave.[272] Such a magnetic field would leave the polarization properties of the scattered light unchanged, since (in the dipolar approximation) Larmor's theorem still applies to the electronic motion in an oscillating electric field *parallel* to the magnetic axis, and the amplitudes $C^{\pm}$, $C^0$ of the $\sigma_{\pm}$ and $\pi$ components of the induced electric moment simply turn into $C^{\pm}e^{\pm 2\pi i \bar{\nu}_L t}$, $C^0$, which does not change the corresponding polarization rates.[273]

Thanks to this new type of spectroscopic stability, the polarization properties of the light scattered by a degenerate system could in general be deduced from the dispersion properties of nondegenerate systems, which Kramers and Heisenberg had already determined in the nonresonant case. For instance, Heisenberg's reasoning immediately implied 100 percent polarization for the light dispersed (far from resonance) by the fundamental state of mercury: the magnetic quantum number in this state can only take the value zero, and the virtual oscillator (RP) responsible for the virtual absorption in the Kramers-Heisenberg formula (220) is a $\pi$-oscillator, so that the magnetic quantum number of the intermediate virtual states

271. Bohr 1924b, 1117.
272. In the case of a circular polarization, B must be taken parallel to the direction of propagation of the exciting light, as appears from a transposition of the following argument.
273. Heisenberg 1925a. The general condition of validity of Larmor's theorem for a system of charged particles is the invariance of the interaction energy under rotations around the magnetic axis.

(R) must be zero, and the corresponding virtual emitters (QR) must all be π-oscillators.[274]

However, Wood's original case of fluorescence appeared to be more problematic. In the vicinity of a resonance, the Kramers-Heisenberg formulae ceased to be valid, and, as stated in the BKS paper and again in Bohr's fluorescence paper, the scattered light came from two sources: from the virtual oscillations in the normal state, and from those in the stationary states ending the resonant transitions.[275] Despite the lack of any classical counterpart to this strange duality, Heisenberg assumed that his stability principle also applied to this case. He further assumed (as others had done before him) that the scattered light's state of polarization was always identical with the state of polarization of the light *spontaneously* emitted from the stationary states excited during the resonant illumination. In this way quantitative predictions could be made whenever the intensities of the various Zeeman components of the resonance lines were known.[276]

For instance, the intensity diagram of the $D_1$ line of sodium is, according to Heisenberg (after Ornstein and Burger),

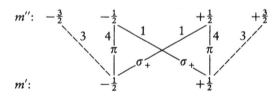

A resonant illumination with a rectilinear polarization parallel to the magnetic field only gives excited states with $m'' = \pm\frac{1}{2}$ (this is why Heisenberg draws dotted lines from $m'' = \pm\frac{3}{2}$). The spontaneous transitions from these states are the ones represented by solid lines on the diagram, with the intensity ratios

$$I_{\sigma_+}/I_\pi = I_{\sigma_-}/I_\pi = \tfrac{1}{4}. \tag{239}$$

For the light observed in a direction perpendicular to the electric vector E of the incident light, I call $I_\parallel$ the intensity of the component polarized in the direction parallel to E, and $I_\perp$ the intensity of the complementary component (see fig. 24). Simple geometric considerations yield

$$I_\perp = \tfrac{1}{2}(I_{\sigma_+} + I_{\sigma_-}), \qquad I_\parallel = I_\pi. \tag{240}$$

274. Heisenberg 1925a, 623.
275. See, e.g., Bohr 1924b, 1115.
276. Heisenberg 1925a, 624.

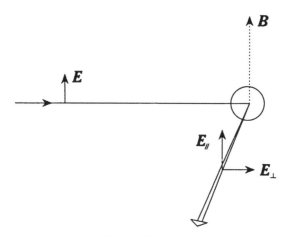

Figure 24.   Diagram for the definition of the parallel and perpendicular components of scattered light.

Combined with (239), this gives the rate of polarization of the fluorescence light:[277]

$$\frac{I_{\parallel} - I_{\perp}}{I_{\parallel} + I_{\perp}} = \frac{4 - 1}{4 + 1} = 60\% \,. \tag{241}$$

The above example shows how sophisticated the sharpening of the correspondence principle could be at the dawn of quantum mechanics. Formal analogies with classical theory operated quantitatively at two stages of the reasoning, in the deduction of the sum rules leading to the intensity diagram, and in the derivation of the stability rule for polarization properties. They even extended to degenerate systems, which had eluded Born's first tries at a discretization of physics. The feeling was growing in Copenhagen that perhaps a general quantitative theory of quantum phenomena was not so far out of reach.

VIRTUALIZATION OR FORMAL TRANSLATION?

In September 1924 Heisenberg reported to Pauli on the new victories of the correspondence principle:

> With Bohr I have . . . come to the conclusion that, against Sommerfeld's opinion, the sum rules do not elude understanding in terms of the correspondence principle; on the contrary, they are a *strict consequence* [*zwangläufige Folge*] of the correspondence principle, and in fact they provide the most beautiful

277. Ibid.

example that the correspondence principle sometimes permits the drawing of unambiguous [eindeutig] conclusions. . . . Since recently the correspondence principle has been criticized so much, it would be good to publish this result "ad majorem Korr. principie [sic] gloriam."[278]

Pauli was not convinced. He had agreed with Sommerfeld that "there was *very little* to be concluded from an application of the correspondence principle to the problem of intensities," which only added to his earlier suspicion of the impotence of the correspondence principle in the problem of the closing of electronic groups. In reply Heisenberg detailed his argument, and commented:

> If by correspondence principle one means, as you do, the wrong assumption that one could arrive at the quantum-theoretical intensity by averaging the classical intensity [an allusion to the procedure introduced in Kramers's dissertation], then *you* are right to state that the correspondence principle does *not* lead to Ornstein's rule; if, instead, one means a natural logical correspondence [sinngemässen logischen Anschluss] with the classical theory, then *I* am right.[279]

After conversations with Heisenberg on this matter, in early 1925 Pauli accepted the considerations of stability (of polarization properties) but condemned the reference to virtual oscillators in this context. Heisenberg reported to Bohr: "He [Pauli] believes in the stability laws but not in the virtual oscillators, and he reviles the 'virtualization' of physics. It is not clear to me what he meant by that."[280] But it will soon be clear, I hope, to my reader.

Four months later, Pauli tried to derive the intensity ratios of Stark components (for which new measurements by Hansen, Takamine, and Werner were available). His strategy in this problem was to find the zero-frequency (static) limit of the Kramers-Heisenberg formula for the "Smekal jumps" induced by an oscillating electric field in the presence of a small parallel magnetic field (to avoid degeneracy). According to Heisenberg's stability principle, the resulting formula had to represent the intensities of the Stark components *without* the magnetic field.[281]

278. Heisenberg to Pauli, 30 Sept. 1924, *PB*, no. 65.
279. Pauli to Sommerfeld, 29 Sept. 1924, *PB*, no. 64; Heisenberg to Pauli, 8 Oct. 1924, *PB*, no. 67.
280. Heisenberg to Bohr, 8 Jan. 1925, *BCW* 5:[357]–[358].
281. Pauli to Heisenberg, 28 Feb. 1925, *PB*, no. 86; Pauli published the full reasoning in November 1925 (Pauli 1925c). Pauli's method gives unambiguous results in the approximation for which a single term (the one with the smallest frequency denominator) is retained in the Kramers-Heisenberg formula for Smekal jumps. Indeed, in this case the (unknown) phases of the virtual oscillators are irrelevant, and one only needs to know the Ornstein-Burger intensities.

In the opinion of Bohr, Kramers, and Heisenberg, all basic sources of this reasoning were consequences of the correspondence principle. Pauli nevertheless refused to admit his capitulation to the Copenhagen views and parodied instead Heisenberg's medieval scholasticism: "In Copenhagen one of course says: 'sharpen the correspondence principle'—*id est imperialismus* of the correspondence principle."[282] One may wonder, as Heisenberg did in the letter earlier quoted,[283] how Pauli could exploit formal and logical analogies with the classical theory and at the same time criticize the sharpeners of the correspondence principle and their virtual oscillators. The explanation is as follows.

Pauli certainly recognized the important progress that Kramers, Born, and Heisenberg had made in the process of formally translating classical relations into quantum-theoretical ones.[284] However, he could not accept the broader conceptual context of the translation, namely, the description of radiation phenomena given by the BKS theory. At the source of the condemned confusion was the sharpeners' (of the correspondence principle) presentation of their results as corroborating this theory, either implicitly through their systematic recourse to the terminology of virtual oscillators, or explicitly: "All our considerations are built on the conception of the relations between atomic radiation and stationary states which is developed in a new work by Bohr, Kramers and Slater; the consequences, if they proved to be verified, would provide an interesting support to this conception."[285]

This is how Kramers and Heisenberg perceived the foundation of their dispersion theory. In reality, as Pauli rightly observed, most of the alleged applications of the BKS theory were essentially independent of the adopted description of radiation processes.[286] One argument for the independence can be found in the historical genesis of the relevant formulae. As seen above, Kramers's dispersion formula, Born's perturbation formula, and Heisenberg's sum rules were originally derived through a procedure of

282. Pauli to Heisenberg, 28 Feb. 1925, *PB*, no. 86.
283. See n. 280.
284. This appears most clearly in Pauli's later approving of an extension of Born's program: "I am very interested in the general formal problem [of the determination] of the transition probabilities, especially in the alteration and extension of Born's formalism about which we spoke in Copenhagen" (Pauli to Kronig, 21 May 1925, *PB*, no. 89).
285. Kramers and Heisenberg 1925, 681 (summary).
286. After referring to the paper by Kramers and Heisenberg in his own paper on intensities, Pauli wrote: "It must be emphasized that the formulae . . . used by these authors are independent of their special theoretical representations [Vorstellungen] concerning the detailed description of radiation processes in the quantum theory" (Pauli 1925c, 5n). In a letter to Kramers of 27 July 1925 (*PB*, no. 97), he ironized on the contrast between this remark and that made at the head of Kramers and Heisenberg 1925.

symbolic translation which was independent of any specific picture of the interaction between radiation and atoms. At every instance, the conceptual framework of the BKS theory was introduced only at a later stage, in an attempt to give a "physical" interpretation of the structure of the formulae (remember in particular that Kramers obtained his dispersion formula *before* Slater arrived in Copenhagen).

This is not to say that the reference to virtual oscillators never occurred at the first stage of symbolic translations; it did. Such reference served to identify the basic terms of the quantum-theoretical language into which classical relations had to be translated: atomic frequencies, intensities, polarizations, and even phases in the case of the Kramers-Heisenberg paper. However, the specific function of the virtual oscillators in the BKS scheme was completely irrelevant at that stage. In essence, this function could hardly be to direct quantitative, constructive reasoning, since *the virtual oscillators did not constitute a model in any sense of the word (mathematical or mechanical)*. As Bohr and Kramers emphasized, these "oscillators" did not react to radiation as classical resonators would do; they were just a metaphoric expression of the sum of their descriptive virtues: frequency, intensity, polarization, sign (positive or negative).[287]

Before the collapse of the BKS theory the sharpeners of the correspondence principle did not seem to be aware of the independence of their quantitative results from the BKS picture of radiation. Clearly, they were unwilling to isolate their results from a theory which, as a funnel for revolutionary energies, provided the psychological incentive for their research strategy. I have already quoted Heisenberg and Bohr attributing spectroscopic stability and the polarization of fluorescence radiation to a somewhat magic effect of the virtual oscillators. In late 1924 Bohr even dreamed of a further extension of the magic:

> Lately, we have entertained a new hope of an essential progress, perhaps with the [virtual] oscillator viewpoint. The experiments about fluorescence radiation have clearly revealed an independence between the electronic motion and the reaction of an atom to external actions; this independence may perhaps be useful in the question of the coupling between electron orbits in atoms.[288]

The suggestion was characteristically ambiguous: it could mean an extension of the formal sharpening of the correspondence principle, or a generalization of the virtual communications described in BKS. As we shall

---

287. See Kramers's comment quoted on p. 228, (n. 245), and Bohr's comment quoted on p. 222, (n. 235).
288. Bohr to Pauli, 22 Dec. 1924, *PB*, no. 77.

now see, the subsequent decline of the BKS theory eliminated the second alternative.

## THE FALL OF THE BKS THEORY

### EINSTEIN'S HOSTILITY

Born and Heisenberg were about the only enthusiastic supporters of the "revolutionary putsch" in Copenhagen, as Pauli stigmatized the BKS theory.[289] Most physicists were either indifferent or hostile. Among the sharpest critics, Einstein naturally came first. Many years earlier, in 1910, he had already envisioned a radiation theory without quanta, .without detailed energy conservation—and without success. His reaction to the BKS paper was therefore very prompt.[290]

First of all Einstein condemned the rejection of the energy principle and causality, which, up to then, had given good service in physics, even in atomic theory. This argument had ambiguous effect: it convinced conservative physicists of the ineptness of the BKS proposal; but lack of energy conservation attracted instead the interest of the more progressive ones. For instance Schrödinger declared: "The most exciting thing about [the BKS theory], so to speak, is the fundamental violation of the energy-momentum law in any radiation process."[291]

Einstein formulated more specific objections to the BKS theory. Some of these were irrelevant, which proves his poor knowledge of the precise contents of the paper. For instance, he overlooked the *secondary* virtual field ("scattered" by the virtual oscillators) and therefore questioned the ability of the BKS theory to reproduce ordinary optical effects. However, most of his specific objections were much harder to dismiss. One considered "very disturbing" by Bohr himself concerned an absurd consequence of the type of independence of quantum jumps in distant atoms implied by the BKS theory: energy fluctuations in a macroscopic sample of matter had to grow linearly and indefinitely in time, which seemed to ruin the basis of statistical thermodynamics.[292]

Kramers argued that an appreciable growth of energy fluctuations would take an extremely long time, so that a small adjustment of the theory was

289. Pauli to Kramers, 27 July 1925, *PB*, no. 97.
290. Einstein to Laub, 4 Nov. 1910, quoted in *BCW* 5:[27].
291. Schrödinger 1924, 720. Einstein's objections to the BKS theory are in Einstein to Ehrenfest, 31 May 1924, *BCW* 5:[26]–[27]; in unpublished notes sent to the *Vossische Zeitung* (French in Einstein 1989b, 166–168); in Pauli to Bohr, 2 Oct. 1924, *PB*, no. 66; and in Heisenberg to Pauli, 8 June 1924, *PB*, no. 62). See Klein 1970a; Stolzenburg 1985.
292. Bohr to Franck, 21 Apr. 1925, *BCW* 5:[350]–[351]. Schrödinger 1924 gave an explicit proof of the linear increase of fluctuations.

likely to remove the absurdity.[293] From Einstein's point of view this would cure, at the very best, only one symptom of a more general disease: the BKS theory required a "preestablished harmony" between continuous and discontinuous aspects of quantum phenomena, which grated his epistemological conscience.[294] For instance, Einstein reproached the BKS theory with giving two different explanations for the natural width of spectral lines: one based on the finite length of the wave trains emitted in the active stationary states; the other drawing from an unsharpness of stationary states, as derived from the correspondence principle (there being no sharp separation between Coulomb and radiation forces in the classical theory). Redundance in theoretical description was precisely what Einstein had striven to eliminate with his relativity theory.[295]

PAULI'S DISGUST

As a champion of relativistic thinking, Pauli was particularly sensitive to this type of argument. After a short period of sympathy for the BKS theory, in the fall of 1924 he became a strong opponent of it and added to the list of Einstein's objections. According to Bohr, the radiation scattered by a resonant vapor (the fluorescence light) had a double origin, from the "forced vibrations" of the virtual oscillators connected with the lower level, and from the spontaneous emission of the virtual oscillators connected with the upper level (which followed the upward quantum jumps induced by the resonant radiation). Such a distinction, Pauli argued, lacked both a classical counterpart and empirical significance. Furthermore, for an excitation within the line (in the absence of a pressure broadening of the absorption line) there would be a nonnegligible contribution of the upper level to the scattered light in addition to that of the lower level. This contradicted the correspondence principle, since, according to the BKS theory, the energy scattered by the lower level was given by Einstein's $B\rho_\nu$, and this contribution alone provided asymptotic agreement with classical theory.[296]

Bohr's reply to this objection was twofold. On the one hand, he did not believe that the correspondence principle could constrain the value of Einstein's coefficients in the case of unbroadened lines. Since, within the

293. Kramers, unpublished manuscript, AHQP, discussed in Stolzenburg 1985, [31].
294. After Pauli to Bohr, 2 Oct. 1924, *PB*, no. 66.
295. BKS 1924, 795–796. After Pauli to Bohr, 2 Oct. 1924, *PB*, no. 66.
296. Pauli to Bohr, 2 Oct. 1924, *PB*, no. 66. A more detailed version of the same argument is in Pauli 1926a, 100. The calculation of resonant scattering in the BKS theory was made by Becker 1924, and the result also criticized by van Vleck 1925 and Slater 1925a. The latter proposed an alternative conception (rejected both by Bohr and Pauli).

natural width of a spectral line, the reaction of radiation could no longer be neglected, no sharp separation could be made between stationary states and transitions, and there was a corresponding ambiguity in the definition of the asymptotic limit of the quantum theory.[297] On the other hand, Bohr believed that the distinction between coherent (connected with the lower level) and incoherent (connected with the spontaneous emission in the upper level) resonance radiation had a definite empirical meaning. In order to prove this, he imagined various thought experiments: none of them convinced Pauli.[298]

Bohr was not completely unstirred by the accumulation of criticisms of the BKS theory. In a letter to Pauli of December 1924 he admitted: "I ought perhaps to have bad conscience with respect to the radiation problem." However, he continued, "even if from a logical point of view perhaps it is a crime, I must confess that I am nevertheless convinced that the swindle of mixing the classical theory and the quantum theory still in many ways will show itself to be fruitful in tracking the secrets of nature."[299] The prototype of such a swindle was of course given by the BKS theory, with its blending of the continuity of Maxwell's equations and the discontinuity of quantum jumps. Contrary to Einstein and Pauli, Bohr did not fear the resulting dual mode of explanation employed in the discussion of radiation phenomena. As he wrote to Slater in January 1925, "[Kramers and I] are rather inclined to be more and more reconciled with the duplicate character of the resonance radiation, in which we see a natural consequence of the dualistic nature of the whole quantum theory in its present state." Bohr meant that the duality involved in the BKS theory just added to already acknowledged dualities, for instance those found in Heisenberg's new quantum principle and in Pauli's fourth electronic quantum number.[300]

297. See Kramers [1925b]. In a comment on Becker's result reported in Becker 1924, 186, Kramers also explained that a dense sequence of energy levels had more resemblance to the classical energy continuum if the energy levels were taken to be unsharp. Previously, Bohr had frequently emphasized that the notion of stationary state was not sharply defined for phenomena involving the reaction of radiation. For example, in BKS 1924, 795, he wrote: "In fact, the postulate of the stability of stationary states imposes an *a priori* limit to the accuracy with which the motion in these states can be described by means of classical electrodynamics, a limit which on our picture is directly involved in the assumption that the virtual radiation field is not accompanied by a continuous change in the motion of the atom, but only acts by its induction of transitions involving finite changes of the energy and momentum of the atom."

298. A first reference to Bohr's thought experiment is in Pauli to Bohr, 2 Oct. 1924, *PB*, no. 66; also Pauli to Kronig, 21 May 1925, *PB*, no. 89, and Heisenberg to Pauli, 16 Nov. 1925, *PB*, no. 105. For details, see Stolzenburg 1985, [60]–[61].

299. Bohr to Pauli, 11 Dec. 1924, *PB*, no. 73.

300. Bohr to Slater, 10 Jan. 1925, *BCW* 5:[66].

In Pauli's opinion the pervasiveness of "dualities" or "duplicities" in the quantum theory marked the final death of classical concepts. The "virtualization" of classical physics was too mild a medicine to cure quantum theory of the disease afflicting it. The formal ghosts of electromagnetic fields and electronic orbits maintained in the BKS theory, Pauli asserted, muddled quantum theory more than they served it. In May 1925, commenting a last time on the duplicate nature of resonance radiation in this theory, he wrote to Kronig: "I would always much rather say that I have so far no complete picture of the phenomena, than even temporarily to put up with a hideousness of this kind which hurts my physical sensibility."[301]

COLLISIONS

Meanwhile, in the first trimester of 1925, Bohr tried to exploit the type of violation of the energy principle implied by the BKS theory to solve a standing paradox in a topic he had worked on earlier, the stopping of swiftly moving particles ($\alpha$ or $\beta$ rays) by atoms. As he constantly remembered, basic information on the atom, like the existence of the nucleus and the density of the electronic swarm, depended on the possibility of a classical treatment of the related collision problem. The paradox was the following. Bohr's classical stopping formula of 1913, which had been well confirmed, gave for the average energy loss per collision a value very inferior to the characteristic energies of atomic transitions. This seemed to contradict Franck and Hertz's observation that, in the case of slower impinging particles, the energy exchange between an atom and a particle was always given by the energy difference between two stationary states (in conformity with the postulate of a supramechanical stability of these states).[302]

Bohr imagined the following solution. In the Franck-Hertz case the target atom returns to a stationary state as soon as its interaction with the traveling electron has stopped: Bohr called such collisions "reciprocal." In contrast, a swiftly moving particle stops interacting with a given target atom before this atom has had the time to accomplish a full transition to its final stationary state (a common estimate of the duration of a quantum jump was $10^{-15}$s, the order of magnitude of an optical period;[303] while the order of magnitude of the interaction time of a $\beta$ particle crossing an atom at nearly the speed of light is $10^{-18}$s): this is what

301. Pauli to Kronig, 21 May 1925, PB, no. 89.
302. Bohr 1925a, 146–147.
303. See Pauli 1926a, 12.

Bohr calls a "nonreciprocal" collision. A similar lack of reciprocity already existed in the BKS theory, where a virtual field could be modified by an atom, without a quantum jump taking place.

On the basis of this analogy, Bohr assumed that energy violations occurred in nonreciprocal collisions, in a way that allowed the target atom to return to a stationary state without the energy loss of the colliding particle being compensated for. Just as in the BKS theory, this assumption would not contradict the large-scale conservation of energy, as long as the probability of a transition of the target atom was taken to be proportional to the energy loss of the colliding particle. Bohr even extended the fundamental dichotomy between the continuous activity of the stationary states and the discontinuous quantum jumps: "The reaction of the atom upon the particle should be regarded as governed essentially by continuous laws [reproducing the classical stopping formula], while the change in the state of the atom, according to our view, can for the present only be described by probability laws."[304]

Bohr had already sent the manuscript containing the above argument for publication, when he started, in early April or late March 1925, to worry about the case of the Ramsauer effect. From a series of experiments started by Ramsauer in 1921, a volume of noble gas was known to be perfectly transparent to a stream of sufficiently slow electrons (around 1 eV). This result sharply contradicted classical electrodynamics, which for this situation predicted a strong deviation—even capture—of electrons coming close to an atom, with a large amount of emitted radiation.[305]

In 1923, Bohr had already discussed the Ramsauer effect as a particular case of the Franck-Hertz effect, corresponding to a situation where the energy of the electronic beam is less than needed to induce a transition to the first excited level. In the same year Hund, following an idea by Franck, had managed to give a first quantitative theory based on the correspondence principle and on the following *ad hoc,* and strongly anti-classical, assumption: those of the impinging electrons which according to classical electrodynamics would be captured by the target atom crossed the atom undisturbed. The correspondence principle came into play in a calculation of the energy loss (through bremsstrahlung) of the rest of the impinging electrons.[306]

The good agreement of Hund's theory with experimental data made clear the necessity of a consequential departure from classical electrody-

---

304. Ibid., 148.
305. Ramsauer 1921a, 1921b, 1923.
306. Bohr. [1923f], [508]; Hund, 1923.

namics, as concluded by Minkowski and Sponer in a review written in 1924:

> To summarize, the collisions between electrons and atoms, in spite of the validity of the energy-momentum principle, proceed in a completely unmechanical way, and seem to obey purely quantum-theoretical laws. This is true not only of the so-called inelastic collisions [of the Franck-Hertz type] but also for the so-called elastic collisions [below the threshold of atomic excitation.][307]

In Bohr's paper on collision, as in the BKS paper, the source of every breakdown of the classical type of description was considered to be located in the quantum jumps. During time lapses in which no quantum jump occurred, the laws ruling the evolution of the various virtual entities were extensions of classical laws, and they could, essentially, be formulated in a space-time framework. This point of view was predicated upon the possibility of sharply separating between stationary motion and transitions. In the case of the Ramsauer effect, however, the large magnitude of the classical reaction of radiation prevented such a separation, as Bohr noted in a footnote to the BKS paper. Consequently, "collisions" of the Ramsauer type eluded his later distinction between reciprocity and nonreciprocity; and the possibility of a space-time description associated with this distinction therefore would have to be abandoned. Bohr drew this dramatic conclusion in April 1925, as seen for instance from a letter to Fowler:

> I shall not wish to publish my little paper [the English version of the work on collisions]. . . . It is still the Ramsauer experiments which is [sic] the chief cause of the trouble. In fact I think that the possibility of describing these experiments without [a] radical departure from an ordinary space-time description is so remote that we may just as well surrender at once and prepare ourselves for a coupling between the changes of states in distant atoms of the kind involved in the light-quantum theory.[308]

THE FINAL WRECK

As appears from the last sentence, Bohr assumed the failure of space-time representations to extend to the radiation theory and was therefore ready

307. Minkowski and Sponer 1924, 84. Hund's theory was soon contradicted by R. B. Brode's electronic absorption measurements on molecules (CO and $N_2$), according to which the Ramsauer transparency occurred only for not-too-slow electrons. See Bohr to Heisenberg, 10 June 1925, BCW 5:[364].
308. BKS 1924, 792; Bohr to Fowler, 21 Apr. 1925, BCW 5:[81]–[82]. At Heisenberg's suggestion, Bohr's collision paper finally appeared in ZP (Bohr 1925a) with an addendum written in July 1925.

to abandon the BKS theory and accept the alternative possibility of "coupling." The Ramsauer effect was not the only source of this radical change of outlook: Pauli the orbit-killer was in Copenhagen for the Easter holiday from 15 April to 25 April, and around that time Geiger concluded with Bothe an experiment proving the conservation of energy in individual Compton scattering events. More specifically, in the interaction between an X-ray beam and quasi-free electrons, the detection of a quantum of radiation in a given direction always coincided with the detection of an electron in the direction expected from the light-quantum explanation of the Compton effect. This implied a coupling between the Compton electrons and the atoms of an X-ray detector of a kind excluded by the BKS theory.[309]

In reply to Geiger's letter announcing these results, Bohr wrote:

> Thank you very much for the great kindness of having informed me of your important results. I was quite prepared to learn that our point of view about the independence of the quantum processes in separate atoms would turn out to be wrong. The whole matter was more an expression of an endeavor to attain the greatest possible applicability of the classical concepts than a completed theory. Not only were Einstein's objections very disquieting, but recently I have also felt that an explanation of collision phenomena, especially Ramsauer's results on the penetration of slow electrons through atoms, presents difficulties to our ordinary space-time description of phenomena to such an extent that, in spite of the existence of coupling, conclusions about a possible corpuscular nature of radiation lack a sufficient basis.[310]

As we can appreciate from this letter, Bohr's reaction to the failure of the BKS theory was just as radical as Pauli's reaction to the failure of multiperiodic models of the anomalous Zeeman effect. With some delay, the two men now agreed about a general collapse of ordinary space-time descriptions. In 1922–23, before the BKS episode, Bohr had already suspected such a failure but had believed it to be restricted to the case of the interaction between atoms and radiation. He now rejected all of the space-time pictures previously used in the quantum theory: electronic orbits in stationary states, trajectories in collision processes, radiation fields, *and* corpuscular light-quanta.

Did the refutation of the BKS theory necessarily imply such a drastic reform of physics? Other physicists involved in the BKS program—

---

309. Geiger to Bohr, 17 Apr. 1925, BCW 5:[352]–[353]. Geiger's letter reached Bohr on 21 April, as appears from the postscript of Bohr to Fowler, 21 Apr. 1925, BCW 5:[81]–[82]. That Bohr had developed strong doubts about the BKS theory even before this date is documented by Bohr to Heisenberg, 18 Apr. 1925, BCW 5:[360]. Bothe and Geiger 1925a, 1925b.
310. Bohr to Geiger, 21 Apr. 1925, BCW 5:[353]–[354].

notoriously, Kramers, Slater, and Born—did not think so. There was a much milder remedy to the absence of the Bothe-Geiger type of correlations in the BKS theory: one just had to return to Slater's original proposal and "hang" light quanta to the virtual fields. With Jordan's help, Born even developed a fairly detailed guiding mechanism that would reproduce optical observations. Under the influence of Bohr's criticism, however, he soon abandoned this attempt.[311]

There is reason to doubt that Bohr's reiterated rejection of the Slater type of theory depended on the specific objections that he made to the Born-Jordan attempt. These objections concerned Born's particular assumptions about the guiding of light quanta, which could be adjusted. Very likely, Bohr had in mind more fundamental defects of this type of theory. As he liked to emphasize, any theory conceding too much reality to light quanta departed from the approach inspired by the correspondence principle. Further, the addition of light quanta did not cure some basic defects of the BKS theory, like the duplicate origin of the resonance radiation denounced by Pauli.

Perhaps Bohr's strongest argument against this type of theory was the general convergence of quantum-theoretical paradoxes toward a renunciation of a detailed description in space-time: on top of the early radiation paradoxes, there were the anomalous Zeeman effect, the Ramsauer effect, and, last but not least, the devastating consequence of strict energy conservation, when applied to Bohr's nonreciprocal collisions. According to the argument in the collision paper, the assumption that swiftly moving particles had a definite course in space-time implied that they lost in each collision an energy much less than the smallest transition energy of the target atoms.[312] In other words, the energy principle, the space-time mode of description, and the quantum postulate could not be simultaneously valid. If, as Geiger's experiment seemed to indicate, the energy principle retained a general validity in the quantum theory, there could be no question of a space-time description. Bohr expressed this viewpoint in an addendum to his collision paper (which he finally decided to publish):

> If one wants to postulate a strict validity of the conservation laws . . . it must be emphasized that, for the collisions called nonreciprocal above, we must not only, as for reciprocal collisions, expect an interaction that is incompatible with

311. Kramers to Urey, 16 July 1925, quoted in BCW 5:[86]; Slater 1925b; Born and Jordan [1925a]; Born to Bohr, 24 Apr. 1925, BCW 5:[308]–[310]; Born to Bohr, 1 May 1925, BCW 5:[310]–[311].

312. In Bohr 1925a, 156, Bohr discussed a similar paradox concerning the capture of electrons by fast particles.

the properties of mechanical models, but we must in fact even be prepared to find behavior that is as alien to the ordinary space-time pictures as the coupling of individual processes in distant atoms is to a wave description of optical phenomena.[313]

Pauli rejoiced over the refutation of the BKS theory, and even more over Bohr's surrender to his views about the failure of space-time descriptions in the atomic realm. He nevertheless lamented about the desolate outlook of quantum theory: "Physics at the moment is again very muddled; in any case, for me it is too complicated, and I wish I were a film comedian or something of that sort and had never heard of physics. Now I do hope nevertheless that Bohr will save us with a new idea."[314] Not too wishful a hope: Bohr's psychology worked counter-Pauli-wise. The more severe the crisis, the higher Bohr's expectations for a quick resolution. In his opinion the climax of quantum paradoxes did not sink the theory into an inextricable chaos. Instead, the past conquests of spectral theory and atomic building had to retain some value: "In spite of all the obscurity, at the moment things are relatively much better with the secrets of the atom than with the general description of the space-time occurrence of quantum processes."[315]

What can a searching physicist base his thinking on when things refuse to be visualized? Bohr had already asked this question two years earlier, in the conclusion of "The fundamental postulates." At that time he had already pointed to the independence of the correspondence principle from the classical mode of description. At the turning point encountered in the spring of 1925, Bohr again placed his ultimate hope in the correspondence principle, now understood as a formal analogy between classical electrodynamics and a purely *symbolic* quantum theory: "We must have recourse

313. Bohr 1925a, 154–157.
314. Pauli to Kramers, 27 July 1925, *PB*, no. 97: "After all I think that we are extremely lucky that the conception of Bohr, Kramers, and Slater has been so quickly refuted by the beautiful experiments of Geiger and Bothe as well as those just published by Compton. . . . *It is not the energy concept that must be modified but the concepts of motion and force.* In cases when interference phenomena occur, one cannot, for sure, define precise [bestimmte] 'trajectories' for light quanta; but also for electrons in atoms one cannot define such trajectories; and there is no more reason to question for that the existence of electrons than there is to question the existence of light quanta because of interference phenomena. For any unprejudiced physicist we may now consider it to be proved that light quanta have just as little (and just as much) physical reality as electrons do. However, one cannot generally apply the classical kinematic concepts to either of them." Pauli to Kronig, 21 May 1925, *PB*, no. 89.
315. Bohr to Heisenberg, 18 Apr. 1925, *BCW* 5:[360].

to symbolic analogies to a still higher degree than before. Just lately I have been racking my brain to imagine [*hineinträumen*] such analogies."[316]

A letter from Pauli to Kronig written after his stay in Copenhagen indicates the source from which Bohr and other Copenhageners were expecting to draw fruitful analogies: "I am very interested in the general formal problem [of the determination] of the transition probabilities, especially in the alteration [*"Ummodelung"; "Umdeutung"* crossed out] and extension of Born's formalism about which we spoke in Copenhagen."[317] Pauli meant a purification of the symbolic procedures introduced by the sharpeners of the correspondence principle through the removal of any allusion to the dead picture of the BKS theory. Any reference to virtual oscillators and fields had to be eliminated, and there had to be imagined an extension of Born's quantum-mechanical relations that would fill the main conceptual gap in that theory: the indetermination of intensities (or $\Gamma$ coefficients in Born's paper).

This program emerged around Easter time in Copenhagen, from discussions involving Bohr, Pauli, Heisenberg, and Kronig. It was precisely the one soon brought to completion by Heisenberg.

SUMMARY

Until 1922 Bohr contented himself with applying the correspondence principle as a formal connection between atomic motion and emitted radiation and refrained from more specific assumptions about the mechanism of radiation processes. His attitude changed after the helium crisis and the empirical "proofs" of Einstein's light quantum in the period 1920–1923 (especially Maurice de Broglie's and Compton's). In his "Fundamental postulates" of 1922 he made up his mind to publish some reflections based on thoughts dating from 1917 at the latest. Einstein's fluctuation argument of 1916, he said, proved that the quantum postulates together with the assumption that conservation laws were strictly valid led necessarily to the conclusion that light quanta existed. Like most of his colleagues, Bohr believed light quanta to be incompatible with the best-verified laws of optics; therefore, he concluded, energy and momentum could not be conserved during individual processes of emission and absorption of radiation. Bohr further suggested that the quantum transitions would elude any detailed description in space and time, even though formal orbits and fields could still be used when no transition occurred.

316. See above, p. 216; Bohr to Born, 1 May 1925, *BCW* 5:[310]–[311].
317. Pauli to Kronig, 21 May 1925, *PB*, no. 89.

Bohr found a more positive inspiration in Ladenburg's dispersion theory of 1921. Classically, optical dispersion is understood as resulting from the interaction between electromagnetic waves and electric oscillators inside atoms. A direct adaptation of this picture to Bohr's orbital model would have given resonances at the *orbital* frequencies, clearly at variance with empirical results. Ladenburg therefore introduced "Ersatz-oscillators" at the *atomic* frequencies (i.e., the frequencies given by Bohr's frequency rule). The effect of light on atoms, Bohr commented, was not only to induce quantum jumps (in the resonant case); there also had to be a continuous action, based on some unknown mechanism in which the atomic frequencies, not the orbital ones, would play a role.

In late 1923 John Slater brought to Bohr and Kramers the key to the development of such a mechanism. Radiation fields, Slater said, had to be continuously emitted or absorbed during the sojourn of an atom in a stationary state, not during the quantum jumps. The fields did not carry any energy; their only function was to guide light quanta that were emitted or absorbed during the quantum jumps. Bohr and Kramers adopted the radiative activity of stationary states but rejected the dual representation of radiation, which conflicted fundamentally with the Copenhagen strategy based on the correspondence principle. In the theory soon published by Bohr, Kramers, and Slater (BKS), the radiation field itself was in charge of inducing quantum jumps, though necessarily in a statistical manner.

The new conception, Bohr commented, greatly refined the "correspondence" between atomic motion and radiation, since the continuous character of radiation now reflected continuous existence of the atom in the stationary states. A space-time description of radiation processes seemed in large part possible, save for the sudden switches of atoms between different stationary states and the corresponding changes of radiative activity. In this sense the BKS theory was the best possible realization of the correspondence idea in terms of space-time pictures. The price to be paid was the relaxation of the energy principle and the "virtualization" of the electromagnetic field: quantum jumps occurred in atoms without energetic compensation (other than statistical), and the fields were emitted without weakening of their source (the motion in stationary states). The only real objects of the BKS theory were the stationary states, for their individual characteristics could be defined through continuous deformation (recall Bohr's comments on the adiabatic principle).

The radiative activity of a given stationary state was attributed to a set of "virtual oscillators" formally connected with the transitions to and from this stationary state, with a frequency given by Bohr's frequency rule and an amplitude approximately given by the "corresponding" harmonic component of the orbital motion. Unlike the virtual fields, which obeyed Maxwell's equations in vacuum, the virtual oscillators did *not* behave like classical oscillators (a point overlooked by several commentators). Indeed, a "positive" ("negative") oscillator—that is, one connected with a downward (upward) transition—reinforced (depressed) a resonant incident wave, whatever the phase of this wave might be; a positive oscillator spontaneously emitted radiation, a negative one did not. In fact, the virtual oscillators were nothing but a condensed expression of their effects, which could be deduced from the correspondence principle piece by piece but could not be synthesized in any classical model.

The BKS paper was mostly qualitative, concerned with the general assumptions of the theory and with the resolution of the conflict between the quantum postulates and specific optical experiments. Quantitative work within this framework was published later in the same year, 1924. Most influential was Kramers's dispersion theory. The resulting formula, although presented in the BKS context, had been obtained before Slater arrived in Copenhagen. Kramers had simply calculated the classical dispersion formula given by a multiperiodic charge system in terms of action-angle variables, and then used the "correspondence" between harmonic components and quantum transitions to translate the result into purely quantum-theoretical terms, that is, quantum numbers, transition probabilities, and atomic frequencies. His formula was more general than Ladenburg's and involved two types of terms, which he associated with positive and negative virtual oscillators.

Kramers's method proved even more important than his result. It involved symbolic rules for translating classical relations into quantum-theoretical ones. Suggested by the combined use of the correspondence principle and action-angle variables, these rules appealed to Bohr's friends, who quickly extended them to other cases. In late 1924 Kramers and Heisenberg generalized the dispersion formula to the case where the frequency of the scattered radiation differs from that of the incident radiation. In this process they introduced "transition amplitudes" indexed by two quantum numbers, and combinations of these amplitudes which we would now recognize to be matrix products and commutators. Even before this episode, Max Born applied Kramers's translation rules to the interaction of electrons within atoms, because he believed such interactions to be of the same nature as the interaction between atoms and radiation, for which Kramers's methods had proven to work. Born interpreted his result, a second-order perturbation formula involving only quantum-theoretical quantities (and identical with the modern one), as a decisive step toward a "quantum mechanics."

As Born himself emphasized, the new quantum-theoretical formulae had an important defect: they involved certain coefficients (transition probabilities in Kramers's case) that could not yet be (exactly) calculated by a priori means. Further progress toward quantum mechanics depended on finding a new access to these coefficients. By a happy coincidence, in the same period a group of Dutch physicists were able to derive semi-empirical relations between transition probabilities (the "Utrecht sum rules"), which were so simple and so well verified that theorists were inclined to regard them as *exact* quantum-theoretical laws. Sommerfeld and Pauli interpreted this progress as one more sign of the impotence of the correspondence principle, which gave only *approximate* values for transition probabilities. To Bohr's pleasure, in Copenhagen Heisenberg quickly inverted the situation by showing that the Utrecht sum rules could be *derived* from the correspondence principle, for they were the natural symbolic translations of corresponding classical relations. With similar methods he soon determined the polarization rates of fluorescence radiation, a problem that had puzzled physicists for a while.

Kramers's dispersion theory, Born's "quantum mechanics," and Heisenberg's sum rules all resulted from what Heisenberg called a "sharpening of the correspondence principle." In previous deductive uses of the correspondence principle, the *magnitude* of the harmonic components of the atomic motion was calculated classically and taken as an estimate of the magnitude of the "corresponding" transition probabilities. In the sharpened version, classical *relations* between harmonic components were translated into relations between the "corresponding" transition probabilities (or amplitudes). As Pauli emphasized, this symbolic translation procedure was rather automatic, and essentially independent of the detailed picture of radiation processes. Nevertheless, since it developed in the context of the BKS program, it was generally "explained" in terms of virtual oscillators and sometimes even presented as a consequence of the new radiation theory.

Meanwhile, severe objections had been raised against the BKS theory. Einstein pointed to several absurd consequences, and condemned the "preestablished harmony" that was assumed in the theory to reconcile continuous and discontinuous aspects of radiation processes. Pauli soon approved of these criticisms. Phenomena that received a simple, unique explanation in classical electrodynamics (for instance, resonance and line width) received *dual* explanations in the BKS theory, so that artificial adjustments were required. Pauli condemned this duality or duplicity, and recommended, in conformity with his analysis of the complex structure of spectra, a complete elimination of classical concepts, even the virtual ones, from the quantum theory. Instead, Bohr and Kramers, in early 1925, welcomed this duality as a universal feature of the quantum theory: there were antecedents in Heisenberg's new quantum principle—and in Pauli's ambiguous electrons.

From then on, however, Bohr's confidence in the BKS theory gradually diminished, and almost totally vanished when in the spring of 1925 he encountered difficulties in a collision theory that had some analogy to the BKS theory. Under Pauli's influence he was already prepared to abandon all space-time descriptions of atomic phenomena, when he received a letter from Geiger reporting strict energy conservation in individual Compton processes, a sharp contradiction of the BKS theory. At that stage Born, Slater, and Kramers decided to go back to Slater's original idea and hook light quanta to the virtual fields. But this would have contradicted the correspondence strategy and would not have solved the paradoxes concerning collision processes. Bohr preferred to erase all visual elements of the quantum theory, the orbits, the waving fields, *and* the light quanta, and started to dream about "symbolic analogies" that would extract from classical laws a purely symbolic content that conformed to the quantum postulates. At the same time a few visitors to Copenhagen—Heisenberg, Kronig, and Pauli—contemplated the possibility of extending the procedure for symbolic translation of classical relations initiated by Kramers and Born.

# CHAPTER X

# Matrix Mechanics

## FROM COPENHAGEN TO GÖTTINGEN

### DUALITY OR SYMBOLIC TRANSLATION?

Before the final collapse of the BKS theory, Heisenberg had been follow-ing two different lines of research, one in "term zoology and Zeeman botany," the other in the problems of dispersion and intensities. Along the first line, under the influence of Born and Pauli, he had reduced classical atomic models to "symbolic pictures" that were related to the observable properties of atomic spectra in an indirect formal way. An unmistakable sign of the symbolic character of these models was the fact that several complementary pictures were needed to describe a given system, each model illuminating only one part of observed regularities. As already mentioned, Heisenberg was not entirely satisfied by this situation, which made the explanation of atomic spectra more complex than the spectra themselves.[318]

Fundamentally, the multimodel type of explanation conflicted with the correspondence principle, according to which the observed simplicity of spectral patterns should have been the reflection of an underlying simplicity of atomic motion. Also, among Heisenberg's various symbolic pictures, only the one providing the radiating electrons with a definite tra-jectory was adapted to the harmonic analysis that sustained the "corre-

---

318. "Term zoology and Zeeman botany" was the title of Heisenberg's talk at the Kapitza club on 28 July 1925, according to the "minute book" of this club (AHQP).

spondence." To summarize, this part of Heisenberg's theoretical activity, as successful as it was, did not comply with Bohr's idea of a complete generality of the correspondence principle; all the same, it adopted only one-half of Pauli's dictum about a new kinematics: visualizable orbits were eliminated, but unobservable quantities proliferated.

Heisenberg's other approach, the one applied to the problems of dispersion and intensities, proceeded directly from the correspondence principle in the sharp form developed within the BKS program. In this strategy the closest formal and logical analogy with classical electrodynamics was supposed to give the best results in the quantum theory. In this line of arguments Pauli distinguished two aspects: a conservative tendency to return to a largely spatiotemporal description of radiation phenomena (which he disapproved) and a subprogram for symbolically translating classical laws into a form expressed in terms of genuine quantum-theoretical concepts (which he approved).

Heisenberg left Copenhagen for Göttingen a little before the results of Geiger's experiment were known. His first reaction appears in a letter to Kronig of 8 May 1925 (AHQP): "To judge from your letter, a terrible confusion about the radiation theory must reign in Copenhagen. If I were there, I would, as in the case of the Zeeman effects, plead for a formal dualistic theory: Everything must be describable both in terms of the wave theory and in terms of light quanta." In other words, Heisenberg suggested a generalization of his symbolic multimodel approach that would encompass the dual aspects of radiation. However, the idea was too vague to fulfill the ambition of a calculating physicist; it would provide only a temporarily tranquilizing medicine. For a more definite solution of an enigma where Bohr himself had erred, Heisenberg preferred to wait: "I myself do not dare to deal with such dangerous problems, of which one cannot be sure at all whether they are really to be solved now."[319]

The "secrets of atoms," as Bohr put it, seemed to be more easily accessible. Naturally, Heisenberg decided to concentrate his efforts on what was left of his correspondence-sharpening approach after the collapse of the BKS theory: Born's program of symbolic translation. In particular, he henceforth proscribed any reference to virtual oscillators or Ersatz-radiators, which he now viewed as a pathetic mark of Bohr's ultimate attempt at retaining a space-time picture of radiation phenomena. As he wrote to Kronig, "The word 'Ersatz-radiator' has come to mean: 'unclean application of the correspondence principle, which one cannot understand'; I

---

319. Pauli to Kronig, 8 May 1925, AHQP.

urge you to eliminate this word, which reminds me of wartime *Ersatz-jam* etc."[320]

When Heisenberg arrived in Göttingen, Born and Jordan were working on a new radiation theory with guided light quanta, as previously mentioned. Heisenberg had "no great faith" in this approach, since it focused on what he had just identified as "dangerous problems." He was more interested in another project of these colleagues, an extension of Born's program of quantum translation, with a special emphasis on eliminating unobservable quantities. In the resulting paper of Born and Jordan one can read: "According to a fundamental principle of great importance and fertility, the only quantities [Grösse] that enter the true laws of nature can in principle be observed [beobachtbar] and determined [feststellbar]," with the following comment in a footnote: "Relativity theory emerged from Einstein's awareness of the fundamental impossibility of determining the absolute simultaneity of two events in different places."[321]

The specific aim of this paper was to give quantum-theoretical formulae that would describe the reaction of atoms to *aperiodic* electromagnetic fields and thus apply to processes like the scattering of white light or radiation damping. To serve this purpose, they first calculated the *phase-averaged* (with respect to atomic phases) reaction of a nondegenerate multiperiodic system to a varying electric field. To justify the phase-averaging operation they turned to the above "fundamental principle," declaring the relative phase between two separate atoms to be unobservable. The resulting formulae contained only symbols like $\bar{v} \cdot \tau$, $\tau \cdot \partial/\partial J$ and $|C_\tau|^2$ (where $C_\tau$ is the $\tau$-component of the electric moment), which translated into the "corresponding" quantum-theoretical symbols. By consideration of the case of radiation damping, the $\Gamma$ coefficients "corresponding" to $|C_\tau|^2$ were now *proved* to be proportional to Einstein's emission coefficients.[322]

Altogether, this ambitious paper failed to bring its authors closer to a new, self-sufficient quantum mechanics: There was still no a priori method to calculate the $\Gamma$ coefficients. In a comparison with Kramers and Heisenberg's anterior work on dispersion, Born and Jordan might retrospectively

320. Bohr to Heisenberg, 18 Apr. 1925, *BCW* 5:[360]; Heisenberg to Kronig, 20 May 1925, AHQP.
321. Heisenberg to Bohr, 16 May 1925, *BCW* 5:[361]–[362]; Born and Jordan 1925b, 493.
322. Born and Jordan 1925b.

even be seen as regressing, since their phase-averaging procedure killed the quantum-mechanical products of the general dispersion formula. Ironically, the main reason of this regression was the very principle of observability!

## UMDEUTUNG

### HYDROGEN

In his first paper on "quantum mechanics" (June 1924) Born had identified the problem of the determination of intensities (or $\Gamma$ coefficients) as "the most important one for the future development of quantum theory." Heisenberg followed this directive better than Born himself. When, in early May 1925, he engaged all his energies in an attempt to further develop Born's program, he started with a consideration of intensities, in particular those of the hydrogen atom's spectral lines. For somebody who had managed a quantitative derivation of the intensity ratios for the lines in Zeeman multiplets, the next natural step was indeed a similar treatment of the simplest of all atoms, as had been agreed upon by Kronig and Heisenberg in Copenhagen.[323]

In his dissertation of 1919 Kramers had already reached semiquantitative expressions for the intensities of hydrogen lines on the basis of the correspondence principle: he assumed the Fourier components of the electric moment of the orbiting electron to give an estimate of the probability of the "corresponding" quantum transitions. He also suggested a certain averaging procedure (over all values of the action variables between the two levels of the transition) that would have made the correspondence more quantitative. As he (or Bohr?) realized, however, the resulting expressions could not be exact since they violated Bohr's principle of spectroscopic stability.[324]

With the experience acquired in the case of Zeeman patterns and in dispersion theory, Heisenberg believed that he could replace Kramers's averages by an exact symbolic translation of the classical Fourier amplitudes $C_\tau$ of the electric moment. On 8 May 1925 he reported to Kronig "a very serious possibility" for the correct translation based on Kramers's

323. Heisenberg to Kronig, 8 May 1925, AHQP. For Kronig's work on multiplet intensities, see n. 260. A profusion of details about the scientific and extrascientific circumstances of Heisenberg's discovery of matrix mechanics can be found in Mehra and Rechenberg 1982b.
324. Kramers 1919, 330n.

explicit formula for the $C_\tau$ of the nonrelativistic Kepler motion (in the complex plane of the motion):[325]

$$C_\tau = \frac{A}{2\tau} \left[ \left(1 + \frac{k}{n}\right) J_{\tau-1}(\tau\varepsilon) - \left(1 - \frac{k}{n}\right) J_{\tau+1}(\tau\varepsilon) \right], \qquad (242)$$

where $n$ and $k$ are the principal and azimuthal quantum numbers; $\varepsilon = \sqrt{1 - k^2/n^2}$, $A = n^2 h^2/4\pi^2 e^2 \mu$ are respectively the eccentricity and half the major axis of the ellipse; and $J_\tau$ is the Bessel function of order $\tau$. In the power-series development of the above formula Heisenberg replaced the powers of quantum numbers by broken factorials; for instance, one replacement was:

$$(n - k)^p \rightarrow (n - k)(n - k - 1) \cdots (n - k - p + 1). \qquad (243)$$

This recipe was an obvious generalization of the $j^2 \rightarrow j(j-1)$ rule for the Zeeman effects, which was itself connected with the $\tau \cdot \partial/\partial J \rightarrow \Delta_\tau$ rule by the identities (175). It had already been used in Copenhagen by Heisenberg and Pauli to guess at the quantum-theoretical intensity formula for a one-dimensional anharmonic oscillator, where it gave[326]

$$C_\tau(n) = \chi(\tau)\sqrt{n^\tau} \rightarrow \chi(\tau)\sqrt{n(n-1) \cdots (n - \tau + 1)}. \qquad (244)$$

The hydrogen case was not so simple. But for a given value of $n$, Heisenberg noticed, only a finite number of terms of the power-series developments survived the quantum translation (since, for instance, the value of $p$ in $(n - k)^p$ must be inferior to $n - k$), as if the result had been derived from the integration of finite difference equations. Conscious of the arbitrariness of his own procedure, Heisenberg commented: "I shall trust the [intensity] formulae only if I can determine which difference equations they satisfy."[327] These words, in spite of their anodyne appearance, marked an essential procedural step toward quantum mechanics. Up to then Born and his collaborators had been satisfied with the application of substitution rules to *final* classical formulae resulting from the integration of the dynamic equations for a generic mechanical problem. Heisenberg now proposed to trace back a quantum-theoretical counterpart of every single classical step leading to these formulae, until he reached some finite difference equations corresponding to the original classical differential equations.

---

325. Ibid., 299.
326. Heisenberg to Kronig, 8 May 1925, AHQP; mention of Pauli's and Heisenberg's past considerations on the anharmonic oscillator is in Heisenberg to Kronig, 5 June 1925, AHQP.
327. Heisenberg to Kronig, 8 May 1925, AHQP.

ANHARMONIC OSCILLATOR

Heisenberg found the mathematical procedure leading from the equation of motion for the Kepler problem to the Fourier components of the motion to be too complicated to serve this purpose. After a few weeks he therefore decided to come back to the simplest nontrivial dynamic problem, the one-dimensional anharmonic oscillator which he had already discussed with Pauli. This oscillator had the equation of motion

$$\ddot{q} + \omega_0^2 q + \lambda q^2 = 0. \qquad (245)$$

In early June 1925, Heisenberg achieved a good part of his aim, as we know from another letter to Kronig.[328] For small $\lambda$ and not too large amplitudes the solutions of the above dynamic equation are periodic (see chapter 6) with a frequency $\omega$ differing from $\omega_0$. But they are not harmonic: at the order $\tau$ of a perturbation calculation there appears a harmonic component with the frequency $(\tau + 1)\omega$.[329] Accordingly, the general form of the solution is

$$q = \lambda a_0 + a_1 \cos \omega t + \lambda a_2 \cos 2\omega t + \cdots + \lambda^{\tau-1} a_\tau \cos \tau \omega t + \cdots, \qquad (246)$$

where $\omega$ and $a_\tau$ are themselves power series of $\lambda$. Substituting this expression in the equation of motion, separating the various harmonics, and retaining only the terms with the lowest order in $\lambda$ gives the following system of equations:

$$\left.\begin{array}{r} \omega_0^2 a_0 + \frac{1}{2} a_1^2 = 0 \\ (-\omega^2 + \omega_0^2) a_1 = 0 \\ (-4\omega^2 + \omega_0^2) a_2 + \frac{1}{2} a_1^2 = 0 \\ (-9\omega^2 + \omega_0^2) a_3 + a_1 a_2 = 0 \\ \cdots\cdots\cdots\cdots\cdots\cdots\cdots\cdots \end{array}\right\} \qquad (247)$$

and the resulting recursive formulae for the zero-order part of $a_\tau$:

$$\left.\begin{array}{r} -3\omega_0^2 a_2 + \frac{1}{2} a_1^2 = 0 \\ -8\omega_0^2 a_3 + a_1 a_2 = 0 \\ \cdots\cdots\cdots\cdots\cdots\cdots \end{array}\right\}. \qquad (248)$$

The general solution of this system has the form

$$a_\tau = k(\tau) a_1^\tau \qquad \text{(for } \tau \geq 1\text{)}. \qquad (249)$$

328. Heisenberg to Kronig, 5 June 1925, AHQP.
329. The classical perturbative discussion of the anharmonic oscillator could be found, for instance, in Born's lectures (Born 1925, par. 12), which Heisenberg had helped to prepare in the winter of 1923–24.

In order to provide the necessary basis for an application of the correspondence principle, these coefficients must be given for a stationary state selected according to the Bohr-Sommerfeld rule $\oint p\,dq = nh$. This is readily done by noticing that $a_1$ represents the amplitude of the solution of the dynamic equation at the zero order of perturbation, which gives $a_1 \propto \sqrt{n}$, and, combined with (249),

$$a_\tau = K(\tau)\sqrt{n^\tau}. \tag{250}$$

Heisenberg then proceeded to construct the quantum-theoretical counterpart of the above calculation, on the basis of the correspondence

$$a_\tau(n) \to a(n, n - \tau)$$

between classical and quantum amplitude that had been introduced in the paper by Kramers and Heisenberg (see p. 230). For the translation of the recursion formulae Heisenberg also needed to know the counterpart of the products $a_\tau a_{\tau'}$. In this respect the same paper by Kramers and Heisenberg suggested a rule: the equality of median indices in the translation,

$$a_\tau a_{\tau'} \quad \text{or} \quad \begin{array}{l} a(n, \underline{n - \tau})a(\underline{n - \tau}, n - \tau - \tau') \\[1em] a(n, \underline{n - \tau'})a(\underline{n - \tau'}, n - \tau - \tau'). \end{array} \tag{251}$$

There was, originally, no fundamental justification for this rule other than that it afforded the highest possible symmetry of intermediate calculations. In his letter to Kronig, Heisenberg filled this gap in the following manner.

He first looked for a physical interpretation of the product $a_\tau a_\tau$ in terms of the radiation properties of the oscillator. This is simply obtained by noticing that the expression $a_\tau a_{\tau'} e^{i(\tau + \tau')\omega_0 t}$ is one of the terms of the Fourier component of $q^2$ with the frequency $(\tau + \tau')\omega_0$; $q^2$ itself has the physical meaning of the quadrupolar moment of a unit charge with the elongation $q$. At the quantum level, in a natural generalization of the correspondence principle, Heisenberg expected the quantum products "corresponding" to $a_\tau a_{\tau'} e^{i(\tau + \tau')\omega_0 t}$ to contribute to a component of the quadrupolar moment with the frequency $\omega(n, n - \tau - \tau')$ connected to the transition $n \to n - \tau - \tau'$. This remark suggests the correspondence

$$a_\tau e^{i\tau\omega_0 t} a_{\tau'} e^{i\tau'\omega_0 t} \to a(n, n - \tau)e^{i\omega(n, n-\tau)t}a(n - \tau, n - \tau - \tau')e^{i\omega(n-\tau, n-\tau-\tau')t}, \tag{252}$$

or the one obtained from this one by permuting $\tau$ and $\tau'$. These are the only two possibilities complying with the combination rule

$$\omega(n, n - \tau - \tau') = \omega(n, n - \tau) + \omega(n - \tau, n - \tau - \tau')$$
$$= \omega(n, n - \tau') + \omega(n - \tau', n - \tau - \tau'). \tag{253}$$

For the sake of symmetry, Heisenberg adopted

$$a_\tau a_{\tau'} \rightarrow \tfrac{1}{2}[a(n, n - \tau)a(n - \tau, n - \tau - \tau')$$
$$+ a(n, n - \tau')a(n - \tau', n - \tau - \tau')]. \tag{254}$$

According to this rule, the recursion formulae (248) translate into

$$\left.\begin{aligned} -3\omega_0^2 a(n, n - 2) + \tfrac{1}{2}a(n, n - 1)a(n - 1, n - 2) &= 0 \\ -8\omega_0^2 a(n, n - 3) + \tfrac{1}{2}[a(n, n - 1)a(n - 1, n - 3) \\ + a(n, n - 2)a(n - 2, n - 3)] &= 0 \\ \dots\dots\dots\dots\dots\dots\dots\dots\dots\dots\dots\dots\dots\dots\dots\dots \end{aligned}\right\} \tag{255}$$

This system can be used to express $a(n, n - \tau)$ in terms of coefficients of the type $a(m, m - 1)$:

$$a(n, n - \tau) = k(\tau)a(n, n - 1)a(n - 1, n - 2) \cdots a(n - \tau + 1, n - \tau). \tag{256}$$

For the pure harmonic oscillator the coefficients $a(m, m - 1)$ are the only nonzero ones, as results from the correspondence principle. Without explicit justification (for the time being), Heisenberg admitted that their values in this case were equal to the classical corresponding values, which gives

$$a(m, m - 1) = a_1(m) \propto \sqrt{m} \tag{257}$$

(a good justification for this could have been that, in Planck's derivation of the blackbody law, the average radiation properties of a resonator are correctly given by classical electrodynamics).[330] The insertion of this expression into (256) gives the general formula for the amplitude of a transition $n \rightarrow n - \tau$ at the lowest order of perturbation ($\tau - 1$) for which it is nonzero:

$$\lambda^{\tau - 1}a(n, n - \tau) = \lambda^{\tau - 1}K(\tau)\sqrt{n(n - 1) \cdots (n - \tau + 1)}, \tag{258}$$

as originally guessed by Heisenberg and Pauli (244).

In the limited case of an anharmonic oscillator, Heisenberg had thereby reached his aim of translating every step of the classical derivation of intensities into a form expressed in terms of quantum-theoretical amplitudes and frequencies. However, his procedure depended on the special simplicity of the unperturbed system, the harmonic oscillator. For in this case the coefficients $a(n, n - 1)$ corresponding to the classical integration constants $a_1(n)$ were known. In the general case, Heisenberg complained to Kronig,

330. See also Pauli 1926a, 51.

the quantum-theoretical integration constants were not a priori known, and he did not yet know how to derive them.[331]

## THE QUANTUM CONDITION

Heisenberg's solution to this difficulty is found in a letter to Pauli of 24 June 1925.[332] There he gave a general translation of the Bohr-Sommerfeld condition in terms of quantum amplitudes and frequencies. This step was to provide for the missing "integration constants." In terms of the usual Fourier development (still for one degree of freedom),

$$q = \sum_\tau C_\tau e^{i\tau\omega t}, \qquad (259)$$

the action variable

$$J = \oint p \, dq = \mu \oint \dot{q}^2 \, dt \qquad (260)$$

reads

$$J = 2\pi\mu \sum_\tau \tau(\tau\omega)|C_\tau|^2, \qquad (261)$$

If written under the form "$J = nh$," the quantum condition has no clear quantum-theoretical counterpart, since, from previous considerations of formal correspondence, only the translations of $\tau\omega$ and $C_\tau$ were known, not that of $\tau$ by itself. Heisenberg therefore took the derivative of both members of (261) with respect to $J$, which gives

$$1 = 2\pi\mu \sum_\tau \tau \frac{\partial}{\partial J} (\tau\omega |C_\tau|^2), \qquad (262)$$

with a straightforward quantum translation dictated by dispersion theory:[333]

$$1 = \frac{2\pi\mu}{h} \sum_\tau \Delta_\tau[\omega(n, n - \tau)|a(n, n - \tau)|^2]. \qquad (263)$$

The final quantum condition reads

$$h = 2\pi\mu \sum_\tau [\omega(n + \tau, n)|a(n + \tau, n)|^2 - \omega(n, n - \tau)|a(n, n - \tau)|^2]. \qquad (264)$$

---

331. In the same letter to Kronig (5 June 1925), Heisenberg gave a new derivation of Kronig's intensity formulae for the Zeeman effect (see n. 260). This calculation (published in Heisenberg 1925c, 892) was based on a quantum-theoretical translation of the classical rule (236) on p. 237, and of the rule $j_z = mh$ (which, in this simple case, could be directly written, without recourse to the general quantum rule later established by Heisenberg).

332. Heisenberg to Pauli, 24 June 1925, PB, no. 93.

333. In his letter, Heisenberg assumed the coefficients $a(n, n - \tau)$ to be real. As he later realized, this was true only in particular cases (for instance, those treated in the letter to Pauli, fortunately!).

Heisenberg immediately tried it on the harmonic oscillator. In this case the restriction $\tau = \pm 1$ leads to

$$h = 4\pi\mu\omega_0[|a(n + 1, n)|^2 - |a(n, n - 1)|^2].$$  (265)

Admitting the existence of a fundamental level $n = 0$ below which no transition can occur, this equation completely determines the intensities as

$$|a(n, n - 1)|^2 = nh/4\pi\mu\omega_0,$$  (266)

in conformity with Heisenberg's earlier assumption (257).


ENERGY CONSERVATION

Just as the "classical" Bohr-Sommerfeld condition did, the new quantum condition also determined the energy spectrum. Heisenberg first tested this by examining the quantum-theoretical counterpart of the classical energy of the harmonic oscillator,

$$E = \tfrac{1}{2}\mu(\dot{q}^2 + \omega_0^2 q^2).$$  (267)

Classically, one has

$$q = a_+ e^{i\omega_0 t} + a_- e^{-i\omega_0 t}$$  (268)

and

$$q^2 = a_+^2 e^{2i\omega_0 t} + a_-^2 e^{-2i\omega_0 t} + 2a_+ a_-.$$  (269)

The general translation rule (254) yields[334]

$$\left.\begin{array}{l} a_+^2 \to a(n, n - 1)a(n - 1, n - 2) \\ a_-^2 \to a(n, n + 1)a(n + 1, n + 2) \\ a_+ a_- \to \tfrac{1}{2}[a(n, n - 1)a(n - 1, n) + a(n, n + 1)a(n + 1, n)] \end{array}\right\}$$  (270)

Heisenberg further assumed the reality of all amplitudes and the equality of emission and absorption probabilities, which simplifies the last of the previous substitutions to

$$a_+ a_- \to \tfrac{1}{2}[a^2(n, n - 1) + a^2(n, n + 1)].$$  (271)

Proceeding in the same way for $\dot{q}^2$ and substituting the results into (267) gives the quantum-theoretical energy

$$E_n = \mu\omega_0^2[a^2(n, n - 1) + a^2(n, n + 1)].$$  (272)

---

334. More exactly, Heisenberg substituted the expression $q(n) = a(n, n - 1) \cos \omega_0 t$ (suggested by the initial cosine Fourier development) into (243) but calculated the products $q^2$ and $\dot{q}^2$ "in a symbolic way" that reproduced the operations given here.

To Heisenberg's satisfaction no oscillating term subsisted, as ought to be the case for the energy of a closed system. Furthermore, replacing the intensities with their value (266) gave[335]

$$E_n = (n + \tfrac{1}{2})h\omega_0/2\pi, \tag{273}$$

as expected from the ordinary quantization of a harmonic oscillator à la Planck, and in conformity with Bohr's second postulate:

$$\frac{h}{2\pi}\,\omega(n, n-1) = E_n - E_{n-1} = h\omega_0/2\pi. \tag{274}$$

Heisenberg managed to repeat the above considerations in the case of an anharmonic oscillator (with a $\lambda q^3$ anharmonicity and to the order $\lambda^2$) and found an energy spectrum identical with the one given by the Bohr-Sommerfeld method (with a half-integral $n$). Nevertheless, he still doubted the generality of the procedure:

> The strongest objection [against the generality of the above considerations] seems to me that the energy expressed as a function of $q$ and $\dot{q}$ in general does not need to become a constant, even if the equations of motion are satisfied; in the last analysis this has to do with the fact that the product of two "Fourier" series is not unambiguously defined.[336]

It is not quite certain what Heisenberg meant by this reference to ambiguity. He might have tried to give a general derivation of energy conservation at the quantum level, by differentiating a generic Hamiltonian

$$H = \tfrac{1}{2}\mu\dot{q}^2 + V(q). \tag{275}$$

This would have given

$$\dot{H} = \mu\dot{q}\ddot{q} + \dot{q}\,dV/dq, \tag{276}$$

where the products $\dot{q}\ddot{q}$ and $\dot{q}\,dV/dq$ then have to be replaced with symbolic counterparts. The rule systematically used by Heisenberg for the translation of $c_\tau c_{\tau'}$ gave the square of a "'Fourier' series" (i.e., the set of quantum amplitudes), but not the product of two such series. At the classical level, if

$$a = \sum_\tau a_\tau e^{i\tau\omega t} \quad \text{and} \quad b = \sum_\tau b_\tau e^{i\tau\omega t}, \tag{277}$$

---

335. The "zero-point" energy $\tfrac{1}{2}\hbar\omega_0$ did not worry Heisenberg: it had already appeared in Planck's "second theory" (1910) and was favored by those who believed in half-integral quantum numbers.
336. Heisenberg to Pauli, 24 June 1925, PB, no. 93.

then

$$c = ab = \sum_\tau c_\tau e^{i\tau\omega t}, \qquad (278)$$

with

$$c_\tau = \sum_{\tau' + \tau'' = \tau} a_{\tau'} b_{\tau''}. \qquad (279)$$

At the quantum level, a possible natural counterpart of the latter rule is

$$c(n, n - \tau) = \sum_{\tau' + \tau'' = \tau} a(n, n - \tau')b(n - \tau', n - \tau' - \tau''). \qquad (280)$$

However, an equally natural one would be obtained by permuting the letters $a$ and $b$ in this expression.

In his final paper Heisenberg pointed to this "significant difficulty" and indicated a way to solve it, for the example of the product $a\dot{a}$. The quantum translation of this product must be the time derivative of the translation of $\frac{1}{2}a^2$. It is therefore obtained by applying the rule (280) to $\frac{1}{2}(a\dot{a} + \dot{a}a)$.[337]

Having made this remark, and another essential one concerning the role of phases in quantum amplitudes, which will presently be discussed, Heisenberg trusted his new scheme enough to publish it, under the title "Quantum-theoretical reinterpretation [Umdeutung] of kinematic and mechanical relations." The contents of this paper will now be summarized briefly.[338]

A NEW CONCEPT OF MOTION

In the introduction Heisenberg acknowledges the failure of the usual quantum theory of atoms. Many of the difficulties, he asserts, are connected with the abundant recourse to unobservable quantities like the position and period of revolution of an electron. His stated alternative strategy then aims to construct a theory formally analogous to classical mechanics, but in which only observable quantities occur.

These quantities are first defined (for one degree of freedom) as those characterizing the radiation emitted by the moving electron. They turn out to be complex amplitudes $a(n, n - \tau)e^{i\omega(n, n - \tau)t}$ "corresponding" to the Fourier components $a_\tau e^{i\tau\omega t}$ of the classical motion $q(t)$ (in the $n^{\text{th}}$ stationary state). In the classical theory the set of Fourier components completely determines the motion; similarly, Heisenberg takes the square table of

337. Heisenberg to Kronig, 24 June 1924, AHQP; also Heisenberg 1925c, 890.
338. Heisenberg 1925c.

characteristic amplitudes to represent the quantum-theoretical motion, that is, the object of a "new kinematics."

The next logical step is the introduction of a new quantum product

$$c(n, n - \tau) = \sum_{\tau' + \tau'' = \tau} a(n, n - \tau')b(n - \tau', n - \tau' - \tau'') \qquad (281)$$

which gives the "table" of the quantity $c$, if one knows the tables of the quantities $a$ and $b$. As in the letter to Kronig of 5 June, this expression is justified on the basis of the correspondence with the analogous expression for the product of classical Fourier series, and by the fact that such a definition groups together terms with the same frequency, as follows from the combination rule (253)

$$\omega(n, n - \tau) = \omega(n, n - \tau') + \omega(n - \tau', n - \tau' - \tau'').$$

The latter condition is necessary, if the frequencies appearing in any quantum-theoretical quantity have to be the observable atomic frequencies.

At this point Heisenberg makes an interesting remark: one might be tempted to introduce sums[339]

$$q = \sum_{\tau} a(n, n - \tau)e^{i\omega(n, n - \tau)t} \qquad (282)$$

analogous to the corresponding classical Fourier series; but such an expression would conflict with the symmetrical role played by $n$ and $n - \tau$ in the quantum theory (a more practical reason could have been advocated: such sums would be useful only if they could be multiplied in the usual manner, but this would introduce frequencies different from those obtained by the combinations (253)).

Heisenberg further maintains the equation of motion

$$\ddot{q} + f(q) = 0 \qquad (283)$$

as a formal relation between the tables representing $\ddot{q}$ and $q$, where $f(q)$ is understood as a series of quantum powers. He derives the quantum rule[340]

$$h = 2\pi\mu \sum_{\tau} [|a(n + \tau, n)|^2 \omega(n + \tau, n) - |a(n, n - \tau)|^2 \omega(n, n - \tau)] \qquad (284)$$

as in his letter to Pauli (with the addition of the moduli) and relates it to a rule already derived by Thomas and Kuhn by taking the high-frequency limit of Kramers's dispersion formula.

---

339. Indeed, Kramers and Heisenberg 1925, 683, had introduced such sums (to no avail, however).

340. I have corrected a trivial mistake made by Heisenberg, regarding the order of the arguments $n$ and $n + \tau$.

There follow two applications of the above general scheme: a lowest-order calculation of the anharmonic oscillator with a $\lambda q^2$ anharmonicity, and a second order calculation in the case of a $\lambda q^3$ anharmonicity (which is simpler), including the derivation of $\omega(n, n-1)$ and $E_n$, and a verification of Bohr's relation:

$$\frac{h}{2\pi} \omega(n, n-1) = E_n - E_{n-1}. \tag{285}$$

Finally, Heisenberg gives two other examples: a planar rotator and a precessing spatial rotator for an account of Zeeman intensities. His brief conclusion calls for a thorough mathematical investigation of the new scheme.

## GUIDING PRINCIPLES

Heisenberg's presentation of this work as "an attempt to establish a theoretical quantum mechanics analogous to classical mechanics, but in which only relations between observable quantities occur" has been a matter of endless debate. Was he really describing his own strategy, or was he trying to seduce Bohr, Pauli, and Born, who, by that time, all advocated a radical elimination of unobservable quantities? The truth seems to lie somewhere in between.[341]

Heisenberg's breakthrough was certainly inscribed in a program of symbolic translation initiated by Born, and approved by Pauli as soon as it was purged of the pseudovisualization brought about by the virtual oscillators. Heisenberg had even joined Kramers in emphasizing that their dispersion formulae eliminated everything reminiscent of the quantum theory of multiperiodic systems. In his letter to Kronig about the nascent quantum mechanics (5 June 1925), he wrote: "What I like in this scheme is that one can really reduce all interactions between atoms and the external world (apart from the problem of degeneracy) to transition probabilities."[342]

Nevertheless, one should note that Heisenberg did not at first refer to the observability principle as the source of his and Kramers's identification of the fundamental quantities of the quantum theory. Not until he recommended to Pauli his already constituted quantum-mechanical scheme, as late as 24 June 1925, in the following words: "The fundamental axiom

341. Heisenberg 1925c, 879.
342. Kramers and Heisenberg 1925, 691, quoted above in n. 249; Heisenberg to Kronig, 5 June 1925, AHQP.

is: In calculating any quantities like energy, frequency, etc., there should occur only relations between quantities that can be controlled in principle," did he emphasize the observability principle. In fact, the historical origin of the focus on transition probabilities and spectral frequencies lay elsewhere: in Bohr's formulation of the quantum postulates.[343]

Deliberately, Bohr had enunciated these postulates in terms of concepts that could be defined in a way independent of the description of electronic motion in terms of classical mechanics; for he regarded the classical electron orbits employed in the theory only as an approximate and provisional representation of atomic motion. In his dispersion theory Kramers emphasized the restriction, in the final formulae, to "such quantities as allow of a direct interpretation on the basis of the fundamental postulates of the quantum theory." Likewise, in his quantum mechanics Heisenberg maintained Bohr's *postulates* and eliminated all other provisional assumptions, replacing them with a procedure for the symbolic translation of classical dynamic equations into quantum counterparts.[344]

In this light Heisenberg's choice of the word *Umdeutung* in his title appears to be a judicious characterization of his genuine endeavor. This word had been used by Sommerfeld in 1922 to mean a re-expressing, in terms compatible with Bohr's postulates, of a line spectrum *directly* obtained as the Fourier spectrum of a classical mechanical model. That is to say, he was providing a set of atomic energy values such that every spectral line could be expressed as a difference $v = \Delta E/h$ of two of these energy values. In Heisenberg's paper the word *Umdeutung* also meant a reformulation in terms of the central concepts of Bohr's theory, stationary states and transition probabilities. The identity of these concepts with those selected by Pauli in the name of the observability principle might have been heartening, but it was not a decisive element in the gestation of Heisenberg's paper.[345]

As essential as it was, the reference to Bohr's postulates was not sufficient to identify the fundamental quantities appearing in the new theory: these were not quite transition probabilities but rather transition *amplitudes,* as in the earlier paper by Kramers and Heisenberg. The complex amplitude included a quantum-theoretical phase, which Heisenberg considered to be necessary even though Born and Jordan had just rejected

343. Heisenberg to Pauli, 24 June 1925, *PB,* no. 93.
344. Kramers, already quoted on pp. 230–231, (n. 249).
345. Sommerfeld 1922. The word *Umdeutung* also appeared in Heisenberg to Kronig, 5 June 1925, AHQP.

this phase, for the very reason that it could not be observed! Presumably to mark his departure from the latter view, Heisenberg emphasized:

> At first sight the phase contained in [the amplitude] would seem to be devoid of physical significance in quantum theory, since in this theory frequencies are in general not commensurable with their harmonics. However, we shall see presently that also in quantum theory the phase has a definite significance that is analogous to its significance in classical theory.[346]

In the examples treated by Heisenberg in his letters to Kronig and Pauli the phase did not appear because all amplitudes could be taken to be real. Nevertheless, as Heisenberg correctly noted, this could not be the case in general.[347] The "Fourier" components $b(n, n - \tau)$ of $q^2$, which are the sources of quadrupolar radiation, depend on the relative phase of the "Fourier" components of $q$ itself as a result of the multiplication law

$$b(n, n - \tau) = \sum_{\tau' + \tau'' = \tau} a(n, n - \tau')a(n - \tau', n - \tau' - \tau''). \qquad (286)$$

At this stage one might still try to save the observability principle by arguing that the phase appears to be *indirectly* observable, as a consequence of the observability of multipolar radiation. But this would hide the fact that the relation between phases and multipolar intensities is not given; it is a new theoretical construct obtained from an analogy between products of classical Fourier series and their quantum-theoretical counterparts. Even in the identification of the fundamental quantities to which the new quantum mechanics had to apply, the correspondence principle appears to have played a more crucial role than the principle of observability.

To summarize in a few words, Heisenberg's breakthrough resulted from an attempt to symbolically translate classical mechanics into a form expressed in terms of genuine quantum-theoretical concepts that were identified in accordance with the two following criteria: to have a direct relation to Bohr's quantum postulates, and to lead to a coherent symbolic scheme. The role of the observability principle appears to have been limited to the elimination of alternative strategies (like the one referring to virtual oscillators) that would have retained more of the visual apparatus of the classical theory. If one still wishes to isolate a single element that contributed more than any other to Heisenberg's quantum mechanics, the only reasonable candidate is the correspondence principle. The idea of a

346. Heisenberg 1925c, 882.
347. Ibid., 883.

symbolic translation of classical mechanics finds its roots in the general context of this principle, namely, the idea that a formal analogy exists between the laws of quantum theory and those of classical theory. The precise expression of this analogy as formulated by Heisenberg must be traced back to a more specific aspect of the same principle: the correspondence between quantum-theoretical spectrum and the harmonics of a classical motion.

Admittedly, Bohr would have preferred a realization of this formal analogy leading to a higher descriptive content, instead of Heisenberg's or Born's utterly symbolic procedures. But he admitted, after the failure of the BKS program, that this was the only strategy left to quantum theorists. And he was quick to recognize the importance of Heisenberg's attempt, which, in an essay published in *Nature* in December 1925, he characterized in the following terms:

> In contrast to ordinary mechanics, the new quantum mechanics does not deal with a space-time description of the motion of atomic particles. It operates with manifolds of quantities, which replace the harmonic oscillating components of the motion and symbolise the possibilities of transitions between stationary states in conformity with the correspondence principle. These quantities satisfy certain relations which take the place of the mechanical equations of motion and the quantisation rules. . . . In brief, the whole apparatus of the quantum mechanics can be regarded as a precise formulation of the tendencies embodied in the correspondence principle.[348]

An echo of the latter comment can be found in Heisenberg's own appreciation of quantum mechanics as a "quantitative formulation of the correspondence principle." In December 1925 the "three men" who developed this mechanics also wrote: "The new theory can be regarded as an exact formulation of Bohr's correspondence considerations . . . in which symbolic quantum geometry goes over into visualizable classical geometry."[349]

## EPILOGUE

In mid-July 1925 Heisenberg handed over his manuscript to Max Born, and before he heard Born's reaction, he left Göttingen for a trip to England. Born immediately perceived an important breakthrough, as may be judged from a letter to Einstein of July 15: "Heisenberg's new work, which appears

348. Bohr 1925c, 852.
349. Heisenberg 1929, 493; Born, Heisenberg, and Jordan 1926, 558.

soon, looks very mystical, but it is certainly right and profound." Four days later he met Pauli in a train from Göttingen to Hannover and asked him to collaborate on the new mechanics.[350]

Pauli was enthusiastic about Heisenberg's paper, as he reported to Kramers on 27 July:

> I have greatly rejoiced in Heisenberg's bold attempts. . . . To be sure, one still is very far from saying something definitive, and we stand at the very beginning of things. However, what has pleased me so much in Heisenberg's considerations is the *method* of his procedure and the *aspiration* that guided him. On the whole I believe that I am now very close to Heisenberg in my scientific views and that our opinions agree in everything as much as is in general possible for two independently thinking men. I was also pleased to notice that Heisenberg has learned some philosophical thinking from Bohr in Copenhagen and takes a sharp turn away from purely formal methods. I therefore wish him success in his endeavors with all my heart.[351]

Not surprisingly, Pauli admired the "aspiration" of the work, which Heisenberg asserted to be the elimination of unobservable quantities. He also seems to have appreciated the way Heisenberg played down the positive part of his paper, as "fairly formal and meager." He nevertheless impertinently declined Born's offer, on the grounds that Göttingen's futile mathematics would "spoil" Heisenberg's physical ideas.[352]

This rejection failed to demoralize Born, who immediately set out to work with a more benevolent collaborator, Pascual Jordan. Progress was so fast that, even before Heisenberg's return from England, the two men had managed to put Heisenberg's ideas on a firm mathematical foundation, including a general proof of energy conservation. The resulting paper was received in late September and published in November 1925.[353]

Born and Jordan first noticed that Heisenberg's multiplication rule, somewhat obscured by the "$n - \tau$" notation derived from the correspondence principle, was nothing but the ordinary matrix product. The rule (281)

$$c(n, n - \tau - \tau') = \sum_{\tau} a(n, n - \tau)b(n - \tau, n - \tau - \tau')$$

350. Born to Einstein, 15 July 1924, in Einstein-Born 1969. On the train story see van der Waerden 1967, 37, and Born 1978, 118.

351. Pauli to Kramers, 27 July 1925, *PB*, no. 97.

352. Heisenberg to Pauli, 9 July 1925, *PB*, no. 96; van der Waerden 1967, 37, and Born 1978, 118. In a letter to Kronig of 9 Oct. 1925 (*PB*, no. 100), Pauli vilified "Göttinger formalen Gelehrsamkeitsschwall" (Göttingen's torrent of erudite formalism).

353. Born and Jordan 1925c. About the relative roles of Born and Jordan, see van der Waerden 1967, 38–39.

just reads c = ab, if a is the matrix corresponding to the table $a(n, m)$ and so on. This prompted them to express every relation in Heisenberg's paper in terms of matrices.

First of all, the quantum rule (284) may be written

$$(pq - qp)_{nn} = h/2\pi i, \tag{287}$$

if q is the matrix corresponding to the table $a(n, m)e^{i\omega(n,m)t}$, and p is the one defined as $p = \mu\dot{q}$. From this form Born guessed the more elegant[354]

$$pq - qp = \frac{h}{2\pi i} 1, \tag{288}$$

where 1 is the unit matrix. For nondegenerate systems, Jordan managed to prove this remarkable relation in the following manner.

The classical equations of motion are first assumed, in the spirit of Heisenberg's paper, to give formal relations between matrices. In the Hamiltonian formulation of one-dimensional mechanics these equations read

$$\dot{p} = -\frac{\partial H}{\partial q}, \qquad \dot{q} = \frac{\partial H}{\partial p}. \tag{289}$$

The partial derivatives must be defined for a specific ordering of p and q in H(q, p), which Jordan managed to identify for any function H admitting a power-series development. For simplicity let us limit our considerations to a Hamiltonian

$$H(q, p) = \frac{p^2}{2\mu} + V(q), \tag{290}$$

for which no ordering is necessary.

In this case the time derivative of $pq - qp$ is easily seen to vanish from the identities

$$\frac{d}{dt}(pq - qp) = (\dot{p}q - q\dot{p}) + (p\dot{q} - \dot{q}p)$$

$$= \left(-\frac{\partial H}{\partial q}q + q\frac{\partial H}{\partial q}\right) + \left(p\frac{\partial H}{\partial p} - \frac{\partial H}{\partial p}p\right), \tag{291}$$

where the two last commutators vanish because $\partial H/\partial p$ depends only on p, and $\partial H/\partial q$ only on q. Now, according to Heisenberg, the time dependence of any quantum-mechanical table $g_{nm}$ is given by

$$g_{nm}(t) = g_{nm}(0)e^{i\omega_{nm}t} \tag{292}$$

354. Born and Jordan 1925c, 880.

or

$$\dot{g}_{nm} = i\omega_{nm} g_{nm}. \tag{293}$$

In the nondegenerate case (for which $\omega_{nm} \neq 0$ if $n \neq m$), in order to be time-independent, g has to be diagonal. Consequently, $pq - qp$ must be a diagonal matrix, the diagonal elements of which are given by Heisenberg's rule (287). This ends the proof of Born's conjecture.

Jordan went on to prove the conservation of energy $dH/dt = 0$, which Heisenberg had shown to hold only in particular cases. From the "strong" quantum condition (288) result the identities

$$Hq - qH = \frac{h}{2\pi i} \frac{\partial H}{\partial p}, \qquad Hp - pH = -\frac{h}{2\pi i} \frac{\partial H}{\partial q} \tag{294}$$

for any function $H(q, p)$ expressible in a power-series development. Combined with Hamilton's equations (289), this gives

$$Hq - qH = \frac{h}{2\pi i} \dot{q}, \qquad Hp - pH = \frac{h}{2\pi i} \dot{p} \tag{295}$$

and, more generally,

$$Hg - gH = \frac{h}{2\pi i} \dot{g} \tag{296}$$

for any function $g(q, p)$. The case $g = H$ gives $dH/dt = 0$, as originally hoped by Heisenberg.

Finally, for the new scheme to be coherent, the above equation of motion (296) must be compatible with the one earlier assumed in (293). This is indeed the case, for (296) is equivalent to

$$\dot{g}_{nm} = \frac{2\pi i}{h} (E_n - E_m) g_{nm}. \tag{297}$$

The latter equation is identical with (293), as soon as the following relation holds:

$$E_n - E_m = \frac{h}{2\pi} \omega_{nm}, \tag{298}$$

which is identical with Bohr's frequency condition.

Born and Jordan commented: "It is in fact possible, starting with the basic premises given by Heisenberg, to build up a closed mathematical theory of quantum mechanics which displays strikingly close analogies with classical mechanics, but at the same time preserves the characteristic features of quantum phenomena."[355]

355. Ibid., 858.

During the following months Heisenberg, Born, and Jordan joined their efforts to further develop the new mechanics. In November 1925 they sent to the *Zeitschrift für Physik* the soon famous "three-men paper," in which they widely extended previous methods and results. They dealt with the case of several degrees of freedom, treated continuous and mixed spectra, developed a perturbation theory analogous to the classical perturbation theory, showed the equivalence of the basic quantum-dynamic problem with Hermite's problem of diagonalizing infinite quadratic forms, and even quantized cavity radiation according to the new mechanics. Moreover, having overcome his initial disgust at Göttingen's formalism, Pauli solved the hydrogen atom with heavy matrix artillery.[356]

Enthusiasm for the new mechanics spread quickly from Göttingen and Germany. Even before the predictive power of the previous quantum theory could really be improved on, a fair number of theoreticians eventually mastered Heisenberg's new scheme and convinced themselves of its essential correctness. Among these pioneers was the young Paul Dirac of Cambridge. I will now turn to his approach to quantum mechanics, for it was the one that drew the best profit from the classical analogy.[357]

## SUMMARY AND CONCLUSIONS

Heisenberg left Copenhagen for Göttingen a little before Bohr received Geiger's letter reporting detailed energy conservation in the Compton process. The failure of the BKS theory led him to reflect on his previous lines of research. One of these lines, the use of complementary symbolic pictures, was empirically successful and independent of the BKS theory, but it made the *explicatio* almost more involved than the *explicandum* and therefore seemed to have already yielded what it could. Heisenberg's other line of research, the "sharpening of the correspondence principle," involved two aspects, speculations on the mechanism of radiation, and Born's program of symbolic translation. The first aspect accompanied the BKS theory to the grave; but the second could be extended, as it was really independent of the notion of virtual oscillators. Heisenberg therefore decided to focus on the determination of transition probabilities, which was the most obvious lacuna in Born's original attempt at a quantum

356. Born, Heisenberg, and Jordan 1926; Pauli 1926b. More comments on these papers will be found in part C.
357. On the early reception of matrix mechanics, see Kozhevnikov and Novik 1987; Mehra and Rechenberg 1982d.

mechanics. Encouraged by his earlier success with the Utrecht sum rules, Heisenberg tried that paradigmatic case, the hydrogen atom, and failed. The classical intensity formula in this case involved expressions whose symbolic translations were unknown (from previous applications of the correspondence principle). Heisenberg therefore switched to a simpler case, the anharmonic oscillator, and decided—this was a crucial step—to "translate" not the final intensity formula but the successive steps of its derivation. He found that this could be done almost unambiguously, if only each classical calculation was expressed in terms of the harmonic components of the motion. Indeed, the correspondence principle provided a natural counterpart for each harmonic component, a "transition amplitude," the square modulus of which gave the transition probability. Moreover, in his contribution to Kramers's dispersion theory, Heisenberg had already encountered such amplitudes and their symbolic products (which Born soon recognized to be matrix products).

In the end Heisenberg succeeded in directly "translating" the general classical equations of motion (for one degree of freedom) and the Bohr-Sommerfeld quantum rule into a form involving just transition amplitudes and atomic frequencies. He could then show that for simple examples the resulting equations correctly determined not only intensities but also the energy spectrum, and, further, that energy had the good taste to be conserved. After this breakthrough of June 1925, in the fall of the same year Born and Jordan analyzed Heisenberg's formal scheme with the tools of matrix calculus and managed to prove its mathematical consistency and completeness. Concrete results were not yet at hand, but, as atomic theorists promptly agreed, quantum mechanics was born.

Heisenberg entitled his seminal paper "On the quantum-theoretical reinterpretation of kinematic and mechanical relations" and presented it as "an attempt to establish a theoretical quantum mechanics analogous to classical mechanics, but in which only relations between observable quantities occur." The term "observable quantities" referred to the atomic frequencies and the transition amplitudes, which, taken together, represented the "motion" of the "new kinematics." Such views coincided with the radical renunciation of classical pictures about which Bohr, Born, and Pauli had come to agree. But they should not be taken as an explanation of how Heisenberg identified the fundamental quantities of his theory. These quantities were just those earlier privileged by Kramers in his dispersion theory, on the ground that they allowed of a "direct interpretation in terms of the fundamental postulates of the quantum theory." Bohr's postulates were deliberately independent of any specific assumptions about atomic motion. Not only were they compatible with the "new kinematics," but they continued to provide the basic notions of the new quantum theory: stationary states, transitions, and the frequency rule.

There is direct evidence that on his way to quantum mechanics Heisenberg departed from Pauli and Born's observability principle. When Heisenberg arrived in Göttingen, Born and Jordan were already working on an extension of the symbolic translation procedure, but one based on transition probabilities instead of transition amplitudes. The reason for this choice was, they declared, that the phases (in the amplitudes) were unobservable and therefore meaningless. Instead, such phases played an essential role in Heisenberg's quantum mechanics, for they were necessary to the consistency of the formal scheme resulting from the symbolic translation. In short, three conditions guided Heisenberg toward quantum mechanics:

● All quantities of the new mechanics had to receive a direct interpretation in terms of Bohr's fundamental postulates.

● The new mechanics had to be formally analogous to classical mechanics, there being the usual "correspondence" between classical harmonics of motion and quantum transitions.

● The new mechanics had to be mathematically closed.

Bohr and Heisenberg regarded the new quantum mechanics as a "quantitative formulation" of the correspondence principle. Indeed, Heisenberg's scheme was built in such a way that it automatically ensured asymptotic agreement between the spectra derived from quantum mechanics and from classical electrodynamics (when applied to the same system). Moreover, Heisenberg's equations of motion were formally identical with their classical counterparts, if only ordinary products were replaced by "quantum products." From a historical point of view, Heisenberg's quantum mechanics may also be seen as the ultimate form of the correspondence principle, this principle having continually evolved in the face of forced changes in the concept of atomic motion, to which it is intimately connected. We may now look back at the circumstances surrounding these changes and at their effect on the use of the correspondence principle and its relation to classical analogies.

In the period 1913–1916 Bohr could quantitatively treat only periodic motions of electrons obeying ordinary mechanics. This was not enough to suggest a detailed analogy between the quantum theory and classical dec-trodynamics. The treatment of multiperiodic systems by Sommerfeld and his followers brought enough generality to inspire in Bohr the full-fledged correspondence principle. The qualitative structure of observed spectra (number of lines in multiplets), Bohr noticed, was better represented by classical electrodynamics than by Sommerfeld's quantum rules. This suggested to him that the quantum theory had to be supplemented by a formal analogy with classical electrodynamics. The product of this analogy was the correspondence principle, which Bohr tended to regard as a principle of the quantum theory, because the quantities which this principle related, harmonic components of the electric moment and transition probabilities, belonged themselves to the quantum theory. If the stationary motions were a priori known, the correspondence principle could be used to *deduce* properties of the emitted spectrum. Unfortunately, this happened only in the case of multiperiodic systems, which covered hardly more than the hydrogen atom. Consequently, Bohr also used the correspondence principle in an inductive way, to infer features of the atomic motion from observed characteristics of spectra. This procedure, despite appearances, increased the predictive power of the theory, because the properties of motion induced from some spectral regularity could be used, in combination with a priori constraints on the motion, to deduce other phenomena, both physical and chemical. In this respect the correspondence principle was similar to Boltzmann's principle (the relation between entropy and probability), which was used both to induce properties of the microcosm and to deduce properties of the macrocosm.

However, the predictive power of Bohr's theory was continually diminished as the a priori constraints on atomic motion were released. This happened most dramatically in early 1923 when Born and his associates proved that ordinary mechanics could not even be used for a proper qualitative description of the helium orbits. Despite this failure Bohr maintained that atomic orbits could be used in a limited way, for they were necessary to his inductions based on the correspondence principle, particularly those found in his second atomic theory. Until the spring of 1925 he resisted pressure from Pauli to completely abandon visual concepts in the atomic domain. Instead, in the BKS theory of 1924, to the orbits he added another space-time picture to depict the coupling of atoms and radiation. This was his ultimate attempt at a space-time implementation of the correspondence principle. The failure of BKS and the concomitant difficulties in imagining a coherent picture of atomic collisions made Bohr surrender to Pauli's dictum.

At that point it seems that the correspondence principle should have lost its guiding role, and in Pauli's opinion this was indeed the case. Previous uses of this principle were bound to a picture of atomic motion, and now no picture was left. Yet the correspondence principle played a crucial role even at that stage. While the orbital model could no longer be used as a (formal) representation of atomic motion, it could still be used in a purely *symbolic* manner, as an analogical basis (outside the quantum theory proper) for calculating transition probabilities or amplitudes. Heisenberg's stroke of genius was to realize that not only the harmonics of the orbital motion but also the underlying dynamic equations could be given a correspondence analogue, leading to a closed mathematical system. He then declared the analogue of the classical motion, namely the set of transition amplitudes, to be the true atomic motion. This left no doubt about the purely symbolic character of the orbital model. At the same time, the formal analogy between the mechanics of these amplitudes and the mechanics of orbits was so close that Bohr's old hope for a rational generalization of the classical theory seemed to be largely justified.

# Dirac's Quantum Mechanics

# Introduction

In Heisenberg's best-informed opinion, the new quantum mechanics contained a quantitative form of the correspondence principle in its very foundation. As a result, further developments could proceed on the basis of this foundation without the need to call on the classical analogy, at least as long as the thorny problem of radiation coupling was deferred. Paul Dirac did not concur with this generally held view. It appeared to him in the fall of 1925 that the analogy between classical and quantum mechanics was not limited to Heisenberg's formal transposition of the Newtonian equations of motion. For him the analogy involved deeper-lying structural properties, those classically expressed in the algebra of Poisson brackets. This remark offered a more direct access to the fundamental equations of quantum mechanics; it also suggested a fruitful adaptation of the canonical methods of resolution from classical dynamics, in which Dirac was already an expert. Dirac thus performed an ultimate transfiguration of the classical analogy into a powerful mathematical heuristics. His impressive success in the winter of 1925–26 was intimately connected to this unorthodox view, and other unusual ideas about theory making.

Dirac's style of quantization was not just a cleverer way of solving other people's equations. It was part of a broader strategy of theory making, one which considered that theories should be articulated in three stages. In the first stage, the fundamental equations of the new theory had to be formulated in the most abstract way, independent of any interpretation. In the second stage, the resulting mathematics had to be developed in a way that would exhibit groups of transformations and conservation properties. In the last stage, the latter properties would be used to inspire a physical interpretation of some of the mathematical quantities employed in the theory.

More specifically, in the first stage of Dirac's quantum mechanics, quantum variables had to be abstracted from their matrix representation and turned into the purely symbolic notion of "$q$-numbers." While Göttingen's theorists required explicit constructions of the mathematical objects they were using, Dirac was satisfied with defining his symbols simply by the equations which they obeyed. As we shall observe, he had been prepared to adopt such a bold attitude by his early exposure to symbolic methods in geometry. He knew that there were "non-Pascalian" geometries in which the "coordinates" of a point did not commute. Conversely, many of the relations between $q$-numbers could be given a geometric interpretation, so that noncommutativity did not have to be feared. This explains why Dirac, in spite of his apparent reveling in algebraic manipulations, could later declare his mind to be an essentially geometric one.[1]

The second stage, the exploration of the mathematical consequences of the fundamental equations, depended much on the previous abstracting stage, which eased the analogy with classical dynamics. Many classical relations remained true in quantum symbols, at least for a specific choice of the order of the terms of quantum products. Dirac worked out the consequences of these equations in an abstract manner, leaving the door open to several possible representations of the quantum symbols in terms of ordinary (measurable) numbers. In his first papers on quantum mechanics, however, the only representation that he deduced and applied was Heisenberg's original matrix representation (for which the energy is diagonal). As Dirac noticed in late 1925, this representation resulted from a transposition of the classical method of action-angle variables, with which he was most familiar.

After his exposure to Schrödinger's wave equation in the spring of 1926, Dirac exploited the freedom of representation inherent in his approach. He studied all other matrix representations of the $q$-numbers (including the ones today called $q$- and $p$-representations, in which $q$ or $p$ is diagonal) and related them to Heisenberg's representation through multilinear

1. Dirac [1962].

transformations. Most important, he could show that Schrödinger's wave function was just one of these transformations. To summarize, Dirac's symbolic formulation of the fundamental equations of quantum mechanics eventually yielded the complete mathematical apparatus of modern quantum mechanics.

The third stage of Dirac's strategy, the identification of the physical content of the theory, was perhaps the most subtle. Properties of conservation or transformation, suggestive as they might be, could not by themselves imply the physical interpretation. As we shall observe, they needed to be completed by a touch of correspondence argument. In Dirac's transformation theory of December 1926, the high quantum-number limit gave a first germ of interpretation in the quasi-classical domain, then the transformation properties generated the whole interpretation of the theory from this germ.

What were the origins of Dirac's immensely powerful methodology? One source appears to have been some philosophical remarks by C. D. Broad and A. Eddington. In their critical presentations of Einstein's general relativity, both thinkers replaced events in space and time with networks of abstract relations; they then recommended that the mathematics of these relations be developed at an a priori level, in a way exhibiting covariance (Broad) and permanence of substance (Eddington). In Broad's opinion the first abstractive step was of an inductive nature, so that the physical interpretation of the theory, at least the metrical meaning of the $ds^2$, was already given. According to Eddington, the physical content of the notions of space and time was lost in the original abstraction; and it was recaptured only in a third stage of "identification" informed by the mind's search for permanence. Although Dirac is not likely to have followed Broad and Eddington very far in their philosophical inquiries, he certainly appreciated their methodological lessons. Not only his approach to quantum mechanics but also much of his later research seem to have proceeded from Eddington's ideal of a physical theory, as instantiated in his (Eddington's) reconstruction of general relativity.

Quite naturally, the great inspirer was pleased with the inspired. In lectures given in early 1927, Eddington judged Dirac's thought to be "highly transcendental, almost mystical," and he saw his prophecy of an ever more abstract condition of the world realized:

> I venture to think that there is an idea implied in Dirac's treatment which may be of great significance. . . . The idea is that in digging deeper and deeper into

that which lies at the base of physical phenomena we must be prepared to come to entities which, like many things in our conscious experience, are not measurable by numbers in any way.[2]

Indeed, where others had already abandoned space-time pictures, Dirac forsook ordinary numbers, a key to his success in exploiting the classical analogy.

2. Eddington 1927, 207–208.

# CHAPTER XI

# Classical Beauty

## NONCOMMUTATIVE GEOMETRY

Originally trained to be an engineer, Dirac was unable to find a job in this field because of the postwar economic depression. He therefore accepted from the mathematics department of Bristol University an opportunity to develop his exceptional scientific skills. During this period (1921–1923) he was influenced by a very good teacher of mathematics, Peter Fraser, who imparted to him a love of projective geometry. For long Dirac admired the magical power of projective methods to justify at a glance theorems otherwise very difficult to prove. Late in his life he remembered having used these methods in much of his work, though only on the backstage of his research.[3]

### WHITEHEAD'S PRINCIPLE

On the subjects he found most attractive Dirac read much on his own. He was presumably exposed to a presentation of projective geometry similar to Whitehead's *Axioms of projective geometry,* published in 1906. As we shall later see, he was at least familiar with some important characteristics of Whitehead's conception of geometry. In line with the general contemporary enthusiasm for the axiomatic approach, Whitehead emphasized the abstract character of the fundamental objects of geometry.

3. For biographical information on Dirac see Kragh 1990; Dalitz and Peierls 1986; Mehra and Rechenberg 1982d; Salam and Wigner 1972; Dirac [1962], 1977.

Points, lines, planes, and so on had to be defined not from an intuition of their inner structure but by their mutual relations, which were raised to the status of axioms; for instance: "Through two distinct points one can draw one and only one line." Moreover, within the limits of mutual compatibility there was freedom in the choice of these axioms. This freedom led to geometries different from Euclid's.[4]

Should one venture so far as to deny any relation between "mathematical" and "physical" points? Whitehead did not believe so. In his earlier *Treatise on universal algebra* (1898), he acknowledged the need for an "existential import" in mathematical definitions. The mutual relations playing the role of definitions were the result of an "act of pure abstraction." In his later philosophical writings, starting with the *Principles of natural knowledge* (1919), he gave central importance to the bridge between roughly perceived objects and mathematically defined concepts which he called the "principle of extensive abstraction." In his opinion, a proper definition of geometric points had to imitate the construction of rational numbers as classes of pairs of integers, or, better, Dedekind's construction of real numbers as classes of interlocked rational intervals. In order to avoid the necessarily finite extension of perceived points, mathematical points had to be conceived as classes of interlocked finite volumes, there being no minimal element in each class.[5]

This limited form of inductivism could perhaps constrain the choice of axioms defining basic geometric objects, but it left other axioms to the taste or interest of the mathematician. In fact, Whitehead's treatise of projective geometry culminated in the proof that the "fundamental theorem" of this geometry could be taken as an independent axiom. This theorem concerns the reduction of chains of perspective, and it is equivalent to Pappus's theorem, which is simpler to enunciate (see fig. 25): If $P_1, P_2, P_3$ are any three points on a line, and $Q_1, Q_2, Q_3$ are any three points on another line, intersecting the former, then the three points of intersection of the cross joints $(P_1Q_2, P_2Q_1)$, $(P_1Q_3, P_3Q_1)$, $(P_2Q_3, P_3Q_2)$ fall on a line.[6]

Whitehead had found the basic idea of the proof of independence in Hilbert's *Grundlagen der Geometrie* (1899). More exactly, Hilbert proved something similar, the independence of "Pascal's theorem" in a geometry employing a notion of parallelism (thereby different from projective geometry, which assumes every two lines to intersect). To this end he first

4. Whitehead 1906.
5. Whitehead 1898, iv; Whitehead 1919, 2d part: "The method of extensive abstraction."
6. Whitehead 1906, 59–64.

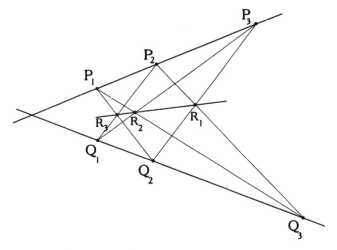

Figure 25.   Figure for Pappus's theorem.

showed that the other axioms of geometry could be represented in terms
of a "system of complex numbers," what we today call a division ring.
A point could then be defined as a triplet $(x, y, z)$ of "coordinates" belonging to the division ring, a plane as a set of triplets satisfying an equation
of the type

$$\alpha x + \beta y + \gamma z + \delta = 0, \tag{1}$$

and a line as an intersection of two planes (in the case of a three-dimensional
geometry). In spite of the superficial resemblance to ordinary analytic geometry, the "complex numbers" $x$, $\alpha$, and so on did not have to be real
numbers; they did not even have to commute. Owing to the latter circumstance, the coefficients in a plane equation (1) had to be kept on the left
of the coordinates.[7]

Pascal's theorem, in the degenerate form used by Hilbert, is a variant
of Pappus's theorem, for which the points $R_1, R_2, R_3$ lie at infinity (see
fig. 26): "If $P_1Q_2$ is parallel to $P_2Q_1$ and $P_2Q_3$ is parallel to $P_3Q_2$, then
$P_3Q_1$ is necessarily parallel to $P_1Q_3$." In the plane of the figure all points
can be represented by two coordinates on the natural axes $OP_1$ and $OQ_1$.
Then,

$$P_i = (x_i, 0), \qquad Q_i = (0, y_i). \tag{2}$$

7. Hilbert 1899, pars. 31–35.

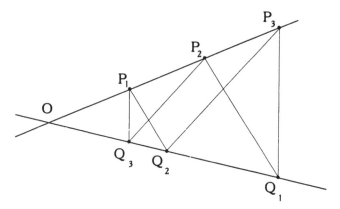

Figure 26.   Figure for a degenerate form of Pascal's theorem.

The equation of the line $P_iQ_j$ can be written:

$$x_i^{-1}x + y_j^{-1}y = 1. \tag{3}$$

That $P_1Q_2$ is parallel to $P_2Q_1$ is determined by the "proportionality" of the coefficients in the corresponding equations, which gives

$$x_1^{-1} = \alpha x_2^{-1}, \qquad y_2^{-1} = \alpha y_1^{-1}. \tag{4}$$

In the same way, for $P_2Q_3$ and $P_3Q_2$ to be parallel, one must have

$$x_2^{-1} = \beta x_3^{-1}, \qquad y_3^{-1} = \beta y_2^{-1}. \tag{5}$$

Combining the two conditions gives

$$x_1^{-1} = \alpha\beta x_3^{-1}, \qquad y_3^{-1} = \beta\alpha y_1^{-1}, \tag{6}$$

while the condition for $P_3Q_1$ and $P_1Q_3$ to be parallel would read

$$x_1^{-1} = \gamma x_3^{-1}, \qquad y_3^{-1} = \gamma y_1^{-1}. \tag{7}$$

Consequently, Pascal's theorem shall hold true if and only if any two $\alpha$, $\beta$ commute (which makes the division ring a field). Since there are noncommutative division rings (skew fields, like the quaternions), Hilbert continued, there exist "non-Pascalian" geometries for which Pascal's theorem is not valid. Whitehead used a similar technique, based on the possibility of noncommutative projective coordinates, to show the independence of Pappus's theorem in projective geometry.[8]

8. Ibid. The notation and presentation are mine. I discuss Pascal's theorem instead of Pappus's because the role of commutativity is more evident in Pascal's theorem. Whitehead 1906.

Even if he did not read Hilbert or Whitehead, Dirac certainly became aware of the existence of non-Pappusian or non-Pascalian geometries, and of their relation to noncommutative algebra. When he arrived in Cambridge in 1923, he was invited to participate in the mathematical tea parties organized by Henry Frederick Baker, a friend of his previous mathematics teacher. Baker's main interest at that time was in geometry. In an axiomatic framework similar to Whitehead's or Hilbert's he developed projective methods, studied extensions in higher dimensions of space, and frequently relied on what he called "symbolic methods." A look at his *Principles of geometry* of 1922 gives a precise idea of the nature of the symbols in question. Like Hilbert's "complex number systems" or Whitehead's projective coordinates, they belonged to a division ring and were used to represent the objects of geometry. More specifically, they provided an extension of Moebius's old barycentric calculus (1827), extracting the algebraic properties of barycentric coefficients from their original identification to real numbers. Every manipulation of the symbols, Baker showed, had a precise geometric meaning, which could be exploited to substitute algebraic for geometric proofs. Moreover, the symbols did not have to commute, again opening the door to non-Pappusian geometries.[9]

Regarding the status of symbolic methods, Baker's attitude was ambiguous. He commended them "to fix ideas and for the purpose of verification" but defended the "purity" of geometry:

> While the view is taken that all the geometrical deduction should finally be synthetic, it is also held that to exclude algebraic symbolism would be analogous to preventing a physicist from testing his theories by experiments; and it becomes part of the task to justify the use of this symbolism.

So that it did not degenerate into a mere algebra of symbols, geometry had to remain clearly connected to its observational roots. Hence follows Baker's epistemological *credo:*

> A science grows up from the desire to bring the results of observation, or the relations of a class of facts that appear to be connected, under as few general propositions as possible. Into these general propositions it is generally found necessary, or convenient, when the science has reached a sufficient development, to introduce abstract entities, transcending actual observation, whose

---

9. See Dirac [1962]; Baker 1922, reference to Moebius on 182. The relation between Baker and Dirac is discussed in Mehra and Rechenberg 1982d.

existence is only asserted by the postulation of their mutual relations. If the science is to be arranged as a body of thought developed deductively, it is necessary to begin by formulating fundamental relations connecting *all* the entities which are to be discussed, from which other properties are to follow as a logical consequence. If this is done we may in the first instance regard all the entities involved in these fundamental propositions as being abstract, even those which we regard as subject to actual observation. The usefulness of the science, for the purpose for which it was undertaken, will depend on the agreement of the relations obtained for the latter entities with those which we can observe.[10]

This is of course reminiscent of Whitehead's principle of extensive abstraction. However, Baker did not mention any source, which shows the pervasiveness of Whitehead's philosophy among contemporary British scientists.

QUATERNIONS

Both in Whitehead's and in Baker's presentation of non-Pappusian geometries, the canonical example of a skew field adduced was Hamilton's quaternions. By the time Dirac studied mathematics and physics, quaternions were not as popular as they had been in nineteenth-century England. While in Maxwell's and Tait's hands they were omnipotent precursors of vector algebra, they disappeared from twentieth-century physics textbooks. Nevertheless, they remained an interesting mathematical curiosity. For this reason the young Dirac read a treatise on quaternions, probably the one by Kelland and Tait, which was the most commonly available. The general emphasis of the book was on the geometric interpretation of quaternions, as given by decomposition into scalar and vector part. Dirac's conception of algebra might have been influenced by this reading, if it was, as he later suggested, his only exposure to algebra.[11]

The main characteristic of quaternions according to Kelland was their noncommutativity:

> About the year 1843 [Hamilton] perceived clearly the obstruction to his progress in the shape of an old law which, prior to that time, had appeared like a law of common sense. The law in question is known as the *commutative* law of multiplication. . . . When it came distinctly into the mind of Hamilton that this law is not a necessity, with the extended signification of multiplication, he saw his way clear, and gave up the law. The barrier being removed,

10. Baker 1922, foreword; 1, 2.
11. See Dirac [1962].

he entered on the new science as a warrior enters a besieged city through a practical breach. The reader will find it easy to enter after him.[12]

This was indeed the way Dirac would enter quantum mechanics. To him the noncommutativity of kinematic variables was not an obstacle but instead the sign of a fundamental advance. In his eyes noncommutativity was not counterintuitive, since it could be understood in geometric terms, and even integrated in a geometric structure, for instance that of non-Pascalian geometry. Neither did it threaten the solidity of the foundation of the new theory, since fundamental entities were sufficiently defined by their mutual relations. According to Kelland, a conqueror of noncommutative extensions would not even have to fear rigor: "It is only by standing loose for a time to logical accuracy that extensions in the abstract sciences—extensions at any rate which stretch from one science to another—are effected." For an engineering student like Dirac, who had learned Heaviside's juggling with derivatives of discontinuous functions and other symbolic methods, the remark hardly needed to be made.[13]

## THE LESSON OF RELATIVITY

### BROAD'S LECTURES

Eddington's eclipse expedition of 1919 verified one of the predictions of Einstein's theory of gravitation and started a wave of enthusiasm for relativity, in England more than anywhere else. Accordingly, Dirac's first love was for the $ds^2 = g_{\mu\nu}\,dx^\mu\,dx^\nu$ of general relativity, though it was not included in the physics curriculum of Bristol University. He actually first learned this theory from a philosopher, Charlie Dunbar Broad, who happened to be teaching a course in philosophy for scientists there in the years 1920–1921. An inspired and methodic thinker, Broad was learned in physics and mathematics and could competently comment on the newest theories. Late in his life he remembered:

> I may compare myself with John the Baptist in at least one respect (though I do not share his taste for an unbalanced diet of locusts and wild honey), viz., that there came to these lectures one whose shoe-latches I was not worthy to unloose. This was Dirac, then a very young student, whose budding genius

12. Kelland and Tait 1882, 5.
13. Ibid., 2.

had been recognised by the department of engineering and was in process of being fostered by the department of mathematics.

Indeed, Broad hardly succeeded in arousing in Dirac an interest in philosophy. But his comments on the nature of relativity thinking may have been heard.[14]

An amplified form of Broad's lectures was published in 1923 as *Scientific thought*. The main focus was on critical philosophy: "The most fundamental task of philosophy is to take the concepts that we daily use in common life and science, to analyse them, and thus to determine their precise meaning and their mutual relations." Half of the lectures were dedicated to a criticism of space and time, culminating in general relativity. Broad founded his analysis on Whitehead's principle of extensive abstraction, which he declared to provide "the essential connection between what we perceive but cannot treat mathematically, and what we cannot perceive but can treat mathematically." When applied to extension and duration, this principle eliminated classical prejudices about space and time and cleared the way to relativity theory. In this process, Broad said, "physicists had been their own philosophers."[15]

For the sake of historical accuracy, let it be mentioned that Whitehead limited his own relativistic enthusiasm to special relativity. He defended the necessity of a uniform space-time and reinterpreted the equations of general relativity in a flat space. "The structure of continuum events," he argued, "is uniform because of the necessity for knowledge that there be a system of uniform relatedness, in terms of which the contingent relations of natural factors can be expressed. Otherwise we can know nothing until we know everything." Fortunately for Dirac, Broad did not share this view, and gave in his course a sympathetic review of general relativity. This is where Dirac remembered having seen $ds^2 = g_{\mu\nu} dx^\mu dx^\nu$ for the first time. Moreover, Broad regarded the tensor calculus of Einstein's theory as providing the general form of any future theory: "The aim of science should be to find general formulae for the laws of Nature, which would ultimately give the special expression of the law in terms of any particular frame, as soon as the defining characteristics are known."[16]

EDDINGTON'S IDEALISM

Once his relativistic appetite had been whetted by Broad's lectures, Dirac devoured two books by the "fountainhead of relativity in England,"

14. See Dirac 1977; Broad 1959, 15.
15. Broad 1923, 16, 49, 85.
16. Whitehead 1922, 29–30; Broad 1923, 216.

Arthur Eddington. The first one, *Space, time and gravitation* (1920), gave
a more popular and literary account,[17] while the second one, *The mathematical theory of relativity* (1923), expounded and even extended the
mathematical apparatus of general relativity.[18] Like Sommerfeld, who
"knew of no book as well written" as *Space, time and gravitation,* Dirac
must have had difficulty deciding what to admire most: clarity of exposition or thoroughness of thought, mathematical elegance, or wit.[19]

In his general conception of relativity, Eddington was close to Broad
and Whitehead in some respects. He highlighted the abstract character of
fundamental physical notions: "The ultimate elements in a theory of the
world must be of a nature impossible to define in terms recognisable to
the mind." He defined the aim of physics as the quest for the "condition
of the world," that is, mathematical symbols comprehending their influence on any possible measurement. The energy-momentum tensor of
general relativity provided the paradigm of such a symbol. Quantum
theory, Eddington speculated (already in 1920!), would require an even
higher degree of abstraction, since, for instance, the symbol connected
with light phenomena would have to encompass antagonistic aspects,
wavelike and corpuscular.[20]

Following Einstein's lead, Eddington regarded general relativity as a
geometrization of physics. In a historico-logical analysis he argued that
geometry had first become analytical (in the Kantian sense), so that it
now dealt with variables of an unknown nature, and could be extended
in various ways. The very diversity of extensions made the geometrization
of physics possible:

> As the geometry became more complex, the physics became simpler; until
> finally it almost appears that the physics has been absorbed into the geometry.
> We did not consciously set out to construct a geometrical theory of the world;
> we were seeking physical reality by approved methods, and this is what
> happened.[21]

Eddington's position was even more radical than suggested by the
above extract. He believed the geometrization to be total, and not only in
the context of gravitation phenomena. Unlike Einstein, he reacted
enthusiastically to Weyl's unified theory of electricity and gravitation, for
it gave the electromagnetic field a geometric interpretation as a connection

17. "Fountainhead" is in Dirac 1977; Eddington 1920.
18. Eddington 1923.
19. Sommerfeld to Einstein, 31 Oct. 1926, in Sommerfeld 1968b, 109.
20. Eddington 1920, 182, 185; Eddington 1923, 3. On Eddington's philosophy see Merleau-
Ponty 1965; on his conception of relativity theory, see Stachel 1986.
21. Eddington 1920, 183.

between local gauges of length. Already in the limited context of gravitation Eddington's views departed from Einstein's. The fundamental equation,

$$T_{\mu\nu} = R_{\mu\nu} - \tfrac{1}{2}g_{\mu\nu}R, \tag{8}$$

relating the energy-momentum tensor $T_{\mu\nu}$ to the curvature tensor $R_{\mu\nu\rho\sigma}$, equated, according to Einstein, two quantities of a different essence: the one on the right side was explicitly given as a function of the metric tensor, while the one on the left side was an empty frame, waiting to be completed by an expression of $T_{\mu\nu}$ in terms of well-defined matter fields. Instead Eddington regarded this equation as a *definition* of energy momentum, hoping that a proper modification of the rest of the theory would provide enough equations to determine the evolution of curvature. In his words the usual conception treated matter as a *cause* of curvature, whereas he proposed to regard it as a *symptom* of curvature; matter was the geometric disturbance itself, not a disturbing factor.[22]

For the sake of pure geometry, Eddington also condemned Einstein's a priori identification of the form $ds^2$ as the metric element, for it presupposed the existence of material rulers and clocks and thereby treated matter as an implicit cause. As a proper alternative, he recommended that one should first conceive a completely abstract geometry, then develop its mathematics in order to construct a "conserved" tensor $T_{\mu\nu}$, and finally "identify" $T_{\mu\nu}$ as representing the flow of energy and momentum. The metrical meaning of $ds^2$ would, he hoped, appear at a later stage. From the perspective of Eddington's influence on Dirac, the most important aspect of this program was the "principle of identification," according to which the mathematics of a physical theory had to be developed at an a priori level before the identification of physically accessible quantities took place.[23]

There were, in addition to the principle of identification, more specific elements in Eddington's methodology, the first of which was the "principle of the permanence of substance." A healthy mind, Eddington thought, could not pretend to any understanding of the world without believing in some kind of permanence; while a predilection for change led one to the asylum, the search for permanence led to the energy-momentum tensor. Nevertheless, a legitimate concept of substance had nothing to do with the naive idea of substance drawn from common experience. Substance had to be related to the "condition of the world," in which process

22. Ibid., 191.
23. Ibid., 191–192; Eddington 1923, 119 and 122 for the "principle of identification."

it was boiled down to an abstract "substratum" of relations: "The relativity theory of physics reduces everything to relations, that is to say it is
structure, not material, which counts."[24]

More strikingly, Eddington believed that there should be only one way
to integrate permanence of substance in a geometry of abstract events:
"Our whole theory has really been a discussion of the most general way
in which permanent substance can be built up out of relations, and it is
the mind which, by insisting on regarding only the things that are permanent, has actually imposed the laws on an indifferent world." This
"despotism of the mind" remained a lasting feature of Eddington's philosophy. It suggested the inspired and often-cited conclusion of *Space,
time and gravitation:* "We have found a strange footprint on the shores
of the unknown. We have devised profound theories one after another,
to account for its origin. At last, we have succeeded in reconstructing the
creature that made the footprint. And Lo! it is our own."[25]

There were interesting by-products of Eddington's viewpoint, for instance the idea of an affine connection without a metric. Nevertheless,
most theoreticians did not see much more in it than a graceful intellectual
exercise. Einstein's judgment displayed a typical mixture of admiration
and suspicion: "[Eddington] has always seemed to me an uncommonly
ingenious but uncritical man. . . . With his philosophy he reminds me of a
*prima ballerina,* who does not herself believe in the justification for her
elegant leaps."[26]

## WAS DIRAC EDDINGTONIAN?

Several comments made while Dirac was a young physicist suggest a positive answer to the question, Was Dirac Eddingtonian? In the manuscript
for a talk which he gave at one of Baker's tea parties one finds:

> The modern physicist does not regard the equations he has to deal with as
> being arbitrarily chosen by nature. . . . In the case of gravitational theory, for
> instance, the inverse square law of force is of no more interest—(beauty)?—
> to the pure mathematician than any other inverse power of distance. But the
> new law of gravitation has a special property, namely its invariance under any

24. Eddington 1920, 195–197; "substratum of the world" is, e.g., on 190.
25. Eddington 1920, 197, 199, 201.
26. Einstein to Besso, 29 July 1953, in Einstein and Besso 1972, quoted in Stachel 1986,
246. Eddington believed that the precise mathematical realization of his program necessitated
a finite value of the cosmological constant, whereas Einstein, after he gave up de Sitter's
static model, favored a zero value of this constant. On this and on the affine connection
without metrics, see Stachel 1986.

coordinate transformation, and being the only simple law with this property
it can claim attention from the pure mathematician.

Here Dirac shows that to some extent he shared Eddington's belief in the
necessity of physical laws, although for a slightly different reason: it is
not the search of the mind for permanence but its predilection for mathematical beauty which enforces the necessity of the laws.[27]
Dirac frequently returned to this theme in his later writings without
giving a precise definition of his aesthetics. Mathematical beauty was no
more subject to definition than beauty in art, but it was obvious to the
connoisseur. However, in its first shy appearance under Dirac's pen, the
word "beauty" (added in parentheses and with a question mark above
"interest") had a more specific meaning: it pointed to rich invariance properties. In other contexts it could also refer to the "magic" of some extensions of ordinary geometry and real analysis, for instance projective
geometry and the theory of functions of a complex variable, which cut
the Gordian knot of theorems difficult to prove in their original context.[28]
Since the mathematical theories suited to the physical world were also
the most beautiful ones, Dirac said to Baker's audience, they were ultimately the ones to be favored by mathematicians themselves: "As more
and more of the reasons why nature is as it is are discovered the questions
that are of most importance to the applied mathematician will become
the ones of most interest to the pure mathematician." Conversely, beautiful parts of mathematics which had not yet received much application
would end up being absorbed by a physical theory. For example, in his
study of permutational symmetry in higher atoms (1929), Dirac treated
group theory as a part of quantum mechanics, for, he said, quantum
mechanics was the general science of quantities that do not commute.[29]
In the foreword of the *Principles of quantum mechanics* (1929), instead
of the above-described revelation of a physical quality of mathematics,
Dirac preferred the more common idea of an increasingly mathematical
nature of physics. Whereas the old physics was based on mental pictures
in space and time, the new physics referred to a "nonpicturable substratum" accessible only through a mathematical description. While this idea

27. P. Dirac, manuscript notes for a talk on relativity, undated (probably 1924), AHQP.
On the relation between Dirac and Eddington, see Kragh 1982. On Dirac's idea of mathematical beauty see Kragh 1979b, 1981; Kragh 1990, chap. 14.
28. See Dirac 1939a.
29. Dirac, n. 27 above; Dirac 1929, 716: "Group theory is just a theory of certain quantities that do not satisfy the commutative law of multiplication, and should thus form a part
of quantum mechanics, which is the general theory of all quantities that do not satisfy the
commutative law of multiplication."

was likely to please Bohr just as much as Eddington, the term "substratum" was specifically borrowed from Eddington's *Space, time and gravitation*. Also like Eddington and Broad, Dirac attached great importance to the existence of groups of transformations relating different "point of views" in the new theory: "The growth of the use of the transformation theory, as applied first to relativity and later to the quantum theory, is the essence of the new method in theoretical physics." And he continued in a very Eddingtonian tone: "This state of affairs is very satisfactory from a philosophical point of view, as implying an increasing recognition of the part played by the observer in himself introducing the regularities that appear in his observations, and a lack of arbitrariness in the ways of nature."[30]

As already remarked, Dirac's concern with philosophy was generally so limited that one may doubt the sincerity of the above-mentioned reflections. He was generally suspicious of any statement that could not be expressed in a mathematical form: what could be said clearly had to be said mathematically. His Eddingtonian utterances might well have been intended as a decorative device, the content of which was soon counterbalanced by crude positivistic statements such as: "The only object of theoretical physics is to calculate results that can be compared with experiments." Or perhaps they pertained to an inaccessible ideal; perhaps they were the mathematical Grail drawing his intellectual energies. In his Scott lecture of 1939 Dirac gave the most extreme expression of this ideal: "We must suppose that a person with a complete knowledge of mathematics would deduce not only astronomical data, but also all the historical events that take place in the world, even the most trivial ones." Such a person, he admitted, did not exist. When under positivistic attack, he would rather profess a balanced mixture of inductive and deductive methods.[31]

Even if Dirac's great admiration for Eddington did not entail a complete adoption of his philosophical stand, he certainly drew important methodological lessons from the Eddingtonian reconstruction of relativity. To imitate the presentation offered in *The mathematical theory of relativity,* a new physical theory had to start with the most abstract development of relations between mathematical symbols possible. Then transformations leading to invariant or covariant properties had to be sought.

30. Dirac 1930, v.
31. Ibid., 7; Dirac 1939a, 129; Dirac 1937. Dirac was here answering Herbert Dingle's positivist assault (Dingle 1937) against cosmological speculations by Eddington, Milne, and himself. See Kragh 1990, 231–233.

Finally, the structures revealed in this process had to suggest the identification of physically observable quantities.

Dirac explicitly enunciated this methodology in 1931, in his paper on quantized singularities (giving birth to the magnetic monopoles):

> The most powerful method of advance that can be suggested at present is to employ all the resources of pure mathematics in attempts to perfect and generalize the mathematical formalism that forms the existing basis of theoretical physics, and *after* each success in this direction, to try to interpret the new mathematical features in terms of physical entities (by a process like Eddington's Principle of Identification).[32]

As explained in the Scott lecture, the notion of mathematical beauty was an integral part of this strategy. One first had to select the most beautiful mathematics—not necessarily connected to the existing basis of theoretical physics—and then interpret them in physical terms. Here also the paragon of beauty was the tensor calculus of general relativity, with its generous transformation properties:

> A powerful new method . . . is to begin by choosing that branch of mathematics which one thinks will form the basis of the new theory. One should be influenced very much in this choice by considerations of mathematical beauty. It would probably be a good thing also to give a preference to those branches of mathematics that have an interesting group of transformations underlying them, since transformations play an important role in modern physical theory, both relativity and quantum theory seeming to show that transformations are of more fundamental importance than equations. Having decided on the branch of mathematics, one should proceed to develop it along suitable lines, at the same time looking for that way in which it appears to lend itself naturally to physical interpretation.[33]

One does not find similar declarations in Dirac's early papers on quantum mechanics. Nevertheless, we shall observe that his spectacular success resulted in good part from an imitation of the model of general relativity as portrayed by Eddington.

## THE ART OF ACTION-ANGLE VARIABLES

Dirac arrived in Cambridge in 1923, at the peak of his love for general relativity. For his supervisor he was hoping to get E. Cunningham, a specialist of this theory. To his disappointment, but also to his advantage, he got instead a specialist in statistical mechanics and quantum theory, Ralph Fowler. Being a friend of Niels Bohr and an occasional visitor at

32. Dirac 1931, 66.
33. Dirac 1939a, 124–125.

the Copenhagen Institute, Fowler was well informed on the developments of atomic theory and taught the main course on this topic at Cambridge. From Dirac's and Thomas's notes on this course one can appreciate how faithful it was to Bohr's ideas and how concerned it was with the latest advances in this field.[34]

As should be the case with any account of the Bohr-Sommerfeld theory, Fowler gave a thorough treatment of the Hamilton-Jacobi method of classical mechanics, with a special emphasis on the "transformation theory of dynamics," which was Whittaker's expression for the theory of canonical transformations. These tools led to an easy quantization of multiperiodic systems and to the Bohr-Kramers theory of perturbations. Fowler also presented the adiabatic principle and the correspondence principle as having great importance—something rather exceptional outside Copenhagen. He treated the BKS theory, the Kramers-Heisenberg dispersion theory, and the sharpened applications of the correspondence principle in detail. Immediately after their publication, and sometimes even before, he reported about Pauli's ambiguous electron, Heisenberg's multimodel theory of multiplets, and the spin hypothesis.[35]

Fowler was also prompt to detect the exceptional qualities of his new student and to encourage his originality. Only six months after his arrival in Cambridge, Dirac started to publish substantial research papers. Whenever the subject had not been imposed on him, he tried to clarify or to generalize in a relativistic way points which he had found obscure in his readings: for instance, the definition of a particle's velocity in Eddington's relativity, or the covariance of the Bohr frequency condition. The main characteristics of his style already showed through: directness, economy in mathematical notation, and little reference to anterior work.[36]

At the end of 1924, after reading Bohr's "Fundamental postulates" (1923) and following suggestions by Fowler and C. G. Darwin, Dirac focused on the more fundamental problem of consolidating and generalizing the adiabatic principle. Burgers's original proof of the adiabatic invariance of action integrals remained incomplete. It strictly required that the fundamental frequencies of the deformed system never become commensurable. But this was clearly impossible, since commensurable frequencies are "dense" among incommensurable ones in the same sense as rational numbers are dense among real ones, and therefore necessarily occur infinitely frequently in a continuous deformation.[37]

34. Dirac's and Thomas's notes on Fowler's course are in AHQP.
35. Fowler's source for classical dynamics was Whittaker 1904, which Dirac read.
36. For more details see Mehra and Rechenberg 1982d.
37. See part B, pp. 176–177, for comments on Bohr's "Fundamental postulates" (Bohr 1923b); Burgers 1917, 169; see part B, pp. 116–118, for Burgers's proof.

Through subtle ε-splitting, Dirac found a rigorous condition of adiabatic invariance, which, fortunately, held in all practical cases. He also touched other problems in the adiabatic register, like the case of varying magnetic fields or, under Darwin's suggestion, the problem of the invariance of the weights (degrees of degeneracy) of stationary states in the degenerate case. Within a few months he had become an expert in the most sophisticated methods of classical dynamics, especially in the art of action-angle variables.[38]

According to a widespread belief, Dirac lacked interest in the other "principle" of Bohr's theory, the correspondence principle. While generally correct, this is not completely true. In one of his unpublished manuscripts, he did investigate an application of this principle to nonperiodic integrable systems. His intention was to provide a more systematic foundation for previous calculations (for instance by Kramers) of the radiation emitted during collision processes. According to a usual procedure of his, he looked for an invariance property, here the independence of "corresponding" radiation intensities with respect to the choice of first integrals of the system. This property ended up holding only in the limit where there would be agreement between quantum-theoretical and classical intensities. The result was too trivial to warrant publication.[39]

In general Dirac (or his adviser) did not believe that the correspondence principle furnished him with a good opportunity to deploy his mathematical skills. The systematic side of this principle—that is, the set of rules used to derive selection rules and approximate intensities—had already been thoroughly studied; the heuristic side, the deep-lying formal analogy between old and new mechanics, he felt to be too vague to be helpful.

## SUMMARY

An important figure of the intellectual milieu in which Dirac grew up was Alfred North Whitehead. Through his studies of the foundations of geometry this philosopher-mathematician was led to the "principle of ex-

---

38. Dirac 1925a. By ε-splitting I mean the current mathematical procedure in which a majoration of each of a series of $n$ terms by $\varepsilon/n$ is sought for, so that the sum of the terms may be smaller than an arbitrary $\varepsilon$ given in advance. M. von Laue obtained similar results with a somewhat different method: see Mehra and Rechenberg 1982d, 103–104. Dirac 1925b; Darwin to Dirac, undated, and 19 Jan. 1925, AHQP; unpublished manuscript concerning the proof that the quantization of a degenerate system is independent of the coordinate choice, in AHQP with Darwin's letters. Dirac also performed lengthy calculations (in AHQP) on a two-particle system, probably with the helium atom in view.

39. Dirac [1924a]; Kramers 1923c.

tensive abstraction," according to which the mathematical concepts of points, lines, planes, and so on are *defined* by mutual relations suggested by experience, while any intuition of their inner essence is meaningless. In spite of this belief in an inductive origin of the objects of geometry, but in conformity with the axiomatic trend at the turn of the century, Whitehead emphasized the freedom left in the choice of axioms used to supplement the definitions. For example, like Hilbert he recognized the possibility of "non-Pascalian" geometries in which the coordinates of a point do not commute (for instance, these coordinates could be quaternions).

Drawn by his love for projective geometry, in his early Cantabrigian years Dirac attended the scholarly tea parties of Henry Frederick Baker, a mathematician who approved of Whitehead's principle of extensive abstraction, and also of his considerations on noncommutative geometries. Moreover, in his multivolume *Principles of geometry* Baker frequently called forth "symbolic methods" in which geometric objects were represented by systems of algebraic relations, and geometric proofs reduced to algebraic manipulations. Since adolescence Dirac had been familiar with another example of noncommutative algebra, Hamilton's quaternions. The standard text on this topic, Kelland and Tait's, praised Hamilton's relaxation of commutativity and the consequent conquest of new mathematical territories. In the spirit of Hamilton and Baker, Dirac would quickly welcome noncommutativity when exposed to Heisenberg's quantum mechanics. We shall also observe that the type of relation which he perceived between classical and quantum mechanics was reminiscent of Whitehead's principle of extensive abstraction. Roughly, Dirac's quantum mechanics could be said to be to ordinary mechanics what noncommutative geometry is to intuitive geometry.

In physics Dirac's first passion was for Einstein's relativity. He was initiated into this theory by his philosophy professor at Bristol in 1920–21, the highly respected Charlie Dunbar Broad. Relativity according to Broad was the result of a systematic criticism of the intuitive notions of space and time, a specific anticipation of Whitehead's principle of extensive abstraction. Impressed by the philosophical foundation of general relativity, Broad presented this theory, with its transformation properties and tensor algebra, as the paradigm of any physical theory to come. The other source of Dirac's knowledge of relativity, Eddington's brilliant essays, gave a similar gloss to the subject and emphasized even more than Broad the abstract character of the fundamental "symbols" of relativity. Unlike Broad, however, Eddington believed that the geometry of abstract events could be reached by a priori means and that its physical content could be "identified" in the end by employing the principle of the permanence of substance (which directed attention to divergence-free tensors).

Like Eddington, Dirac frequently expressed a belief in the a priori necessity of physical laws. But his interest in philosophy was generally so limited that one can only speak of his sympathy with the methodological implications of Eddington's views. In his approach to quantum mechanics, as well as in most of his later work, he tended to first work out the mathematics and *then* "identify" the physical content. In the first abstract stage, the ultimate guide was his "principle of mathematical beauty," which meant, essentially, that he emphasized and searched for rich transformation properties (as found in Riemannian geometry and in Hamiltonian mechanics). These properties also helped in the second stage, the identification of the physical content of the theory.

Under the influence of his supervisor in Cambridge, Ralph Fowler, Dirac shifted his interest toward quantum theory. Fowler's lectures in this field were exceptionally clear and thorough. They exposed the most sophisticated analytic tools, including action-angle variables. They discussed, in a Bohrian manner, the applications of the adiabatic and correspondence principles and reported the latest advances in the field. In the best of his early work, Dirac deployed his exceptional mathematical skills in extending the most formal aspects of the quantum theory and thus became an expert in the handling of action-angle variables. But he paid little attention to the correspondence principle and did not appreciate its constructive value.

# CHAPTER XII

# Queer Numbers

In July 1925 there came to Cambridge a visitor who thought differently about the power of the correspondence principle and had just drawn from it the first elements of a new mechanics. In fact Heisenberg lectured on "Term zoology and Zeeman botanics" at an informal club of young physicists created by Kapitza. The title of this talk referred to the multimodel approach of multiplet theory (see part B, on pp. 205–207). We do not know for sure whether Heisenberg mentioned his more recent inspiration or whether Dirac was present. However, Fowler almost certainly heard of the new kinematics in private conversations, and asked to be kept informed.[40]

## POISSON RESURRECTED

In late August Dirac received from Fowler the proofs of Heisenberg's seminal paper. Even before he was able to judge the relevance of the new scheme, he tried his favorite game, finding a relativistic extension. While this premature attempt fell short, it revealed what Dirac considered to be the essence of Heisenberg's new ideas. First there was a substitution of "Heisenberg's product" for the ordinary product, then an endeavor to maintain as much as possible of the structure of classical dynamics: "The

40. "Minutes of the Kapitza Club," AHQP, entry of 28 July 1925. In Dirac 1977, Dirac remembered having heard Heisenberg's talk, in contradiction of his earlier statement in Dirac [1962]. The source of Heisenberg's talk was Heisenberg 1925b. See Kragh 1990, 14, 317 n. 1.

main point in the present dynamics is that when we have to choose a quantum coefficient, we do so in such a way as to make as many classical relations as possible still true between the quantum quantities."[41]

Another characteristic of Heisenberg's paper, the organic relation between the new kinematics and the structure of the emitted radiation, initially diverted Dirac's attention from more essential features of the theory. In his tentative relativistic extension, he invoked the unidirectional character of the emitted radiation to justify the introduction of the atomic momentum in the labeling of stationary states. In another manuscript he tried to explain the absence of radiation in the fundamental state by introducing a new distinction between two types of "virtual oscillators." The "$i$-type" with an amplitude $q = ae^{i\omega t}$ was unable to radiate by itself, if only the general expression of radiated energy was assumed to be $A^2 + B^2$, where $A$ and $B$ are defined by

$$q = A \cos \omega t + B \sin \omega t. \tag{9}$$

Not to radiate, the fundamental state had to be a pure $i$-oscillator; the possibility of emission in the other states was then to be attached to a corruption of the $i$-oscillators by "$j$-oscillators," with an amplitude $be^{j\omega t}$, wherein $j = -i$.[42]

As Dirac quickly realized, this strange idea had every chance to be irrelevant, since it connected subsets of virtual oscillators to definite levels, in the naive fashion

$$q(n) = \sum_m q_{nm} e^{i\omega_{nm}t}, \tag{10}$$

which is not compatible with Heisenberg's product. The manuscript ends with the words: "We cannot, however, put $xy(n) = x(n)y(n)$, so that coordinates associated with a stationary state can have only a very restricted meaning."[43]

The title and introduction of the above-mentioned manuscript, "Virtual oscillators," clearly indicates that Dirac originally interpreted Heisenberg's new scheme as a modification of the BKS theory. In this modification the distinction between positive and negative oscillators was erased, but an alternative distinction, that between $i$- and $j$-oscillators, was needed to give some insight into the mechanism of radiation. Heisenberg's own

41. Heisenberg 1925c; Dirac [1925d], [1925e], undated but almost certainly written before the discovery of the connection between Poisson brackets and commutators. The quotation is from the second manuscript.

42. Dirac [1925f].

43. Ibid.

emphasis on radiation properties—the only observable things—probably suggested this misinterpretation. Nevertheless, his careful elimination of the term "virtual oscillator" indicated a fundamental departure from the BKS approach: radiation properties could no longer be connected with a given stationary state, as reflected by the interlocked character of quantum products. After his failed distinction between $i$- and $j$-oscillators Dirac also emphasized this impossibility: "The components of a varying quantum quantity are so interlocked . . . that it is impossible to associate the sum of certain of them with a given state."[44]

## THE BRACKETS

Having given up on trying to gain insights into the mechanism of radiation, Dirac turned to the more formal side of Heisenberg's scheme, first to the new quantum rule. Since Heisenberg presented this rule as deducible from the high-frequency limit of Kramers's dispersion formula, Dirac naturally went back to the Kramers-Heisenberg paper for a full derivation. On the one hand, he found that Heisenberg's new product already appeared in the dispersion formulae for the incoherent case (see (220) of part B).[45]

On the other hand, he knew well that in Hamiltonian dynamics the first-order perturbation $P_1$ of a quantity $P_0$ (like the electric moment that was responsible for classical dispersion) could be expressed in the form

$$P_1 = \varepsilon \sum_i \frac{\partial f}{\partial w_i} \frac{\partial P_0}{\partial J_i} - \frac{\partial f}{\partial J_i} \frac{\partial P_0}{\partial w_i}, \tag{11}$$

wherein $\varepsilon f$ is the generating function of the first-order canonical transformation connecting old and new action-angle variables. He probably had learned this from Whittaker's *Analytical dynamics,* or from Fowler's lectures, which used this type of expression in the perturbative treatment of the Stark effect, and in the classical dispersion formula leading to the Kramers-Heisenberg formulae (see (202) of part B). Poisson brackets also occurred in several of Dirac's early manuscripts, even though he might not have remembered that they were named so. Now, according to the Kramers-Heisenberg procedure for translating from the classical dispersion formula, the Poisson bracket had to be translated into a commutator.[46]

44. Dirac 1925g, 652.
45. Kramers and Heisenberg 1925.
46. Whittaker 1904, 302 of the 1960 ed. for the Poisson brackets in perturbation theory; see Dirac's notes and Thomas's more detailed notes on Fowler lectures, AHQP. Poisson brackets (without the name) appeared in Dirac [1924?b] and in the manuscript kept with Darwin's letter (see n. 38 above).

This explanation of Dirac's first important discovery in the new quantum mechanics is not unfounded reconstruction; it may be surmised from a rough calculation found on a back page of a recycled manuscript. The following transcription is the closest possible.[47]

$$\frac{\partial x_\tau}{\partial J}\,\tau' y_{\tau'} - \frac{\partial y_{\tau'}}{\partial J}\,\tau x_\tau = \tau'\,\frac{\partial x_\tau}{\partial J}\,y_{\tau'} - \tau\,\frac{\partial y_{\tau'}}{\partial J}\,x_\tau$$

$$\{x_\tau(J + \tau'h) - x_\tau(J)\}y_{\tau'}(J) - x_\tau(J)\{y_{\tau'}(J + \tau h) - \ldots$$

$$\{x(n, n - \tau) - x(n - \tau', n - \tau - \tau')\}y(n, n - \tau')$$
$$- x(n, n - \tau)\{y(n, n - \tau') - y(n - \tau, n - \tau - \tau')\}$$

$$(x_3 - x_1)y_2 - (y_2 - y_4)x_3 = x_3 y_4 - x_1 y_2$$

$$a(n, m) = \frac{2\pi}{ih}\sum_k \{x(n, k)y(k, m) - y(n, k)x(k, m)\}$$

The diagram was obviously taken from the Kramers-Heisenberg paper. In fact, the whole calculation is very similar to that of Kramers and Heisenberg (which is discussed in the equations (214–220) of part B). The second line results from the prescription[48]

$$\tau\,\frac{\partial}{\partial J} \to \frac{1}{h}\,\Delta_\tau. \tag{12}$$

The factor $2\pi/ih$ in the expression of $a(n, m)$ enables us to reestablish its meaning as the quantum amplitude "corresponding" to the harmonic $n - m$ of the classical bracket

$$B = \frac{\partial x}{\partial w}\frac{\partial y}{\partial J} - \frac{\partial x}{\partial J}\frac{\partial y}{\partial w}. \tag{13}$$

Indeed, if

$$x = \sum_\tau x_\tau e^{2\pi i \tau w} \qquad \text{and} \qquad y = \sum_\tau y_\tau e^{2\pi i \tau w}, \tag{14}$$

47. Back page of Dirac [1925c]. Mehra and Rechenberg 1982d discuss Dirac's back-page calculation.
48. I remind the reader that $\Delta_\tau$ is defined by $\Delta_\tau f(J) = f(J + \tau h) - f(J)$, and that $x_\tau(J)$ "corresponds" to a quantum amplitude $x(n', n'')$ with $n' - n'' = \tau$. The choice of $n'$ is directed by considerations of symmetry (for Kramers and Heisenberg) or by the desire to make matrix products appear in the final formula (in Dirac's case). The diagram is in Kramers and Heisenberg 1925, 694.

then

$$B = \frac{2\pi}{i} \sum_{\tau\tau'} e^{2\pi i(\tau+\tau')w} \left( \frac{\partial x_\tau}{\partial J} \tau' y_{\tau'} - \frac{\partial y_{\tau'}}{\partial J} \tau x_\tau \right), \tag{15}$$

where the quantity in parentheses is the exact starting point of Dirac's note. Finally, the $h$ in $2\pi/ih$ comes from the translation rule (12).

Most important, Dirac's discovery of the relation between commutators and Poisson brackets appears to have been based on Kramers's procedure of symbolic translation. Therefore, it was directly connected with the previous sharpening of the correspondence principle. Here lies the secret of Dirac's revelation of a structural analogy between old and new mechanics—one more significant than Heisenberg's formal transposition of classical dynamic equations.

In his final paper, however, Dirac adopted a different presentation of the relation between classical and quantum brackets. There he used the correspondence principle backward, from the commutator to the Poisson bracket, and in its narrower but safer acceptance as an asymptotic convergence of quantum relations toward classical ones. The resulting calculation looks artificial, since it is nothing but the original one, read from bottom to top:[49]

$$(xy - yx)_{nm} = \sum_{\substack{\tau\tau' \\ \tau+\tau'=n-m}} [x(n, n - \tau) - x(n - \tau', n - \tau - \tau')]y(n, n - \tau')$$
$$- x(n, n - \tau)[y(n, n - \tau') - y(n - \tau, n - \tau - \tau')], \tag{16}$$

which is asymptotically equal to

$$\sum_{\substack{\tau\tau' \\ \tau+\tau'=n-m}} [x_\tau(J + \tau'h) - x_\tau(J)]y_{\tau'}(J) - x_\tau(J)[y_{\tau'}(J + \tau h) - y_{\tau'}(J)],$$

or

$$\sum_{\tau+\tau'=n-m} h\left( \tau' \frac{\partial x_\tau}{\partial J} y_{\tau'} - \tau \frac{\partial y_{\tau'}}{\partial J} x_\tau \right).$$

The latter expression is, as we saw, $ih/2\pi$ times a Fourier coefficient of the Poisson bracket[50]

$$\{x, y\} = \frac{\partial x}{\partial w} \frac{\partial y}{\partial J} - \frac{\partial x}{\partial J} \frac{\partial y}{\partial w}. \tag{17}$$

49. Dirac 1925g, 647–648.
50. Dirac used the notation [x, y] instead of {x, y} for the Poisson brackets. In order to avoid confusion, I will conform to the modern usage, which reserves [x, y] for the commutator.

As immediately noticed by Dirac, the first attractive feature of the Poisson brackets is their canonical invariance: for any choice $q$, $p$ of the canonical coordinated, they can be expressed as

$$\{x, y\} = \sum_r \left( \frac{\partial x}{\partial q_r} \frac{\partial y}{\partial p_r} - \frac{\partial x}{\partial p_r} \frac{\partial y}{\partial q_r} \right). \tag{18}$$

Moreover, they have the same simple algebraic properties as commutators: antisymmetry, bilinearity, distributivity, and Jacobi's identity, which respectively read:

$$\{x, y\} = -\{y, x\}, \tag{19}$$

$$\{x + y, z\} = \{x, z\} + \{y, z\}, \tag{20}$$

$$\{xy, z\} = \{x, z\}y + x\{y, z\}, \tag{21}$$

$$\{\{x,y\}, z\} + \{\{y, z\}, x\} + \{\{z, x\}, y\} = 0. \tag{22}$$

All of this suggested to Dirac the following assumption:[51] "*The difference between the Heisenberg product of two quantities is equal to $ih/2\pi$ times their Poisson bracket expression*. In symbols,

$$xy - yx = (ih/2\pi)\{x, y\}." \tag{23}$$

In the case of a canonical pair $q$, $p$, this rule gave

$$q_r p_s - q_s p_r = (ih/2\pi)\delta_{rs}. \tag{24}$$

In this way Dirac reached the canonical form of the new quantum rule independently of Born and Jordan, and in a more profound way, one showing the intimate structural analogy between classical and quantum mechanics. He concluded: "The correspondence between the quantum and classical theories lies not so much in the limited agreement when $h \to 0$ as in the fact that the mathematical operations on the two theories obey in many cases the same laws." What Heisenberg had judged to be an "essential difficulty" of his new scheme, the noncommutativity of the quantum product, Dirac viewed as having a natural classical counterpart in the Poisson bracket algebra. As Dirac could not have failed to notice, it also had antecedents, even geometrically meaningful ones, in the algebra of quaternions or in Baker's symbols. This prompted him to develop a "quantum algebra," abandoning commutativity but saving associativity and distributivity.[52]

51. Dirac 1925g, 648.
52. Ibid., 649, italicized by Dirac; Heisenberg 1925c, 883.

For the sake of homogeneity of quantum operations, Dirac required every classical operation to have a counterpart in the quantum algebra. Consequently, he introduced a "quantum differentiation" $d/dv$, with the characteristic property that

$$\frac{d}{dv} xy = \left(\frac{d}{dv} x\right)y + x\left(\frac{d}{dv} y\right). \tag{25}$$

Linear realizations of this property, he showed, could always be expressed under the form

$$\frac{d}{dv} x = xa - ax. \tag{26}$$

For example, the partial derivatives of Hamilton's equations could be represented as commutators in the equations

$$\dot{p} = \frac{2\pi}{ih}(pH - Hp), \qquad \dot{q} = \frac{2\pi}{ih}(qH - Hq) \tag{27}$$

resulting from the corresponding classical equations

$$\dot{p} = \{p, H\}, \qquad \dot{q} = \{q, H\}. \tag{28}$$

In this elegant manner Dirac dispensed with the awkward mixture of differential and algebraic operations that was being developed with great pain in Göttingen. As Fowler wrote to Bohr: "I think it is a very strong point of Dirac's that the only differential coefficients you need in mechanics are really all Poisson brackets, and that the direct redefinition of the Poisson brackets is better than the invention of formal differential coefficients."[53]

ACTION-ANGLE VARIABLES

On the basis of the extended analogy between classical and quantum mechanics, Dirac hoped to be able to transpose classical methods of resolution of dynamic problems. One method, the introduction of new canonical variables, received an immediate counterpart through the canonical criterion: The variables $Q$, $P$ shall be canonical if and only if

$$[Q_r, Q_s] = 0, \qquad [P_r, P_s] = 0, \tag{29}$$

$$[Q_r, P_s] = (ih/2\pi)\delta_{rs}. \tag{29'}$$

53. Dirac 1925g, 645; Fowler to Bohr, 22 Feb. 1926, AHQP; see also penetrating comments in Birtwistle 1928, 77.

For systems that were multiperiodic at the classical level, there would presumably be something like quantum action-angle variables (which Dirac rather called uniformizing variables).[54]

In a first exploration of this notion, Dirac found it convenient to introduce the canonical variables (similar to the modern creators and annihilators)

$$\zeta_r = (2\pi)^{-1/2} J_r^{1/2} e^{2\pi i w_r}, \qquad \eta_r = -i(2\pi)^{-1/2} J_r^{1/2} e^{-2\pi i w_r}. \tag{30}$$

In the light of the correspondence principle he requested that the corresponding quantum variables have vanishing matrix elements, except for the elements $\zeta_r(n, \bar{n})$ and $\eta_r(\bar{n}, n)$ with $\bar{n}_s = n_s - \delta_{rs}$. This condition implies the identities

$$\zeta_r \eta_r(n, n) = \zeta_r(n, \bar{n}) \eta_r(\bar{n}, n) = \eta_r(\bar{n}, n) \zeta_r(n, \bar{n}) = \eta_r \zeta_r(\bar{n}, \bar{n}). \tag{31}$$

In order to be canonical the variables have to verify another identity:

$$\zeta_r \eta_r(n, n) - \eta_r \zeta_r(n, n) = ih/2\pi. \tag{32}$$

Hence,

$$\zeta_r \eta_r(n, n) = \zeta_r \eta_r(\bar{n}, \bar{n}) - ih/2\pi, \tag{33}$$

and

$$\zeta_r \eta_r(n, n) = -n_r(ih/2\pi) + \text{const.}, \tag{34}$$

wherein the constant must be taken to be zero in order that all amplitudes may vanish when $n_r \leq 0$.

Granted that the classical relation

$$\zeta_r \eta_r = (-i/2\pi) J_r \tag{35}$$

still holds at the quantum level, the (diagonal) values of the action variables are restricted to $J_r = n_r h$. Implicitly assuming the classical expression of the energy in terms of the $J$'s, Dirac commented: "This is just the ordinary rule for quantising the stationary states, so that in this case [when relation (35) is true] the frequencies of the system are the same as those given by Bohr's theory." This was too simple to be true, as we shall presently see. Nevertheless, the general tendency to adapt classical methods in the new mechanics proved to be very productive in Dirac's subsequent work.[55]

"The fundamental equation of quantum mechanics" was received in early November 1925 by the editors of the *Proceedings of the Royal Society*

54. Dirac 1925g, 651–653.
55. Ibid., 653.

and was hurried to publication by Fowler. The introduction expressed Dirac's personal view of quantum mechanics:

> Heisenberg puts forward a new theory, which suggests that it is not the equations of classical mechanics that are in any way at fault, but that the mathematical operations by which physical results are deduced from them require modification. *All* the information supplied by the classical theory can thus be made use of in the new theory.

Dirac contrasted this outlook with the one associated with the correspondence principle, which confined the validity of classical equations to the asymptotic case of high quantum numbers and to "certain other special cases." In reality the discovery of the connection between commutators and Poisson brackets was inspired by the conception of correspondence as formal translation earlier developed by Kramers, Born, and Heisenberg under Bohr's guidance. The concomitant formal analogy between classical and quantum mechanics was, though Dirac did not know it, the most perfect expression of the "general tendency" expressed in the latest form of the correspondence principle.[56]

## THE CANONICAL METHOD

How did Dirac's conception differ from that developed in Göttingen? Before seeing Dirac's fundamental paper, Born, Heisenberg, and Jordan had already been aware of the connection between Poisson brackets and commutators. As Kramers communicated to the Dutch Academy in November 1925, Pauli had encountered this relation in the same way as Dirac had, through the classical dispersion formula. However, once the "three men" knew the fundamental equations of quantum mechanics, they stopped referring to their classical origin. They felt that they had in hand an essentially new and self-contained theory, which should be developed from its own axioms with the suitable tools of matrix theory. Instead, Dirac tried his best to transpose the classical methods of solution and apply them to quantum problems. The correspondence between the two theories, he believed, was not limited to the form of the fundamental equations; it concerned mathematical *structures,* in the modern sense of the word.[57]

From this perspective, the transformation theory of Hamiltonian dynamics and the action-angle variables, suitably adapted, were likely to be

56. Ibid., 642.
57. Kramers 1925c, 376. According to Kragh 1990, 21, 319, J. C. Slater also discovered the connection between Poisson brackets and commutators, and *used* it to develop quantum formalism in an unpublished work.

useful in the new theory. As Dirac explained at the Solvay congress of 1927, the operator $\partial/\partial J$ of the Hamilton-Jacobi theory even anticipated the interlocked character of stationary states in matrix theory, since it connected *two* infinitely close orbits, in the same way as matrices connected two stationary states.[58]

The analogy, as profound as it might be, was full of traps. As Heisenberg wrote to Dirac in December 1925, quantum mechanics, did not simply result from a reinterpretation of the equations of classical mechanics. The very concept of motion had to be changed. Moreover, the formal correspondence between the two mechanics was not as close as Dirac imagined. One could not simply identify *all* Poisson brackets with commutators without getting into trouble. For instance, the commutator $[q^2, p^2]$ could be evaluated in two contradictory ways, through an algebraic reduction in terms of the commutator $[q, p]$:

$$\begin{aligned}
[q^2, p^2] &= q[q, p^2] + [q, p^2]q \\
&= qp[q, p] + q[q, p]p + p[q, p]q + [q, p]pq \\
&= \frac{ih}{2\pi}(2qp + 2pq),
\end{aligned} \tag{36}$$

or directly through the corresponding Poisson brackets:

$$[q^2, p^2] = \frac{ih}{2\pi}\{q^2, p^2\} = \frac{ih}{4\pi}4qp. \tag{37}$$

Fortunately, the correspondence still held for canonical pairs, which was all that Dirac needed for his developments.[59]

Heisenberg also reproached Dirac with a more important overestimation of the classical analogy. Contrary to the assumption made at the end of "The fundamental equations," the expression $H(J)$ of the Hamiltonian in terms of action variables could not be simply adapted from the classical theory; the quantum-mechanical spectrum of multiperiodic systems had to differ, in general, from that given by the Bohr-Sommerfeld theory. Indeed, as a consequence of the noncommutativity of quantum variables, the quantum-mechanical expression of the original configuration variables $(q, p)$ in terms of the action-angle variables $(w, J)$ generally differs from the classical one; so does the expression of $H$ as an implicit function of $(w, J)$ through $(q, p)$. For instance, for a rotator the action variable is the

58. P. Dirac, manuscript notes for the Solvay congress of 1927, AHQP, and *Electrons et photons*, proceedings of the Solvay congress of October 1927 (Paris, 1928), 182.

59. Heisenberg to Dirac, 1 Dec. 1925, AHQP (Berkeley copy), quoted in Dirac 1977, 128.

angular momentum $J$ around the axis; as a function of this variable, the classical Hamiltonian is $\frac{1}{2}\alpha J^2$ while the quantum-mechanical one is $\frac{1}{2}\alpha J(J + 1)$.[60]

These limitations did not discourage Dirac. In his subsequent papers he adapted more correctly the technique of canonical transformations and action-angle variables, taking properly into account the modifications required by noncommutativity. His general strategy was to find the explicit form of a classical transformation, say $(q, p) \rightarrow (Q, P)$, and then modify this expression in such a way that $(Q, P)$ would remain canonical in the quantum-mechanical sense $[Q, P] = ih/2\pi$, and so on. Take for instance the transformation from plane Cartesian coordinates and momenta to polar coordinates and momenta. Classically,

$$x = r \cos \theta, \qquad y = r \sin \theta \tag{38}$$

$$p_\theta = xp_y - yp_x, \qquad p_r = p_x \cos \theta + p_y \sin \theta. \tag{39}$$

Quantum-mechanically, the latter relation must be modified to

$$p_r = \tfrac{1}{2}(p_x \cos \theta + \cos \theta \, p_x) + \tfrac{1}{2}(p_y \sin \theta + \sin \theta \, p_y). \tag{40}$$

With this type of consideration, Dirac solved the hydrogen atom, discussed the composition of angular momenta in atoms with several electrons—imitating the classical procedure of "the elimination of the nodes" in celestial mechanics—and even derived the Compton scattering probability (his result being that later obtained from the Klein-Gordon equation). With the exception of the latter result, which was almost the only one not available from the old quantum theory, similar progress had been made in the Göttingen group, but by three or four men instead of one, and only with a great amount of heavy mathematics and some hints from Hilbert and Courant.[61]

## QUANTUM ALGEBRA

Dirac's successful adaptation of the canonical methods of classical dynamics depended much on his conception of "quantum algebra." Several of the symbolic operations which he performed on quantum variables were, indeed, meaningless from the point of view of Göttingen's authorities. For instance, as Jordan explained in a letter to Dirac, there could be

60. Dirac 1925g, 653; Heisenberg to Dirac, 23 Nov. 1925, quoted in Dirac 1977, 127–128.
61. Dirac 1926a. Pauli's method (Pauli 1926b), based on Lenz's invariant vector ($\sigma \times v + e^2 r/r^3$ where $\sigma$ is the angular momentum), was equally ingenious but perhaps less general. Dirac 1926b; Dirac 1926c. With the methods of Bohr's theory Breit 1926 obtained the same scattering formula around the same time.

no matrix (even a continuous one) representing an angle variable, since there is no conjugate operator to operators like the action variables, which have a discrete spectrum and no accumulation point.[62]

In "The fundamental equations" Dirac had adopted Heisenberg's original definition of quantum variables as arrays of ordinary numbers, and also the interpretation of the polarization matrix as giving radiation intensities. In his next article he adopted a more abstract stance. The quantum variables were "magnitudes of a kind that one cannot specify explicitly." They had to be defined only by the fundamental equations which they obeyed, while their representation in terms of infinite matrices, *if any existed*, had to be *deduced* from these equations. To capture the essence of his position in one word, Dirac introduced the "$q$-numbers," which were defined by their algebraic properties alone: they could be added and multiplied as in a ring; only some of them commuted with all other $q$-numbers, in which case they were called "$c$-numbers." Apparently, $c$ stood for "classical," while $q$ stood for "quantum"; but later Dirac suggested that they respectively stood for "commutative" and "queer."[63]

In a spectacular illustration of his strategy, Dirac subsequently derived the existence of a matrix representation for most $q$-numbers in the case of multiperiodic systems.[64] He first modified his definition of quantum action-angle variables in such a way that they no longer presupposed matrices. Just as in the classical theory, $(w, J)$ would be action-angle variables if and only if the Hamiltonian was a function of $J$ only, and any $q$-number (save the multiple-valued ones) could be expressed in the form[65]

$$q = \sum_{\tau} C_{\tau}(J)e^{2\pi i \tau \cdot w}. \tag{41}$$

Consider now two $q$-numbers $x$ and $y$ and their product $xy$. We have

$$x = \sum_{\tau} x_{\tau}e^{2\pi i \tau \cdot w}, \qquad y = \sum_{\tau} y_{\tau}e^{2\pi i \tau \cdot w}, \tag{42}$$

---

62. Jordan to Dirac, 14 Apr. 1927, AHQP.

63. Dirac [1926d], 5; Dirac 1926a, 562; Dirac 1977, 129.

64. Dirac 1926a, 567–569.

65. At this stage it would be convenient to multiply all angle variables by $2\pi$, and divide all action variables, $h$ as well, by $2\pi$, as Dirac did. For the sake of homogeneity I maintain instead the earlier convention. But I introduce Dirac's later notation $\hbar = h/2\pi$. The dot product $\tau \cdot w$ is $\sum_r \tau_r w_r$.

and

$$xy = \sum_{\tau\tau'} x_\tau e^{2\pi i \tau \cdot w} y_{\tau'} e^{2\pi i \tau' \cdot w} \qquad (43)$$

or

$$xy = \sum_{\tau\tau'} x_\tau (e^{2\pi i \tau \cdot w} y_{\tau'} e^{-2\pi i \tau \cdot w}) e^{2\pi i (\tau + \tau') \cdot w}. \qquad (44)$$

In order to transform the latter expression we first prove the identity

$$e^{2\pi i \tau \cdot w} f(J) e^{-2\pi i \tau \cdot w} = f(J - \tau h), \qquad (45)$$

which is valid for any function $f$ expressible as a power series of $J$.

The relation of commutation

$$[w_r, J_s] = i h \delta_{rs} \qquad (46)$$

implies

$$[e^{2\pi i \tau \cdot w}, J] = i h \frac{d}{dw} e^{2\pi i \tau \cdot w} = -\tau h e^{2\pi i \tau \cdot w} \qquad (47)$$

or

$$e^{2\pi i \tau \cdot w} J e^{-2\pi i \tau \cdot w} = J - \tau h. \qquad (48)$$

Equating the $n^{\text{th}}$ powers of the two members of the latter equation produces

$$e^{2\pi i \tau \cdot w} J^n e^{-2\pi i \tau \cdot w} = (J - \tau h)^n. \qquad (49)$$

Then, linearly superposing powers of $J$ and composing the results justifies the identity (45) for power series.

The expression (44) for the product $xy$ now becomes

$$xy = \sum_{\tau\tau'} x_\tau(J) y_{\tau'} (J - \tau h) e^{2\pi i (\tau + \tau') \cdot w}, \qquad (50)$$

or

$$(xy)_{\tau''} = \sum_{\tau} x_\tau(J) y_{\tau'' - \tau}(J - \tau h). \qquad (51)$$

For the sake of transparency change the notation $C_\tau(J)$ into $C(J, J - \tau h)$. Then

$$xy(J, J - \tau''h) = \sum_{\tau} x(J, J - \tau h) y(J - \tau h, J - \tau''h). \qquad (52)$$

This symbolic relation, noticed Dirac, becomes a matrix product as soon as the $J$'s are given $c$-number values $nh$ (i.e., $J_r = n_r h$). Therefore, any $q$-number may be represented by a matrix $q_{mn}$, wherein $n$ and $m$ refer to two possible values of the action variables $J$. The action-variables $J$ themselves and the energy $H(J)$ are represented by diagonal matrices with diagonal elements corresponding to $J = nh$. Naturally, the different values of $J$ are assumed to characterize stationary states in Bohr's sense.

Thus defined, the matrices do not yet exhibit the time dependence implied by the fundamental relation

$$i\hbar\dot{g} = [g, H]. \tag{53}$$

Dirac remedied this by studying the time derivative of the quantum Fourier exponentials:

$$\frac{d}{dt} e^{2\pi i t \cdot w} = \frac{2\pi}{ih} [e^{2\pi i t \cdot w}, H(J)]$$

$$= \frac{2\pi i}{h} (H(J) - e^{2\pi i t \cdot \omega}H(J)e^{-2\pi i t \cdot w})e^{2\pi i t \cdot w}. \tag{54}$$

Through the identity (45) this transforms into

$$\frac{d}{dt} e^{2\pi i t \cdot w} = 2\pi i \frac{H(J) - H(J - \tau h)}{h} e^{2\pi i t \cdot w}. \tag{55}$$

Taking the derivative of a $q$-number with respect to time therefore amounts to multiplying its $C(J, J - \tau h)$ by $2\pi i$ times the Bohr frequency $\Delta_\tau H/h$. In this magic way Dirac recovered Heisenberg's matrix form and Bohr's frequency condition.[66]

Dirac still had to show that the polarization matrix in this scheme provided transition probabilities, as originally asserted by Heisenberg. He did this in the following manner. The harmonic development of the quantum electric polarization P is essentially ambiguous, for it can be written in two equally justified forms:

$$P = \sum_\tau C_\tau(J)e^{2\pi i t \cdot w} \quad \text{or} \quad P = \sum_\tau e^{2\pi i t \cdot w}C'_\tau(J). \tag{56, 56'}$$

According to identity (45), however, the coefficients $C_\tau$ and $C'_\tau$ are related by

$$C_\tau(J) = C'_\tau(J - \tau h). \tag{57}$$

This shows that $C_\tau(J)$ is naturally connected to *two* stationary states, $J = nh$, and $J = (n - \tau)h$, whereas it was connected only to one stationary state in the classical Fourier development. This suggests, in conformity with Bohr's postulates, that radiation is related to a transition $J \rightarrow J - \tau h$

---

66. Dirac 1926a, 567.

between two stationary states and that the matrix $C(J, J - \tau h)$ represents the amplitude of the oscillations connected with this transition.[67]

This reasoning of Dirac's reflected a strategy reminiscent of Eddington's principle of identification. It first introduced abstract entities defined only by their mutual relations, the $q$-numbers, then developed the formal consequences of these relations in such a way as to suggest an identification of their physical meaning. There were, however, some differences. According to Eddington, the primitive relations were dictated by the mind, whereas Dirac obtained them through the classical analogy or, better, through some kind of "extensive abstraction" of the structure of Hamiltonian dynamics. Moreover, the identification of observable quantities was not completely dictated by the mind; it relied on Bohr's postulates and also on the privileging of action-angle variables, which was a remnant of the old form of the correspondence principle.

A MATHEMATICAL DIGRESSION

The essence of Dirac's approach was to leave the properties of $q$-numbers open to the needs of future developments that might occur in quantum mechanics. Nevertheless, his interest in the purely mathematical side of his theory led him to introduce supplementary axioms that would enrich the algebra of $q$-numbers and make it closer to the algebras which he already knew, namely, quaternions and Baker's symbols. For instance, he occasionally admitted that all $q$-numbers had inverses, and he excluded divisors of zero (i.e., numbers such that $qq' = 0$ with $q \neq 0$ and $q' \neq 0$). In a mathematical paper of 1926 he added another axiom that was supposed to be necessary for a proper definition of $q$-number functions: for any two $q$-numbers $x$ and $y$ there had to exist a $q$-number $b$ such that $y = bxb^{-1}$.[68]

As Léon Brillouin noted in a letter to Dirac, none of these axioms was suited to quantum variables. An operator introduced by Pauli in 1926, the spin-raising operator $S_+ = S_x + iS_y$, furnishes a simple counterexample to the two first axioms. It divides zero since $S_+^2 = 0$, and it cannot be inverted since a relation $S_+ q = 1$ leads to an absurdity once multiplied by $S_+$ on the left. Finally, if the last axiom were true, any two quantum variables would have the same spectrum—patently untrue. The algebraic

---

67. Dirac 1926c, 406.
68. Dirac 1926a, 562 for the first two properties; Dirac 1926e, 413.

properties of $q$-numbers, Brillouin concluded, could not differ from those of arbitrary matrices.[69]

Fortunately, Dirac's attempts to axiomatize the $q$-numbers did not interfere with their practical use. Despite Brillouin's claim, the $q$-numbers proved to be more general than Heisenberg's original matrices, since they could cover both discrete and continuous spectra and allowed quantum angle variables that had no matrix representation, and since their applicability was not limited to stationary systems, as exemplified in the calculation of the Compton effect. Above all, Dirac wanted flexibility:

> One can safely assume that a $q$-number exists that satisfies certain conditions whenever these conditions do not lead to an inconsistency, since by a $q$-number one means only a dummy symbol appearing in the analysis satisfying these conditions. . . . One is thus led to consider that the domain of all $q$-numbers is elastic, and is liable at any time to be extended by fresh assumptions of the existence of $q$-numbers satisfying certain conditions, and that when one says that all quantum numbers satisfy certain conditions, one means it to apply only to the existing domain of $q$-numbers, and not to exclude the possibility of a later extension of the domain to $q$-numbers that do not satisfy the condition.[70]

Dirac thus set forth a general program by which arbitrary physical situations might be analyzed with $q$-numbers, the properties of the $q$-numbers being tailored to fit the physical situations as well as the fundamental equations.

STAGNATION

In May 1926 Dirac put together in his dissertation the first fruits of his conception of quantum mechanics. By then he had found nearly all that could be learned from the $q$-number adaptation of the method of uniformizing variables. There were obvious signs that the magic of this method was being exhausted. Even a problem that was simply treated on the basis of the old quantum theory, the H-atom, received a fairly complicated treatment within the $q$-number theory, regardless of the high mathematical skills deployed. The very problem that motivated Heisenberg's discovery of matrix mechanics, the calculation of the intensities of hydrogen lines, was no more accessible to Dirac than it was to the Göttingen group.

69. Brillouin to Dirac, March 1926, original unlocated, discussed in Dirac 1977, 130; see also Brillouin 1926, 154 n.
70. Dirac [1926d], 24.

In general (with the questionable exception of the Compton effect), Dirac's methods could not be used to treat more problems than the old quantum theory, precisely because they were nothing but a noncommutative reformulation—should we say complication?—of the methods of this theory.

There was a more fundamental obstacle which Dirac disclosed in the late spring of 1926, either before or right after his first use of the Schrödinger equation: if, in the spirit of Heisenberg's theory, matrices refer only to observable processes, there cannot be any action-angle representation of them in the case of atoms with more than one electron. Let indeed $m$ and $n$ be two similar quantum numbers referring to two electrons in a given atom. According to a natural extension of Heisenberg's observability principle, stationary states differing only by a permutation of $m$ and $n$ should be identified, since there is no observable difference between the transitions $mn \to m'n'$ and $nm \to m'n'$. Consider now the Fourier exponential $e^{2\pi i \tau \cdot w}$ corresponding to the transition $mn \to m'n'$. Then the Fourier exponential $e^{2\pi i(2\tau \cdot w)}$ corresponds to a transition $mn \to m''n''$, with

$$\left. \begin{array}{l} m - m' = m' - m'' \\ n - n' = n' - n'' \end{array} \right\}. \tag{58}$$

If $m'n'$ is to be identified with $n'm'$, one might as well have written

$$\left. \begin{array}{l} m - n' = n' - m'' \\ n - m' = m' - n'' \end{array} \right\}. \tag{59}$$

Here comes the absurdity: the values of $m''n''$ deduced from each system cannot refer to the same stationary state since they are neither identical nor related through a permutation. With this ingenious argument Dirac closed a first chapter of his involvement in the history of quantum mechanics.[71]

## SUMMARY

In the fall of 1925, Dirac scrutinized Heisenberg's fundamental papers and perceived three essential elements: the new quantum product, the endeavor to maintain as many classical relations as possible, and the direct connection between the quantum amplitudes and the properties of the emitted radiation. Misled by the latter point, Dirac originally interpreted Heisenberg's theory as a modification of the BKS theory and tried to draw

---

71. Dirac 1926f, 668.

from it a new conception of virtual oscillators. But he quickly abandoned this line of thought and addressed a more fruitful question: Where did Heisenberg's new quantum rule come from? In his paper, Heisenberg pointed to the possibility of deducing his quantum rule from the high-frequency limit of Kramers's dispersion formula. Consequently, Dirac went back to the Kramers-Heisenberg paper (or to Fowler's account of it) and observed that the dispersion formula was the symbolic translation of a Poisson bracket (Poisson brackets are differential expressions involving two dynamic quantities: they appear when considering infinitesimal trans-formations in Hamiltonian mechanics, and they enjoy remarkably simple algebraic properties). Together with Heisenberg's remark, this led him to postulate that quantum mechanics was obtained by expressing the fundamental equations of mechanics in terms of Poisson brackets, and by replacing the brackets by purely algebraic expressions, the commutators (divided by $ih/2\pi$).

This conception implied a deep structural analogy between classical and quantum mechanics, from which Dirac drew maximum profit. First of all, his "fundamental equations" were expressed in a very homogeneous form, one involving only algebraic operations (except for time differentia-tion), whereas the mechanics developed in Göttingen awkwardly mixed algebraic and differential operations (with respect to *matrix* coordinates!). On the practical side, Dirac imagined a quantum-mechanical analogue of the canonical methods for solving mechanical problems, particularly an analogue of the powerful technique of action-angle variables. This led him, within a semester, to results comparable, and sometimes superior, to those obtained in Göttingen. At the end of 1925 (a little after Pauli) he solved the hydrogen atom, and, soon after, he derived the algebra of angular momenta and made a relativistic calculation of Compton scattering probabilities.

The superiority of Dirac's method lay in his personal appraisal of the classical analogy in the new mechanics. While Göttingen's physicists judged that this analogy had been integrated, once for all, into the *foundation* of the theory, Dirac believed that it could still be used profitably in the *development* of the theory. Accordingly, Dirac exaggerated the analogy between classical and quantum mechanics. He initially underplayed the revolutionary character of quantum mechanics and asserted that only the physical interpretation, but not the equations of classical mechanics, was at fault. Heisenberg corrected him: the revolution affected the very concept of motion (kinematics); furthermore, the formal analogy between the two mechanics was not quite as close as Dirac first thought. One could not,

without contradiction, replace all Poisson brackets of the classical theory
with commutators, and the energy expression in terms of action variables
was not the same in the classical and the quantum cases, contrary to what
Dirac suggested in his first paper on quantum mechanics. But to Dirac
these were only points of rigor, which did not affect his general view or
strategy.

The success of Dirac's adaptation of classical methods depended on
another unique aspect of his approach, namely his notion of quantum
algebra. In an Eddingtonian manner, Dirac formulated the fundamental
equations of quantum mechanics in a purely abstract way, without having
formerly interpreted the symbols entering these equations. The symbols,
or "$q$-numbers," were defined only by their mutual relations, which for
him constituted a "quantum algebra." The physical interpretation of these
symbols occurred in two steps. The introduction of a quantum analogue
of action-angle variables first suggested a matrix representation of the
symbols; then the matrices were identified with collections of transition
amplitudes, as suggested by some formal properties and a touch of
"correspondence." This strategy was reminiscent of Whitehead's extensive
abstraction, insofar as the relations defining the symbols were abstracted
from ordinary mechanics; and it was akin to Eddington's principle of
identification insofar as it purported to deduce the interpretation of the
symbols from their formal properties. Dirac's symbols, however, in contrast
with Whitehead's geometric objects, were not interpreted on the basis of
their empirical origin; and contrary to Eddington's symbols of the world,
they could not be interpreted without comparing the theory with an already
interpreted theory of the same phenomena, Bohr's old quantum theory.

Dirac did some purely mathematical work on the quantum algebra, in
the course of which he ventured to subject $q$-numbers to axioms similar
to those found in Baker's *Principles* and Kelland and Tait's *Quaternions*.
Some of these axioms did not suit quantum mechanics, as quickly noticed
by Jordan and Brillouin. But Dirac already knew that the guilty axioms
were not necessary to his practical calculations. In general, he wished to
maintain a certain flexibility in his notion of $q$-number: the algebra had
to be adapted to the needs of the developing theory. Also he did not
require rigorous mathematical definitions of the objects he was manipula-
ting; it was sufficient for him that the symbolic operations performed on
these objects would not lead to contradiction.

By the spring of 1926 Dirac's progress had amazed all his colleagues;
yet it seemed to have reached a peak. The method of transposing classical
methods indeed had a defect: essentially, it could only solve problems that

had a solvable counterpart in the old quantum theory based on classical orbits. To make it worse, through a very ingenious argument Dirac proved that action-angle variables could not exist for quantum-mechanical systems containing two or more indistinguishable particles (this case includes all atoms beyond hydrogen!). A new method had to be found to solve the fundamental equations of quantum mechanics.

CHAPTER XIII

# Quantum Beauty

## SCHRÖDINGER'S EQUATION

### THE NEWS

From March 1926 on, Erwin Schrödinger published a series of memoirs on a new theory of quantization based on his famous equation. In a conception derived from de Broglie's, stationary states were identified with stationary modes of electron waves in atoms. The corresponding calculations required none of Göttingen's transcendent algebra; they rested on a mathematical technique well known to anyone versed in the classical theory of waves. Within this new framework Schrödinger could very quickly and simply solve many of the standard problems of quantum theory. Surprisingly, the results appeared identical with those given by matrix mechanics, whenever comparison could be made.[72]

Dirac's first reaction to this spectacular invention was essentially negative. A wave theory of matter, he thought, had to be just as inconsistent as the wave theory of light. Moreover, there was no need for a new quantum mechanics, since there already was one, the foundation of which he did not question. A letter from Heisenberg of May 1926 changed his opinion. It explained how the Schrödinger equation could be used as a tool to derive matrices satisfying the fundamental equations of quantum mechanics. One just had to solve this equation (the time-independent version,

---

72. Schrödinger 1926a, 1926b, 1926d, 1926e.

which was the only one available at that time) for eigenfunctions $\psi_n$ with the energies $E_n$,

$$H\left(q, -i\hbar \frac{\partial}{\partial q}\right)\psi_n = E_n\psi_n, \tag{60}$$

and form the matrix

$$g_{mn} = g^0_{mn}e^{i(E_m - E_n)t/\hbar} \tag{61}$$

associated with the quantum variable $g(q, p)$ according to the rule

$$g\left(q, -i\hbar \frac{\partial}{\partial q}\right)\psi_n = \sum_m g^0_{mn}\psi_m. \tag{62}$$

While such a development is possible only if the functions $\psi_n$ span the space of $\psi$-functions (that is, if the condition of "completion" is met for the original wave equation), Heisenberg, unencumbered by this type of consideration, immediately proceeded to prove his assertion.[73]

First of all, the matrix associated with the product $xy$ of any two quantum variables is the product of the matrices respectively associated with $x$ and $y$. This is a trivial result for the modern reader since the equation (62) just says that $g^0_{mn}$ is the matrix representation of an operator in the base of the $\psi_n$'s. In this scheme the relation

$$[q, p] = i\hbar \tag{63}$$

immediately follows from (62) and

$$\left[q, \frac{\partial}{\partial q}\right] = -1. \tag{64}$$

Finally, the time dependence of matrices introduced *ad hoc* in equation (61) warrants, as usual, the result that

$$i\hbar\dot{g} = [g, H]. \tag{65}$$

The above consideration is so simple that one may wonder why the Schrödinger equation was not derived from the original quantum mechanics before it was inferred from de Broglie's notion of matter waves. At

73. Heisenberg to Dirac, 26 May 1926, AHQP. Dirac's first reaction to wave mechanics is inferred from Heisenberg's comment in this letter: "I quite agree with your criticism of Schrödinger's paper with regard to a wave-theory of matter. This theory must be inconsistent, so just like the wave-theory of light." The connection between matrix and wave mechanics was simultaneously discovered by several theoreticians, among whom were Pauli (letter to Jordan of 12 Apr. 1926, *PB*, no. 131) and Schrödinger 1926c. Unlike Pauli and Heisenberg, Schrödinger noted the role of the condition of completion.

Göttingen, theoreticians had been for some time inhibited by Heisenberg's doctrine of observability, which confined quantum mechanics to the methods of matrix algebra. In early 1926 Born, Wiener, and Lanczos attempted to remove this restriction, and would probably have reached the Schrödinger equation (they came very close to it) if only there had been enough time before Schrödinger's publication. On the other side of the Channel, Dirac would have been in the best position to discover the new wave equation since his conception of $q$-numbers was not bound to any specific representation. For instance, he could have noticed that a differential operator made a good representation of the momentum quantum variable. However, he did not, because his intellectual adventure remained restricted by the analogy with the method of uniformizing variables, as we just saw.

THE CROP

In compensation for not having found it, both Heisenberg and Dirac made faster progress in exploiting the Schrödinger equation than did Schrödinger himself. In his letter to Dirac, Heisenberg announced that he already knew how to solve the helium atom. Within the three following months Dirac reached no less important results. As usual, he put a touch of relativity in the new equation, that is to say, he extended the substitution

$$p \rightarrow -i\hbar \partial/\partial q \tag{66}$$

to

$$E \rightarrow i\hbar \partial/\partial t. \tag{67}$$

This gives instead of equation (60) the so-called "time-dependent Schrödinger equation"

$$H\left(q, -i\hbar \frac{\partial}{\partial q}\right)\psi = i\hbar \frac{\partial \psi}{\partial t}, \tag{68}$$

which Schrödinger obtained at the same time through less straightforward means.[74]

74. Heisenberg to Dirac, 26 May 1926, and Heisenberg 1926a, 1926b; Dirac 1926f, 661–663. Schrödinger had only an a posteriori justification for this form of the time-dependent equation: it led to the correct dispersion formula. He had earlier used another equation,

$$\Delta \psi - 2\mu \frac{E - V(\mathbf{r})}{E^2} \frac{\partial^2 \psi}{\partial t^2}.$$

The stationary solutions leading to the energy spectrum are the ones for which

$$\left. \begin{array}{l} i\hbar \dfrac{\partial \psi}{\partial t} = E\psi \\[2mm] H\psi = E\psi \end{array} \right\}. \tag{69}$$

Consequently, they evolve in time according to

$$\psi_n(t) = \psi_n(0)e^{-iE_nt/\hbar}. \tag{70}$$

The functions $\psi_n$ now directly engender Heisenberg's time-dependent matrix according to a rule similar to (62):

$$g\psi_n = \sum_m g_{mn}\psi_m. \tag{71}$$

Dirac published this most adequate presentation of the relation of Schrödinger's equation to his quantum mechanics in the first part of a paper entitled "On the theory of quantum mechanics." In the second part he gave his argument about the impossibility of action-angle variables for several indistinguishable particles, together with a reference to Heisenberg's observability principle:

> In Heisenberg's matrix mechanics it is assumed that the elements of the matrices that represent the dynamical variables determine the frequencies and intensities of the components of the radiation emitted. The theory thus enables one to calculate just those quantities that are of physical importance, and gives no information about quantities such as orbital frequencies that one can never hope to measure experimentally. We should expect this very satisfactory characteristic to persist in all future development of the theory.[75]

Along the same line, Dirac argued that for a pair of (noninteracting) electrons a *single* wave function had to be associated with a given pair of individual stationary states $m$ and $n$, in order that no distinction could be made between the transitions $mn \to m'n'$ and $nm \to m'n'$. Accordingly, the simplest type of wave function for the compound system had to be

$$\psi_{nm}(q^{(1)}, q^{(2)}) = \psi_m(q^{(1)})\psi_n(q^{(2)}) \pm \psi_n(q^{(1)})\psi_m(q^{(2)}). \tag{72}$$

In general, Dirac took the wave function of a system of identical particles to be a totally symmetrical or antisymmetrical function of the positions of the particles. The symmetrical case led to the Bose-Einstein statistics, while the antisymmetrical one led to Pauli's exclusion principle and to the Fermi gas, according to procedures that have now become standard.

75. Dirac 1926f, 666–667.

Up to that point, Dirac's recourse to the Schrödinger equation was simply a new way to achieve a specific matrix representation of his $q$-numbers. In a subsequent paper on the Compton effect he noted: "The wave equation is used merely as a mathematical help for the calculation of the matrix elements, which are then interpreted in accordance with the assumptions of matrix mechanics." The last part of "On the theory of quantum mechanics" contained, however, a first departure from this limited viewpoint. There Dirac considered the case of a time-dependent Hamiltonian, more specifically the one corresponding to an atom subjected to an electromagnetic field. He calculated the function $\psi$ through a now standard perturbative method, the first step being the development of the wave function in terms of the stationary solutions of the unperturbed Schrödinger equation:

$$\psi = \sum_n c_n \psi_n. \tag{73}$$

Then he interpreted the squared modulus $|c_n|^2$ as the number of atoms to be found in the stationary state $n$, when a large assembly of atoms is subjected to the perturbation. "We take $|c_n|^2$ instead of any function of $c_n$," he commented, "because ... this makes the total number of atoms remain constant." Indeed, the hermiticity of the Hamiltonian operator implies the constancy of $\sum_n |c_n|^2$.[76]

With this rule Dirac derived Einstein's $B$ coefficients for induced atomic transitions. Quantum mechanics, once equipped with Schrödinger's equation, was now able to say something about radiation processes (although not yet anything about spontaneous emission). The procedure leading to this progress once again had some resemblance to Eddington's principle of identification. A new equation, the time-dependent Schrödinger equation, was first introduced by formal considerations; then a permanence property, the conservation of the norm $\sum_n |c_n|^2$ of the wave function, oriented the discussion of the physical meaning of the solutions.

## TRANSFORMATIONS

After this quick and rich crop of fundamental results, Dirac pondered about the general interpretation of quantum mechanics. There were basically two ways to draw observable information from the fundamental

76. Dirac 1926g, note added in proof (to compare with Gordon's approach), 507; Dirac 1926f, 674. Dirac sent this paper for publication in late August 1926, that is, a month after Born gave a talk at the Kapitza club (29 July 1926) on "collisions in wave mechanics," with the same interpretation of $|c_n|^2$ (see further on pp. 334–335). The absence of any reference to Born's considerations in Dirac's paper suggests that he reached his conclusions independently.

equations of quantum mechanics. One could either construct matrices that would then be interpreted à la Heisenberg, or one could try to guess a direct interpretation of Schrödinger's $\psi$. As we just saw, Dirac initially favored the first approach, although with his interpretation of the $c$'s he already had started using the second.

## PARTIAL INTERPRETATIONS

In the last semester of 1926, Dirac became aware of several new points of contact between theory and observation, derived either from the matrix camp or from the wave camp. Originally, Schrödinger regarded the $\psi$-function as describing some substantial oscillation within the atom. He quickly realized, however, that this "heuristic viewpoint" could not be taken too literally, since for a quantum system involving more than one particle, the oscillations no longer occurred in the ordinary physical space but in the $3n$-dimensional configuration space associated with the $n$ particles. In his fourth installment of wave mechanics, completed in June 1926, he therefore reinterpreted the wave function in the following way:

> $|\psi|^2$ is a kind of *weight-function* in the configuration space of the system. The *wave-mechanical* configuration of the system is a *superposition of many*— strictly speaking, of *all*—configurations allowed in the mechanics of a point. . . . For the ones who like paradoxes, the system can be said to occupy all kinematically conceivable positions at the same time, but not "in the same degree."[77]

Nevertheless, Schrödinger still regarded the fluctuation of the electric density (in ordinary space) calculated through $\psi$ as real, certainly more real than the one attached to the corpuscular picture. There was in his opinion more truth in the continuous evolution of the $\psi$ than in the quantum leaps of the matrix theory.[78]

In the same month, Max Born compromised between particles and waves. His purpose was to give a quantum-mechanical treatment of collisions between atoms and particles. As we saw in part B (p. 253), after the Bothe and Geiger experiment disproving the BKS theory he had tried with Jordan to come back to Einstein's and Slater's original idea of a wave guiding light quanta. Born had given up this attempt for about a year when he decided to use it as a heuristic analogy for the wave-mechan-

---

77. Schrödinger 1926e, 135.
78. Schrödinger's expression for the electric density in the physical space was (in Dirac's style of notation)

$$\rho(\mathbf{r}) = \int |\psi(\mathbf{r}_1, \mathbf{r}_2, \ldots, \mathbf{r}_n)|^2 \left( \sum_i e\delta(\mathbf{r} - \mathbf{r}_i) \right) d^{3n}r.$$

ical collision problem. Just as free light quanta were guided by a plane monochromatic "ghost" wave, the free asymptotic motion of colliding particles had to be represented by a plane monochromatic Schrödinger wave; the distribution of scattering angles was obtained by developing the scattered wave in terms of such plane waves:

$$\psi_{out} = \sum_{k} a(k)e^{ik \cdot r}. \tag{74}$$

After some hesitation Born decided that $|a(k)|^2$ (not $|a(k)|$) would give the scattering *probability* in the direction of k. In a slight generalization, if $\psi$ is developed in a basis of stationary states $\psi_n$ according to

$$\psi = \sum_{n} c_n \psi_n, \tag{75}$$

then $|c_n|^2$ determined the relative probability of the state $n$ (Born used the word *Häufigkeit*, which refers to the statistical conception of probability), as in Dirac's perturbation theory. Accordingly, quantum mechanics gave no deterministic prediction of an *observable* quantity, the scattering angle. Born believed this feature to be fixed and central, in conformity with his earlier intuition that the world was essentially a kind of lattice (see part B, p. 196): "I myself am inclined to give up determinism in the atomic world." For the consolation of deterministic thinkers he then introduced the notion of statistical determinism: "The motion of particles is ruled by probability laws, but the probability itself propagates in accordance with the causal laws."[79]

In a letter to Heisenberg of 19 October 1926, Pauli combined Born's idea of a probability wave with Schrödinger's "weight functions" in configuration space: the expression

$$|\psi(q_1, q_2, \ldots, q_n)|^2 \, dq_1 \, dq_2 \ldots dq_n,$$

he wrote, had to give the *probability* for the system to be found in the configuration $q_1, q_2, \ldots, q_n$ (actually within a little volume $dq_1 \, dq_2 \cdots dq_n$). Closer to Born's collision case, one could also build a probability density $|\tilde{\psi}(p_1, p_2, \ldots, p_n)|^2$ in the momentum space by taking the multiple Fourier transform of $\psi$. The latter step was naturally suggested by the existence of a dual form of the Schrödinger equation enjoying the same conservation property as the original equation:[80]

$$H\left(i\hbar \frac{\partial}{\partial p}, p\right)\tilde{\psi}(p) = i\hbar \frac{\partial}{\partial t} \tilde{\psi}(p), \tag{76}$$

79. Born 1926a, 867 for the first quotation; Born 1926b, 804 for the reference to Einstein's *Gespensterfeld*, 827 for the second quotation. See Konno 1978.
80. Pauli to Heisenberg, 19 Oct. 1926, *PB*, no. 143.

for which

$$\frac{d}{dt} \int |\tilde{\psi}(p)|^2 \, d^n p = 0. \tag{76'}$$

Pauli further remarked that Born's scattering probabilities were intimately connected with a special matrix already introduced in the Born-Heisenberg-Jordan paper. Up to a proportionality coefficient they were the matrix elements of the interaction potential with respect to the stationary states of the unperturbed Hamiltonian. This suggested a deep-lying connection between Heisenberg's original interpretation of matrices and the new probabilistic interpretation of Schrödinger's waves.

During the following month Heisenberg made some progress in clarifying the connection. His motivation was to show that Schrödinger's continuum theory was unsuited for giving a correct intuitive understanding of the internal energy fluctuations taking place when two atoms are interacting. Suppose two identical atoms originally in the stationary states $m$ and $n$ to be weakly coupled. A resonant interaction takes place, which according to Heisenberg corresponds to discontinuous exchanges of the energy values $E_m$ and $E_n$ between the two atoms. In this view the energy of one of the atoms can take only the values $E_m$ and $E_n$, whereas in Schrödinger's view it can take all intermediate values given by the continuous $\psi$-evolution.[81]

There was a way, Heisenberg argued, to decide between the two conceptions. Matrix mechanics could certainly not give the energy $H_1$ of one of the atoms as a function of time, but it was nonetheless able to determine the probability for this energy to take a given value. One just had to assume that, for any dynamic variable, the diagonal elements of the corresponding matrix gave its average value in the various stationary states of the global system; according to this rule, one could derive the average value of any function of $H_1$, for example the moments $\bar{H}_1, \overline{H_1^2}, \overline{H_1^3}, \ldots$ and therefore the probability distribution of $H_1$. The result read

$$\overline{f(H_1)} = \tfrac{1}{2}[f(E_m) + f(E_n)] \tag{77}$$

in conformity with Heisenberg's intuition of discontinuous switches.

Moreover, the probability distribution of $H_1$ was explicitly given by the squared moduli of the elements of a matrix $S$ that had been introduced in the three-men paper and related the matrices $g$ of the dynamic variables of the unperturbed system (no coupling between the two atoms) to the

81. Heisenberg 1926c.

matrices $G$ of the corresponding variables of the perturbed system according to

$$G = SgS^{-1}. \tag{78}$$

There is no need here to give the details of Heisenberg's argument, since they will result from Dirac's much more general investigation of relations of the above type.

THE FORMAL APPARATUS

When he set out to elaborate his own interpretation of quantum mechanics, Dirac was aware of Schrödinger's fourth memoir, had heard Born speak on collisions at the Kapitza club (on 29 July 1926), and had seen the manuscript of Heisenberg's fluctuation paper. He was dissatisfied with the multiplicity of partial, disconnected, and sometimes contradictory interpretations. But he saw an important advance in Heisenberg's considerations, for they indicated how to derive probability distributions from the original matrix formulation of quantum mechanics.[82]

However, Heisenberg had limited his interpretative inquiry to simple examples, a result of his search for an intuitive understanding of quantum mechanics. Dirac, following his Eddingtonian bent, returned to his fundamental equations of quantum mechanics and formulated the preliminary interpretative problem very generally: as the search for $c$-numbers connected with the $q$-numbers satisfying these equations. Since the action-angle variables could no longer help in this problem, one needed to consider *all* matrix systems capable of representing the $q$-numbers, without including the restriction that action variables should be diagonal. Even Heisenberg's limitation to a diagonal (total) energy matrix had to be avoided, since it introduced a premature interpretative element.

The fundamental equations read

$$[q, p] = i\hbar, \tag{79}$$

$$[g, H] = i\hbar\dot{g} \tag{79'}$$

(in order to present the equations more compactly, they will be written for the case of one degree of freedom, even though the discussion will refer to the general case). Dirac first noticed that one goes from one matrix

82. Dirac 1927a, 621. Dirac's transformation theory is discussed in Jammer 1966, 300–305, and Kragh 1990, 38–42.

representation of these equations to another through a transformation of the type

$$G = bgb^{-1}. \tag{80}$$

Interestingly, Dirac had learned much earlier from Göttingen's theoreticians that a relation of this type existed between any two sets of canonical coordinates (in Heisenberg's representation). But at that time he believed the remark to be of no practical use, for it had no obvious classical counterpart.[83] In late 1926 he was completely freed from the inhibitory effects of his desire to maintain a close classical analogy, and regarded instead the form (80) of transformation as most essential.

Representations are likely to be of physical interest, Dirac went on, only if they make a given set $\xi$ of dynamic variables diagonal. The set $\xi$ is said to be complete if there is only one representation for which it is diagonal. Dirac found it convenient to represent the transformations from one representation to another by symbols $(\xi'/\alpha')$ with $c$-number values, wherein $\xi'$ and $\alpha'$ represent the diagonal elements of two complete sets $\xi$ and $\alpha$. In this notation the equation (80) becomes

$$g_{\alpha'\alpha''} = \int (\alpha'/\xi')g_{\xi'\xi''}(\xi''/\alpha'') \, d\xi' \, d\xi''. \tag{81}$$

(to the mathematicians' horror, $g_{\alpha'\alpha''}$ was *not* to be regarded as the same function of $\alpha'\alpha''$ as $g_{\xi'\xi''}$ is of $\xi'\xi''$).

Although this type of formula seems to be limited to the case of continuous spectra, Dirac took it to also cover discrete and mixed spectra, assuming the integral to represent a sum in the discrete case. The continuous case itself called for a few mathematical tricks. First of all, in order to be able to write the transformation $(\xi'/\xi'')$ corresponding to the identity $(b = 1)$, Dirac introduced the now famous "$\delta$-function," some kind of limit of sharply peaked functions such that

$$\int f(x) \, \delta(x) \, dx = f(0) \tag{82}$$

for any (regular) function $f$. Then the choice

$$1_{\xi'\xi''} = (\xi'/\xi'') = \delta(\xi' - \xi'') \tag{83}$$

makes (81) an identity, as required. As usual, Dirac did not worry about a rigorous mathematical construction (Laurent Schwarz's distributions later provided such a construction). To him as to his precursor Heaviside,

---

83. Dirac 1926a, 564–565: "These formulae do not appear to be of great practical value."

it was sufficient that the symbolic manipulations of the $\delta$-function did not lead to contradictions.[84]

The next easier matrix after the identity is the one representing $\xi$ in its own scheme. It must be a diagonal matrix with elements $\xi'$, which leads to the expression

$$\xi_{\xi'\xi''} = \xi'\delta(\xi' - \xi''), \tag{84}$$

assuming the $\delta$-function to generalize Kronecker's symbol $\delta_{ij}$ in the continuous case. Dirac further showed that the matrix $\eta_{\xi'\xi''}$ when $\eta$ is canonically conjugate to $\xi$ is given by

$$\eta_{\xi'\xi''} = -i\hbar\delta'(\xi' - \xi''). \tag{85}$$

The proof required a few $\delta$-gymnastics performed on the commutator $[\xi, \eta]$:[85]

$$\begin{aligned}
[\xi, \eta]_{\xi'\xi''} &= \int \xi_{\xi'\xi'''}\eta_{\xi'''\xi''}\, d\xi''' - \int \eta_{\xi'\xi'''}\xi_{\xi'''\xi''}\, d\xi''' \\
&= -i\hbar \int \xi'\delta(\xi' - \xi''')\delta'(\xi''' - \xi'')\, d\xi''' \\
&\quad + i\hbar \int \delta'(\xi' - \xi''')\xi'''\delta(\xi''' - \xi'')\, d\xi'''. \tag{86}
\end{aligned}$$

Performing an integration by part on the last integral gives

$$i\hbar \int [\delta(\xi' - \xi''')(\delta(\xi''' - \xi'') + \xi'''\delta'(\xi''' - \xi''))]\, d\xi'''.$$

The second term of this integral cancels the first integral in (86), while the first provides $i\hbar\delta(\xi' - \xi'')$, that is, $i\hbar$ times the identity matrix, which completes the proof.

On the basis of the general transformation formula Dirac could calculate the matrices $\xi_{\xi'\alpha'}$ and $\eta_{\xi'\alpha'}$ in a representation where the rows refer to $\xi$ and the columns to $\alpha$:

$$\xi_{\xi'\alpha'} = \int d\xi'' \xi_{\xi'\xi''}(\xi''/\alpha') = \xi'(\xi'/\alpha'), \tag{87}$$

$$\eta_{\xi'\alpha'} = -i\hbar \int d\xi'' \delta'(\xi' - \xi'')(\xi''/\alpha') = -i\hbar \frac{d}{d\xi'}(\xi'/\alpha'). \tag{87'}$$

84. Heaviside introduced discontinuous functions like "Heaviside's step function" (which he designated as $Q$), and also its time derivative, that is, a $\delta$-function (which he designated as $pQ$, with $p = d/dt$), to help the solution of linear inhomogeneous differential equations, for instance in Heaviside 1893, 133 of the Dover reprint (1950). Dirac had used $\delta$-functions in several anterior publications, but without a specific notation.

85. Dirac's multiplications of singular distributions can be justified from the modern point of view, because they occur only within convolutions.

More generally, any function $g(\xi, \eta)$ expressible as a sum of products of $\xi$ and $\eta$ allowed the mixed representation:

$$g(\xi, \eta)_{\xi'\alpha'} = g\left(\xi', -i\hbar\frac{\partial}{\partial\xi'}\right)(\xi'/\alpha').$$  (88)

In order to prove this identity, it is sufficient to show that when it holds for $f$ and $g$, it also holds for $f + g$ and $fg$. The first part of the latter assertion being trivial, we are left with the second one:

$$
\begin{aligned}
(fg)_{\xi'\alpha'} &= \int d\xi'' f_{\xi'\xi''} g_{\xi''\alpha'} \\
&= \int d\xi'' \left[ f\left(\xi', -i\hbar\frac{\partial}{\partial\xi'}\right)\delta(\xi' - \xi'')\right] g\left(\xi'', -i\hbar\frac{\partial}{\partial\xi''}\right)(\xi''/\alpha') \\
&= \int d\xi'' f\left(\xi', -i\hbar\frac{\partial}{\partial\xi'}\right)\left[\delta(\xi' - \xi'')g\left(\xi'', -i\hbar\frac{\partial}{\partial\xi''}\right)(\xi''/\alpha')\right] \\
&= f\left(\xi', -i\hbar\frac{\partial}{\partial\xi'}\right)g\left(\xi', -i\hbar\frac{\partial}{\partial\xi'}\right)(\xi'/\alpha').
\end{aligned}
$$  (89)

The identity (88) played a central role in Dirac's transformation theory. As a first outstanding application of it, let us choose $g$ to be the Hamiltonian $H$, $\xi$ the position coordinates, and $\alpha$ a complete set of dynamic variables commuting with $H$:

$$H_{q'\alpha'} = H\left(q', -i\hbar\frac{\partial}{\partial q'}\right)(q'/\alpha').$$  (90)

Using a relation similar to (87), we also have

$$H_{q'\alpha'} = E_{\alpha'}(q'/\alpha'),$$  (91)

and, therefore,

$$H\left(q', -i\hbar\frac{\partial}{\partial q'}\right)(q'/\alpha') = E_{\alpha'}(q'/\alpha').$$  (92)

In this way Dirac could have discovered the Schrödinger equation before Schrödinger, if only he had earlier exploited the freedom of representation of his fundamental equations. At least he was able to realize a posteriori that Schrödinger's $\psi$ was nothing but the transformation from a scheme in which the position variables are diagonal to one in which the energy is diagonal. It further appears that there are as many equations

$$H\left(\xi', -i\hbar\frac{\partial}{\partial\xi'}\right)(\xi'/\alpha') = E_{\alpha'}(\xi'/\alpha')$$  (93)

as there are choices of $\xi$, since the above deduction is not limited to the case of position coordinates.

Finally, the Dirac-Schrödinger evolution for a transformation $(\xi'/\alpha')$, when $\alpha$ is a constant of the motion $(\dot{\alpha} = 0)$,

$$i\hbar \frac{\partial}{\partial t} (\xi'/\alpha') = H\left(\xi', -i\hbar \frac{\partial}{\partial \xi'}\right)(\xi'/\alpha'), \qquad (94)$$

follows from the fundamental equation

$$i\hbar \frac{d}{dt} f(\xi) = [f(\xi), H]. \qquad (95)$$

This is easily proved by calculating the matrix elements of the two members of this equation in the $\alpha$-scheme:[86]

$$i\hbar \dot{f}_{\alpha'\alpha''} = i\hbar \frac{d}{dt} \int f(\xi') \, d\xi' (\alpha'/\xi')(\xi'/\alpha'')$$

$$= i\hbar \int f(\xi') \, d\xi' \left[ (\alpha'/\xi') \frac{\partial}{\partial t} (\xi'/\alpha'') + (\xi'/\alpha'') \frac{\partial}{\partial t} (\alpha'/\xi') \right], \qquad (96)$$

while

$$[f(\xi), H]_{\alpha'\alpha''} = \int f(\xi') \, d\xi' [(\alpha'/\xi')H_{\xi'\alpha''} - H_{\alpha'\xi'}(\xi'/\alpha'')]. \qquad (97)$$

The function $f$ and the values of $\alpha'$ and $\alpha''$ being arbitrary, the expressions are identical if and only if

$$\left. \begin{array}{l} i\hbar \dfrac{\partial}{\partial t} (\xi'/\alpha') = H_{\xi'\alpha'} \\[2mm] -i\hbar \dfrac{\partial}{\partial t} (\alpha'/\xi') = H_{\alpha'\xi'} \end{array} \right\}. \qquad (98)$$

These two equations are equivalent in any Hermitian scheme for which

$$(\xi'/\alpha')^* = (\alpha'/\xi') \qquad \text{and} \qquad H^*_{\xi'\alpha'} = H_{\alpha'\xi'}. \qquad (99)$$

Combined with the identity (88), they lead to the time-dependent Schrödinger equation (94), as stated.

The formal apparatus of quantum mechanics was now thoroughly unified, Schrödinger's equation being harmoniously blended with the fundamental equations ruling $q$-numbers. According to Dirac's own criteria, the whole theory was impressively beautiful, for it displayed a transformation apparatus as elegant and powerful as those of relativity or Hamiltonian dynamics.

86. The following proof differs slightly from Dirac's.

Dirac was now ready to attack the interpretation problem proper. He started with the following words:

> To obtain physical results from the matrix theory, the only assumption one needs make is that the diagonal elements of a matrix, whose rows and columns refer to the $\xi$'s say, representing a constant of integration, $g$ say, of the dynamical system, determine the average values of the function $g(\xi, \eta)$ over the whole of $\eta$ space for each particular set of numerical values for the $\xi$'s in the same way in which they certainly would in the limiting case of large quantum numbers.[87]

There is an unfortunate obscurity in what Dirac meant by the "limiting case" here. The following is a very plausible interpretation.

As a first remark, the term "constant of integration" in the above extract is misleading. Dirac just means that he is considering a dynamic variable *at a given instant of time*, as made clear by an earlier footnote in the paper. If so, the matrix element $g_{\xi'\xi'}$ is independent of the choice of the Hamiltonian $H$. Its interpretation is obtained by exploiting this independence and provisionally considering the Hamiltonian to commute with all the $\xi$'s. Then $g_{\xi'\xi'}$ represents the time average of $g$, according to Heisenberg's previous interpretation of matrices, or from a comparison with the high quantum-number limit in the Bohr-Sommerfeld theory; besides, the conjugate variables $\eta$ are, in the Bohr-Sommerfeld theory, phases (or "angles" in the action-angle formulation) varying linearly in time, so that the time-averaging is here identical with an $\eta$-averaging. In the general case for which $H$ and $\xi$ do not necessarily commute, it is therefore natural to assume that $g_{\xi'\xi'}$ represents the average of $g$ when $\xi = \xi'$ and $\eta$ is uniformly spread.

Very cleverly, Dirac deduced a complete interpretation of the quantum formalism from this seemingly limited assumption. His magic wand was the identity

$$\delta(g - g')_{\xi'\xi'} = |(g'/\xi')|^2, \qquad (100)$$

which simply results, for any complete set $g$ of dynamic variables, from

$$\delta(g - g')_{\xi'\xi'} = \int dg''(\xi'/g'')\delta(g'' - g')(g''/\xi'),$$
$$= (\xi'/g')(g'/\xi') = |(g'/\xi')|^2. \qquad (100')$$

The function $\delta(g - g')$ is nonzero only for $g \sim g'$. Therefore its "$\eta$-average" for $\xi = \xi'$ is nothing but the fraction of the $\eta$-space for which $g = g'$ when

---

87. Dirac 1927a, 637.

$\zeta = \zeta'$. In other words $|(g'/\zeta')|^2$ is the relative probability that $g = g'$ knowing that $\zeta = \zeta'$. According to Dirac, this answered all the questions "to which the quantum theory [could] give a definitive answer." These questions, he added, were "probably the only ones to which the physicist could give an answer."[88]

## WELCOME TO COPENHAGEN

Dirac completed his transformation theory in December 1926 in Copenhagen, where he was treated like a hero. There he learned that Pascual Jordan had formulated a similar theory, though from a different point of view. Instead of studying transformations from one matrix representation of the fundamental equations to another, Jordan examined transformations from one canonical pair, say $(\zeta, \eta)$, to another, say $(\alpha, \beta)$, in a given representation. His theory was similar to Dirac's insofar as it led to unitary operators $b$ generating the canonical transformations according to

$$\alpha = b\zeta b^{-1}, \qquad \beta = b\eta b^{-1}. \tag{101}$$

Superficially, this point of view might have seemed closer to the transformation theory of classical dynamics, which also related canonical pairs. In reality Jordan departed much more from the classical model than did Dirac: he defined canonical conjugation not by commutation rules transposed from the Poisson algebra but by broader axioms at the quantum level. For instance, anticommuting variables like the spin operators $S_x$, $S_y$ were conjugate in Jordan's sense.[89]

Dirac could not be sympathetic to such a wild deviation from classical canons. Later he even rejected some interesting products of Jordan's conception like the notion of anticommuting quantized fields.[90] One might wonder, however, why he did not view transformations as relating canonical pairs, since this was, after all, the conception that dominated his own work before the advent of Schrödinger's equation. The reason might well have been psychological: to somebody who had just discovered that his dear uniformizing variables failed to solve the problem of atoms with more than one electron, other canonical transformations had little chance to stand at the foreground of a fundamental exposition of quantum mechanics. That Jordan somehow succeeded in adopting this conception did not persuade Dirac to change his position. In his first lectures on quantum

88. Ibid., 641.
89. Jordan 1927a, 1927b; Dirac compares the two approaches in Dirac to Jordan, 24 Dec. 1926, AHQP.
90. See Darrigol 1986.

mechanics he did try to lay out his competitor's theory, but he quickly returned to his own, which he found simpler and more elegant.[91]

There is one feature of Dirac's original transformation theory that is likely to surprise the modern quantum physicist: the notion of state vector is completely absent. It was in fact introduced later by Weyl and von Neumann, and subsequently adopted by Dirac himself. In 1939 Dirac even split his original transformation symbol $(\xi'/\alpha')$ into two pieces $\langle \xi' |$ and $| \alpha' \rangle$, the "bra" and the "ket" vectors.[92] The mathematical superiority of the introduction of state vectors is obvious, since it allows—albeit not without difficulty—an explicit construction of mathematical entities (rigged Hilbert spaces) that justify Dirac's symbolic manipulations. There was also a more physical advantage to the notion of state vector: it placed the superposition principle in the foreground, which pleased Bohr, who set wave-particle duality at the core of complementarity. Perhaps modern-day interpreters of quantum mechanics should nevertheless remember that there exists a formulation of quantum mechanics without state vectors, and with transition amplitudes (transformations) only.

In this original conception Dirac had nothing to call the state of a system, except the old $(q, p)$ configuration. This state of affairs conditioned his conclusion to "The physical interpretation of the quantum dynamics":

> It may be mentioned that the present theory suggests a point of view for regarding quantum phenomena rather different from the usual ones. One can suppose that the initial state of a system determines definitively the state of the system at any subsequent time. If, however, one describes the state of the system at an arbitrary time by giving numerical values to the coordinates and momenta, then one cannot actually set up a one-one correspondence between the values of these coordinates and momenta initially and their values at a subsequent time. All the same one can obtain a good deal of information (of the nature of averages) about the values at the subsequent time considered as functions of the initial values. The notion of probabilities does not enter into the ultimate description of mechanical processes; only when one is given some information that involves a probability (e.g. that all points in $\eta$-space are equally probable for representing the system) can one deduce results that involve probabilities.[93]

Such a view was still too conservative to please Copenhagen authorities. Once again Dirac was charged with having overplayed the classical anal-

91. Dirac [1927c].
92. Dirac 1939b. The *Principles* (Dirac 1930), already in the first edition, introduced the state vectors before the transformations.
93. Dirac 1927a, 641.

ogy. Nevertheless, the success of the transformation theory was immediate, with respect to both interpretation and application of quantum
mechanics. From the transformation connecting two conjugate variables,

$$(q/p) = \frac{1}{\sqrt{2\pi\hbar}}\, e^{ipq/\hbar}, \tag{102}$$

Heisenberg deduced the uncertainty relations; and he showed that the resulting limitations in the definition of conjugate quantities exactly corresponded to the concrete limitations of double measurement processes. On
the more practical side, the transformation theory gave a general method
of quantizing everything, since, contrary to Heisenberg's original matrix
mechanics, it was completely independent of the nature of the dynamic
system under consideration. Dirac's radiation theory, published in early
1927, was the first of a long list of spectacular successes resulting from
this method.[94]

As Dirac would have said, Nature was being seduced by mathematical
beauty. The transformation theory equaled the aesthetic qualities that he
had earlier contemplated in classical theories. In the fall of 1927 Dirac
explained this to his first students in quantum mechanics:

> The quantum theory has now reached a form . . . in which it is as beautiful,
> and in certain respects more beautiful than the classical theory. This has been
> brought about by the fact that the new quantum theory requires very few
> changes from the classical theory, these changes being of a fundamental nature,
> so that many of the features of the classical theory to which it owes its attrac-
> tiveness can be taken over unchanged into the quantum theory.[95]

## SUMMARY AND CONCLUSIONS

In March 1926 Schrödinger published the first of a series of memoirs in
which he tried to reduce atomic theory to a mechanics of matter waves
à la de Broglie. Dirac's first reaction was negative, for he already had
placed his hopes in another quantum mechanics. Yet, taking a suggestion
of Heisenberg, he promptly exploited the Schrödinger equation, if only
as a "mathematical help in calculating matrix elements." In the summer
of 1926 he thus reached some of the basic notions of modern quantum

94. Heisenberg 1927b. Equations (87) and (88) for $\xi = q$ and $\alpha = p$ give together a differential equation leading to equation (102). Dirac 1927b. On subsequent applications of
transformation theory see Kragh 1990 and Darrigol 1984, 1986.

95. Dirac 1927c, introduction.

mechanics. Most important, in the name of Heisenberg's principle of observability he introduced symmetrical and antisymmetrical wave functions of the configuration of a set of identical particles, and proceeded to connect respective symmetry classes with Bose's statistics on the one hand, and Pauli's exclusion principle on the other. This provided the general basis both for quantum statistics and for the calculation of properties of atoms with several electrons (but Heisenberg was the one who first solved the helium atom within the new mechanics).

In another important innovation, Dirac presented a method for treating time-dependent perturbations; with it he could derive general expressions for Einstein's transition probabilities. In this case he gave the Schrödinger waves a more direct interpretation (not via Heisenberg's matrices), as a means to calculate the (statistical) probability of the system to be in a given stationary state. Characteristically, he presented this interpretation as being suggested by an invariance property (the invariance of the norm of the wave function).

In the summer of 1926 there were several other contributions to the interpretation of the two or three distinct quantum formalisms that had arisen, proposed both by Schrödinger and by the Göttingen group. Schrödinger, retreating somewhat from his original mechanistic conception of matter waves, now regarded the wave function as a (nonstatistical) sort of "weight function" in the configuration space of the system. While treating the problem of particle scattering, Born related the (outgoing) wave function to the scattering probability. Pauli, blending Schrödinger's notion of a weight function with Born's statistical conception, interpreted the squared modulus of the wave function as giving the probability for the system to be found in a specified configuration, and he vaguely suggested a connection between these probabilities and the transition probabilities of Heisenberg's theory. Finally, Heisenberg made the latter connection entirely explicit within the context of a suggestive example, the energy fluctuation of an atom when coupled with an identical atom (quantum-mechanical resonance): the relevant probability was obtained by taking the squared modulus of the elements of the unitary matrix connecting the stationary states of the coupled system to those of the uncoupled system.

In the fall of 1926 Dirac decided to bring some order to this proliferation of partial interpretations. In accordance with Eddingtonian methodology, and in contrast with Heisenberg, he did not start from specific physical examples but explored instead the transformation properties of the fundamental equations of his quantum mechanics (still the ones he had devised in the fall of 1925). He was now in a position to fully exploit the freedom of representation of $q$-numbers inherent in his conception of quantum algebra. Dropping Heisenberg's restriction to a matrix scheme in which the energy matrix is diagonal, he studied the general set of bilinear transformations mutually connecting all possible matrix schemes and proved by symbolic means that Schrödinger's wave function was just a particular case of transformation. This showed that both matrix mechanics and wave mechanics were implicitly contained in his fundamental equations.

For the interpretation of his general formalism, the only assumption Dirac made was that there existed a limited "correspondence" with classical theory. Through an extremely ingenious argument based on transformation properties, he could show that this assumption was sufficient to derive a general interpretation of matrices and transformations. The standard question Dirac's theory could answer, in the simple case of one degree of freedom, was: What is, for a given value of a dynamic coordinate $q$, the relative probability of the values which a given dynamic quantity can take if the conjugate coordinate $p$ is uniformly distributed? Dirac believed this was the only type of question physicists could answer. In this sense his interpretation was a statistical one; influenced, however, by the classical analogy, he still imagined the state of the system to be represented (at a given time) by definite coordinates $q$ and $p$. In this view, the theory of transformations just implied that it was fundamentally impossible to predict unambiguously the state of the system at a subsequent time.

Bohr and Heisenberg soon persuaded Dirac to give up the "fiction" of a definite $p$ and $q$. Accordingly, in his later presentations of quantum mechanics Dirac abandoned the notion of a $(q, p)$ state, and adopted the formal notion of "state vector" proposed by the Göttingen mathematicians (Weyl and von Neumann).

A transformation theory partly and formally similar to Dirac's was simultaneously invented by Jordan in Göttingen. In general, Jordan tended to build quantum mechanics on autonomous axioms, without reference to classical theory. As a result, his notion of canonical conjugation, in contrast with Dirac's, did not necessarily correspond to the classical one. However, his idea of a transformation was in some respects closer to the classical idea of canonical transformation than Dirac's was. In this case, Dirac distanced himself from the classical analogy, presumably because his early attempts to adapt canonical transformations in quantum mechanics had become stagnant in the spring of 1926. At any rate, Dirac's transformation theory was more elegant than Jordan's; it was easier to apply (at least in its creator's hands), and as a result of the classical analogy it involved a more restrictive concept of canonical conjugation. These virtues explain to a large extent the "miracles" Dirac subsequently performed in the contexts of radiation theory and relativistic quantum mechanics.

That Dirac's success owed much to the classical analogy was obvious and explicit. It remains to be seen to what extent his use of analogies was related to the correspondence principle. Dirac discovered the connection between Poisson brackets and commutators by examining Kramers and Heisenberg's procedure of symbolic translation, which itself derived from a sharpening of the correspondence principle. And his use of classical mechanics as a template for the construction of the new theory can be seen as a mathematical version of Bohr's continual appeal to formal analogies between classical and quantum theory. In this sense Dirac fulfilled Bohr's old prophecy of a "rational generalization" of the classical theory.

However, some essential aspects of Dirac's method were foreign to Bohr's strategy of correspondence. For instance, the way Dirac connected the algebra of Poisson brackets with the algebra of commutators was more similar to the mathematicians' notion of isomorphism than to the formal or symbolic analogies cultivated in Copenhagen. Further, in his approach to the transformation theory Dirac was inspired by another type of classical analogy, one in which he tended to imitate the relativistic strategy of theory building. As we saw, he first developed, at an abstract level, the transformation properties of his fundamental equations and then used these properties in identifying the physical content of the formalism. Since to him this was the royal road to fundamental theories, his greatest satisfaction as a theoretical physicist was obtained in the creation of the transformation theory. More generally, he relished the thought that quantum mechanics, in its genesis and expression, could compete in beauty with the greatest classical monuments, Hamiltonian mechanics and general relativity.

# Bibliographical Guide

As a history of quantum theory, the present book covers only a modest part of its subject. For instance, very little is said of the experimental and theoretical developments concerning quantum radiation and the quantum gas, which ultimately led to Louis de Broglie's matter waves and Schrödinger's wave mechanics. Those who are entering the rich field of the history of quantum theory for the first time should read a general overview, such as Max Jammer's *Conceptual development* (1966), which remains the best to date. While later research has improved on various aspects of the subject, Jammer's book offers a clear, concise, well-wrought conceptual history.

In a diametrically opposed style, Jagdish Mehra and Helmut Rechenberg have published, starting in 1982, five volumes of an extremely detailed history. This collection saves trips to the library and to the AHQP archive (description below) and sometimes reveals important connections. For example, the fourth volume relates Dirac's $q$-numbers to Baker's principles of geometry (the connection is only vaguely suggested in Dirac's AHQP interview). As I have endeavored to show in part C, this is an index of a more general shaping of Dirac's methods by his early exposure to a Whiteheadian mathematical tradition.

For anyone who wishes to do research on the history of quantum theory, there exists a primary source of exceptional quality and convenience, the Archive for the History of Quantum Physics (AHQP) put together in the early sixties by Thomas Kuhn, Paul Forman, and John Heilbron. Copies of this archive are available in various places, including Berkeley, Chicago, New York, Washington, Copenhagen, London, Rome, and Paris (Musée de la Villette). They contain manuscripts, correspondence, and interviews of the main early quantum physicists (description in Kuhn-Heilbron-Forman-Allen 1967).

In general the most illuminating studies in the history of quantum theory are those which have adopted a relatively narrow focus. The subject of part A of this book, Planck's radiation theory, has been well covered, notably by Martin Klein

(1962, 1963a), Hans Kangro (1970), Thomas Kuhn (1978), Allan Needell (1980, 1988), and John Heilbron (1986). Heilbron's *Dilemmas* provides an insightful analysis of the biographical roots of Planck's goals. Kangro describes the experimental and theoretical context in which Planck's program took place. Klein explains how Planck's innovation at the close of the year 1900 was inscribed in this program. Kuhn contests the traditional interpretation of Planck's energy elements as an intrinsic energy discontinuity. Finally, Needell unveils the inner motivation of most of Planck's work between 1895 and 1914, that is, his persistent belief in absolute thermodynamic irreversibility. Here I have brought Needell's insight to bear on Kuhn's thesis and have shown that any ambiguity in Planck's idea of the energy elements disappears as soon as these elements are understood as an outgrowth of analogies guided by the belief in absolute irreversibility. However, Klein's thesis that Planck did introduce a quantum discontinuity in 1900 should not be too hastily dismissed. One should distinguish Planck's own understanding of the nature of his innovation from the interpretation of his readers at the beginning of the century. In at least two important cases, concerning Ehrenfest and Lorentz, the latter interpretation best fits Klein's account. In short, if Klein misinterpreted Planck's intentions, he is in excellent company. And no one would deny the historical significance of misinterpretations.

No reading will better accompany the second part of this book than Bohr's own writings, which have been conveniently united in his *Collected works,* under the competent direction of Leon Rosenfeld and Erik Rüdinger. The eight volumes published to date contain most of the texts quoted in this book, a selection of letters, and highly valuable introductions and comments. Some of the manuscripts I have used are unpublished, but they can be found in the AHQP archive (or in the Bohr archive in Copenhagen).

Strangely enough, the correspondence principle has not been studied as extensively as other aspects of the history of quantum theory. Perhaps historians were long-term victims of the common prejudice against this principle, which is often seen as an irrational piece of heuristics or, worse, as an a posteriori simulation of such heuristics. Notable exceptions are Meyer-Abich (1965), Jammer (1966), Stolzenburg (1977, 1985), Wassermann (1981), Mehra and Rechenberg (1982), Petruccioli (1988). Klaus Meyer-Abich's essay is hardly historical, but it contains deep philosophical insights into the nature of the correspondence principle and its versatility. Jammer had little space to analyze the roots and the applications of this principle, but he does properly acknowledge its historical significance. Stolzenburg (1977) is mainly concerned with the relation between correspondence and complementarity. His account of the BKS episode (1985) is, together with Klein's (1970a), one of the best available. Wassermann examined some of the earlier history of the correspondence principle up to its extension in the BKS theory. Mehra and Rechenberg seem in general to carry little sympathy for Bohr's way of thinking, but, in conformity with Heisenberg's own recollections, they emphasize the role of the correspondence principle in the genesis of matrix mechanics.

Of all historians Sandro Petruccioli seems to be the one who pays the greatest attention to this principle. He does not detail the origins and technical applications of the "correspondence," but he senses the rationality of Bohr's strategy. Also his parallel with the notion of scientific metaphor recently developed by philosophers like Max Black and Richard Boyd is instructive, though, in my opinion, partially misleading. Bohr's correspondence principle did not share one of the commonly acknowledged virtues of constitutive metaphors, namely the furnishing of terms without definition. As I have aimed to show, Bohr tended to ascribe the power of yielding definitions to another principle, the adiabatic principle, or to another metaphor, the orbital picture. Bohr's provisional recourse to a partially classical picture of electronic motion in atoms should not be confused with the correspondence principle, which stated a relation between emitted radiation and atomic motion, whatever this motion could be.

In order properly to situate the context in which the correspondence principle originated, thrived, and evolved, I have recounted aspects of the quantum theory which have already been treated in the existing literature, although with different aims and perspective. The first phase of the Bohr atom was the object of a penetrating study by Heilbron and Kuhn (1969), to which I had nothing to add. John Heilbron dealt with earlier and later developments of theories of atomic structure in a set of fundamental studies: his dissertation (1964), a series of articles (republished in Heilbron 1981), a biography of Moseley (1974), and finally, a history of the exclusion principle (1983) from which I had nothing to subtract. On Bohr's second atomic theory there also exists an insightful study by the Danish historian Helge Kragh (1979a). However, one essential aspect of this mysterious theory of Bohr's has been generally overlooked: the paradigmatic role of the helium atom, especially regarding the use of the correspondence principle in determining the length of chemical periods.

As is well known, the two foremost schools of quantum theory around 1920 were Bohr's in Copenhagen and Sommerfeld's in Munich. Proper accounts of Sommerfeld's research and teaching are found in Jammer's general history, and in Forman's and Heilbron's more specialized studies on related subjects (see above for Heilbron, and below for Forman). But, with the exception of Ulrich Benz's biography (1975) and Shigeko Nisio's "The formation of the Sommerfeld quantum theory" (1973), no historian has yet chosen Sommerfeld's role as his or her central topic. Perhaps this is because Sommerfeld's style of physics is usually regarded as less problematic and foundational than others'. Let us hope that this illusion will be soon dispelled.

The crisis of the old quantum theory that started in the winter of 1922–23 has been the subject of much historical research. Most helpful are Daniel Serwer's "Unmechanischer Zwang" (1977) and Paul Forman's articles on Landé and on the doublet riddle (1968, 1970). Forman's considerations of Landé's psychology and of his relation to Sommerfeld's seminar are a nice antidote to my dry conceptual account of the "term zoology." The least-studied aspect of the crisis is the failure of the Bohr-Kemble model of helium, although it was, from Born's and Bohr's points of view, and according to an argument developed in this book, the most crucial event of the stormy winter of 1922–23.

Pauli's role in clarifying complex spectral issues and in criticizing existing theories has been especially well documented. Karl von Meyenn's exemplary edition of Pauli's scientific correspondence (1979, 1985) enriches us with accurate annotation and valuable comment. John Hendry's *Bohr-Pauli dialogue* (1984) draws an interesting conceptual thread between Pauli's early instrumentalism and his ultimate rejection of atomic orbits. The title notwithstanding, his vision of the origins of quantum mechanics is more akin to Pauli's than to Bohr's; he says relatively little about the origins and importance of Bohr's views on the correspondence principle. I just hope that my book may offer the complementary perspective.

Other key personalities of the history of quantum theory were Werner Heisenberg, Paul Ehrenfest, and Bohr's assistant of the early years (from 1916 to 1925), Hendrik Kramers. Biographical information on Heisenberg is found in Hermann (1976) and in Mehra and Rechenberg (vol. 2). Some of his early work has been well studied by David Cassidy (1976, 1979), who is now writing a full biography. Martin Klein's biography of Ehrenfest and his "Great connections" (between Bohr, Ehrenfest, and Einstein, 1986) are essential works, which convey to us in a most elegant manner the benefits of Ehrenfest's own insights into Bohr's and Einstein's singularities. Regarding Kramers, Max Dresden's biography (1987) must also be recommended for the clear representation of physical issues, and for the vivid description of the psychosociology of Bohr's guests in Copenhagen.

In recent years several philosophers have studied Bohr's thought, and they have generally recognized the necessity of tracing it back to its origins in the practice of atomic physics. This is especially the case for works bordering between history and philosophy, like those of Miller (1978), Beller (1983), Hendry (1984), and Petruccioli (1988). But it is also true of mostly philosophical essays, like those of MacKinnon (1982), Folse (1985), Honner (1987), Murdoch (1987), and Chevalley (1991a). None of these philosophers has really taken up the task of filling the existing gaps in the history of Bohr's theory. However, the questions they asked about the roots of complementarity—which is of course their main subject of interest—have helped to guide my investigation of the early Bohr. This is especially true of Catherine Chevalley's work; in her contacts with me and with other historians, she has made the communication between history and philosophy a pleasant reality.

An interesting historicophilosophical question is posed by the relation between Bohr and his philosopher friend Harald Høffding. In this book I have barely touched upon this interesting topic. In the past the nature of Bohr's debt to Danish philosophy has sometimes been misconceived. Most Bohr scholars now agree that the main source of Bohr's complementarity was his physics, not the ambient philosophy. Yet Bohr's emphasis on language and his constant recourse to psychological analogies seem to bear the stamp of Høffding's influence, whatever the modalities of this connection (discussed, unfavorably, by David Favrholdt and, favorably, by Jan Faye) might be. Norton Wise (1987) gave the most convincing evidence that Bohr actually drew on Høffding in his old quantum theory (before quantum mechanics).

Among the creators of quantum mechanics, Dirac was for long the one most neglected by historians. This state of affairs changed in 1982 with the pioneering study by Michelangelo de Maria and Francesco La Teana, and with the publication of volume 4 of the Mehra and Rechenberg saga, which contains a great deal of information on Dirac's background and his early career (up to 1925). I have already mentioned some of the merits of this work. More recently, Helge Kragh wrote a remarkable scientific biography of Dirac (1990), which includes information on family background and on school and university training, penetrating accounts of the genesis of his most important theories, and sensible renderings of his psychology and implicit philosophy. I had not yet seen this book when I wrote part C; but I was certainly inspired by some of Kragh's earlier writings, in which he noted the affinities between Eddington and Dirac (1982) or gave an illuminating characterization of important aspects of Dirac's methodology (1979b).

To close this rapid survey, I would like to mention that my omission of some references is not necessarily deliberate.

# Abbreviations Used in Citations and in the Bibliography

AHES     *Archive for history of exact sciences.*

AHQP     Archive for the History of Quantum Physics.

AP     *Annalen der Physik.*

BB     Akademie der Wissenschaften, Berlin, Physikalische-mathematische Klasse, *Sitzungsberichte.*

BCW     Niels Bohr, *Collected works,* ed. L. Rosenfeld and E. Rüdinger, 11 vols. (Amsterdam, from 1972).

BMSS     Niels Bohr, Scientific manuscripts, in AHQP.

BWA     Ludwig Boltzmann, *Wissenschaftliche Abhandlungen,* 3 vols. (New York, 1968).

DSB     *Dictionary of scientific biography,* ed. C. C. Gillispie, 16 vols. (New York, 1970–1980).

FT     *Fysisk Tidsskrift.*

HSPS     *Historical studies in the physical (and biological) sciences.*

JSHS     *Japanese studies in the history of science.*

KDM     Det Kongelige Danske Videnskabernes Selskab, *Matematisk-fysiske Meddelser.*

MB     Akademie der Wissenschaften, Munich, Physikalische-mathematische Klasse, *Sitzungsberichte.*

NW     *Die Naturwissenschaften.*

PAV     Max Planck, *Physikalische Abhandlungen und Vorträge,* 3 vols. (Braunschweig, 1958).

| | |
|---|---|
| *PB* | Wolfgang Pauli, *Wissenschaftlicher Briefwechsel,* vol. 1, ed. A. Hermann, K. von Meyenn, V. Weisskopf (New York, 1979), and vol. 2, ed. K. von Meyenn (New York, 1985). |
| *PCPS* | Cambridge Philosophical Society, *Proceedings.* |
| *PGV* | Deutsche Physikalische Gesellschaft, *Verhandlungen.* |
| *PM* | *Philosophical Magazine.* |
| *PRS* | Royal Society of London, *Proceedings,* series A. |
| *PR* | *Physical review* |
| *PZ* | *Physikalische Zeitschrift* |
| *RHS* | *Revue d'histoire des sciences.* |
| *WB* | Akademie der Wissenschaften, Vienna, *Sitzungsberichte,* Abteilung II. |
| *ZP* | *Zeitschrift für Physik.* |

# Bibliography of Secondary Literature

Literature directly or indirectly used in this book (includes physicists' correspondence and collected works)

Aaserud, Finn
  1990      *Redirecting science: Niels Bohr, philanthropy and the rise of nuclear physics.* Cambridge: Cambridge University Press.
Agassi, Joseph
  1967      "The Kirchhoff-Planck radiation law." *Science* 156:30–37.
Agostino, Salvo d'
  1985      "The problem of the link between correspondence and complementarity in Niels Bohr's papers." *Rivista di storia della scienza* 2:369–390.
Balibar, Françoise
  1985      "Bohr entre Einstein and Dirac." *RHS* 38:293–307.
Beller, Mara
  1983      "Matrix theory before Schrödinger: Philosophy, problems, consequences." *Isis* 74:469–491.
Benz, Ulrich
  1975      *Arnold Sommerfeld: Lehrer und Forscher an der Schwelle zur Atomzeitalter, 1868–1951.* Stuttgart: Wissenschaftliche Verlagsgesellschaft.
De Boer, Jorrit, et al., eds.
  1986      *The lesson of quantum theory: Niels Bohr centenary symposium, October 3–7, 1985.* Amsterdam: North-Holland.

Bohr, Niels
1972–        *Collected works.* Ed. L. Rosenfeld and Erik Rüdinger. Amsterdam:
             North-Holland.
             Vol. 1, *Early work (1905–1911).* Ed. J. Rud Nielsen. 1972.
             Vol. 2, *Works on atomic physics (1912–1917).* Ed. U. Hoyer. 1981.
             Vol. 3, *The correspondence principle (1918–1923).* Ed. J. Rud
             Nielsen. 1976.
             Vol. 4, *The periodic system (1918–1923).* Ed. J. Rud Nielsen. 1977.
             Vol. 5, *The emergence of quantum mechanics.* Ed. K. Stolzenburg.
             1984.
             Vol. 6, *Foundations of quantum physics I (1926–1932).* Ed.
             J. Kalckar. 1985.
             Vol. 8, *The penetration of charged particles through matter (1912–
             1954).* Ed. J. Thorsen. 1987.
             Vol. 9, *Nuclear physics (1929–1952).* Ed. R. Peierls. 1986.
Bohr, Niels
             (articles containing historical accounts)
1925         "Atomic theory and mechanics." Supplement to *Nature* 116:845–
             852. Also in *BCW* 5:269–280.
1929         "Die Atomtheorie und die Prinzipien der Naturbeschreibung."
             *NW* 18:73–78. Eng. trans. in *BCW* 6:236–255.
Boltzmann, Ludwig
1968         *Wissenschaftliche Abhandlungen.* 3 vols. Leipzig: Barth.
Born, Max
1963         *Ausgewählte Abhandlungen.* 2 vols. Göttingen.
1978         *My life: Recollections of a Nobel laureate.* New York.
Boyd, Richard
1979         "Metaphor and theory change: What is 'metaphor' a metaphor
             for?" In *Metaphor and thought,* ed. A. Ortony. Cambridge: Cam-
             bridge University Press.
Broda, Engelbert
1955         *Ludwig Boltzmann: Mensch, Physiker, Philosoph.* Vienna:
             Deuticke.
Brown, Laurie, and Helmut Rechenberg
1987         "Paul Dirac and Werner Heisenberg, a partnership in science." In
             *Paul Adrien Maurice Dirac: Reminiscences about a great physicist,*
             ed. B. N. Kursunoglu and E. Wigner, 117–162. Cambridge: Cam-
             bridge University Press.
Brush, Steven G.
1965         *Kinetic theory: Selected readings in physics.* 2 vols. Oxford:
             Pergamon.
1976         *The kind of motion we call heat: A history of the kinetic theory
             of gases in the 19th century.* 2 vols. Amsterdam: North-Holland;
             New York: Elsevier.
Buchwald, Jed Z.
1985         *From Maxwell to Microphysics: Aspects of electromagnetic theory*

*in the last quarter of the 19th century.* Chicago: University of Chicago Press.

Campbell, Norman R.
1920        *Physics: The elements.* Cambridge: Cambridge University Press.

Carazza, Bruno, and Helge Kragh
1989        "Adolfo Bartoli and the problem of radiant heat." *Annals of Science* 46:183–194.

Cassidy, David C.
1976        "Werner Heisenberg and the crisis of quantum theory." Diss., Purdue University.
1979        "Heisenberg's first core model of the atom: The formation of a professional style." *HSPS* 10:187–224.

Chevalley, Catherine
1985        "Complémentarité et langage dans l'interprétation de Copenhague." *RHS* 38:251–292.
1989        "Histoire et philosophie de la mécanique quantique." *Revue de synthèse,* 469–481.
1991a       "Le dessin et la couleur," introduction to Niels Bohr, *Physique atomique et connaissance humaine,* 19–140. Paris: Gallimard. Also the glossary of Bohrian concepts, ibid., 345–567.
1991b       "Complémentarité et représentation: Bohr et la tradition philosophique allemande." In *Lezioni della scuola superiore di storia della scienze della Domus Galileana di Pisa, sezione storico-epistemologica,* ed. S. Petruccioli. Rome.

Dalitz, R. H., and Rudolf Peierls
1986        "Paul Adrien Maurice Dirac, 1902–1984." *Biographical Memoirs of Fellows of the Royal Society* 32:138–185.

Darrigol, Olivier
1984a       "La genèse du concept de champ quantique." *Annales de physique* 9:433–501.
1984b       "A history of the question: Can free electrons be polarized?" *HSPS* 15(1):39–79.
1986        "The origin of quantized matter waves." *HSPS* 16:197–253.
1988a       "Statistics and combinatorics in early quantum theory." *HSPS* 19(1):18–80.
1988b       "The quantum electrodynamical analogy in early nuclear theory or the roots of Yukawa's theory." *RHS* 41:225–297.

Daub, E. E.
1970        "Maxwell's demon." *HSPS* 1:213–227.

De Maria, Michelangelo, and Francesco La Teana
1982        "Schrödinger's and Dirac's unorthodoxy in quantum mechanics." *Fundamenta Scientiae* 3:119–148.
1983        "Dirac's unorthodox contribution to orthodox quantum mechanics (1925–1927)." *Scientia* 118:595–611.

D'Espagnat, Bernard
1971        *Conceptual foundations of quantum mechanics.* New York: Addison Westley/Benjamin.

1985        "Niels Bohr et l'étrangeté du monde." *La recherche* 171:1402–
            1403.
Dirac, Paul Adrien Maurice
1977        "Recollections of an exciting era." In *History of twentieth century
            physics,* ed. C. Weiner, 109–146. New York: Academic Press.
Dresden, Max
1987        *H. A. Kramers: Between tradition and revolution.* Berlin: Springer.
Dugas, René
1955        "Einstein et Gibbs devant la thermodynamique statistique."
            Académie des sciences, *Comptes rendus* 241:1685–1687.
1959        *La théorie physique au sens de Boltzmann et ses prolongements
            modernes.* Neuchâtel: Griffon.
Ehrenfest, Paul
1959        *Collected scientific papers.* Ed. M. J. Klein. Amsterdam.
Einstein, Albert
1949        "Autobiographisches." In *Albert Einstein: Philosopher-scientist,*
            ed. P. A. Schilpp, 9–94. Evanston, Ill.: The Library of Living
            Philosophers.
1987–       *The collected papers of Albert Einstein.* Princeton: Princeton Uni-
            versity Press.
            Vol. 1, *The early years, 1879–1902.* Ed. J. Stachel. 1987.
            Vol. 2, *The Swiss years: Writings, 1900–1909.* Ed. J. Stachel. 1989.
1989        *Oeuvres choisies.* Vol. 1, *Quanta.* Ed. F. Balibar, O. Darrigol,
            B. Jech. Paris: Seuil/CNRS.
Einstein, Albert, and Michele Besso
1972        *Correspondance 1903–1955.* Ed. P. Speziali. Paris: Hermann.
Einstein, Albert, and Hedwig and Max Born
1969        *Albert Einstein–Max Born: Briefwechsel 1916–1955.* Munich:
            Nymphenburger.
Enz, Charles P.
1973        "W. Pauli's scientific work." In *The physicist's conception of
            nature,* ed. J. Mehra. Dordrecht: Reidel.
Everitt, C. W. Francis
1974        "Maxwell, James Clerk." In *Dictionary of scientific biography,*
            ed. C. C. Gillispie, 9:198–230. New York: Scribners.
Ezawa, Hiroshi
1979        "Einstein's contributions to statistical mechanics." *JSHS* 18:
            27–72.
Favrholdt, David
1976        "Niels Bohr and the Danish philosophy." *Danish Yearbook of
            Philosophy* 13:206–220.
1979        "On Høffding and Bohr: A reply to Jan Faye." *Danish Yearbook
            of Philosophy* 16:73–77.
1985        "The cultural background of the young Niels Bohr." *Rivista di
            storia della scienzia* 2:445–461.
Faye, Jan
1979        "The influence of Harald Høffding's philosophy on Niels Bohr's

interpretation of quantum mechanics." *Danish Yearbook of Philosophy* 16:37–72.

1988        "The Bohr-Høffding relationship reconsidered." *Studies in History and Philosophy of Science* 19:321–346.

Folse, Henry

1985        *The philosophy of Niels Bohr: The framework of complementarity.* Amsterdam: North-Holland.

Forman, Paul L.

1968        "The doublet riddle and the atomic physics *circa* 1924." *Isis* 59: 156–174.

1970        "Alfred Landé and the anomalous Zeeman effect, 1919–1921." *HSPS* 2:153–261.

Forman, Paul L., John L. Heilbron, and Spencer Weart

1975        "Physics *circa* 1900." *HSPS* 5:5–185.

Galison, Peter

1981        "Kuhn and the quantum controversy." *British Journal for the Philosophy of Science* 32:71–85.

1987        *How experiments end.* Chicago: University of Chicago Press.

Garber, Elizabeth

1976        "Some reactions to Planck's law, 1900–1914." *Studies in History and Philosophy of Science* 7:89–126.

Gibbs, Josiah Willard

1928        *The collected works of J. Willard Gibbs.* 2 vols. New York.

Goldberg, Stanley

1976        "Max Planck's philosophy of nature and his elaboration of the special theory of relativity." *HSPS* 7:125–160.

Heilbron, John L.

1964        "A history of atomic structure from the discovery of the electron to the beginning of quantum mechanics." Ph.D. diss., University of California, Berkeley.

1966        "The work of H. G. J. Moseley." *Isis* 57:336–364.

1967        "The Kossel-Sommerfeld theory and the ring atom." *Isis* 58: 451–482.

1968        "The scattering of $\alpha$ and $\beta$ particles and Rutherford's atom." *AHES* 4:247–307.

1974        *H. G. J. Moseley: The life and letters of an English physicist.* Berkeley, Los Angeles, London: University of California Press.

1977a       "J. J. Thomson and the Bohr atom." *Physics Today* 30: 23–30.

1977b       "Lectures on the history of atomic physics, 1900–1922." In *History of twentieth century physics,* ed. C. Weiner, 40–108. New York: Academic Press.

1981        *Historical studies in the theory of atomic structure.* New York.

1982        "Fin-de-siècle physics." In *Science, technology, and society in the time of Alfred Nobel,* ed. C. G. Bernhard et al. New York: Pergamon Press.

1983        "The origins of the exclusion principle." *HSPS* 13:261–310.

1985    "Bohr's first theories of the atom." In *Niels Bohr: A centenary volume,* ed. A. P. French and P. J. Kennedy. Cambridge: Harvard University Press.

1986    *The dilemmas of an upright man: Max Planck as spokesman for German science.* Berkeley, Los Angeles, London: University of California Press.

Heilbron, John L., and Thomas S. Kuhn

1969    "The genesis of the Bohr atom." *HSPS* 1:211–290.

Heilbron, John L., and Bruce R. Wheaton

1981    *Literature on the history of physics in the 20th century.* Berkeley: Office for History of Science and Technology, University of California.

1982    *An inventory of published letters to and from physicists, 1900–1950.* Berkeley, Los Angeles, London: University of California Press.

Hendry, John

1981    "Bohr-Kramers-Slater: A virtual theory of virtual oscillators and its role in the history of quantum mechanics." *Centaurus* 25: 189–221.

1984    *The creation of quantum mechanics and the Bohr-Pauli dialogue.* Dordrecht: Reidel.

Hermann, Armin

1969    *Frühgeschichte der Quantentheorie (1899–1913).* Mosbach in Baden: Physik.

1973    *Max Planck in Selbstzeugnissen und Bilddokumenten.* Hamburg: Rowohlt.

1976    *Werner Heisenberg in Selbstzeugnissen und Bilddokumenten.* Hamburg: Rowohlt.

Hermann, Armin, and Karl von Meyenn

1976    "Wolfgangs Pauli Beitrag zur Göttinger Quantenmechanik." *Physikalische Blätter* 32:145–150.

Hesse, Mary B.

1966    *Models and analogies in science.* Notre Dame: University of Notre Dame Press.

Hiebert, Erwin N.

1968    *The conception of thermodynamics in the scientific thought of Mach and Planck.* Wissenschaftlicher Bericht Nr. 5/68, Ernst Mach Institut. Freiburg.

Hirosige, Tetu, and Sigeko Nisio

1964    "Formation of Bohr's theory of atomic constitution." *JSHS* 3: 6–28.

1970    "The genesis of the Bohr atom model and Planck's theory of radiation." *JSHS* 9:35–47.

Holton, Gerald

1970    "The roots of complementarity." *Daedalus* 99:1015–1055.

1973    *Thematic origins of scientific thought.* Cambridge: Harvard University Press.

1978        *The scientific imagination: Case studies.* Cambridge: Harvard University Press.

Høffding, Harald
1923        *Begrebet Analogi.* Copenhagen. French translation: *Le concept d'analogie* (Paris: Vrin, 1931).

Hon, Giora
1989        "Franck and Hertz versus Townsend: A study of two types of experimental error." *HSPS* 20(1):79–106.

Honner, John
1987        *The description of nature: Niels Bohr and the philosophy of quantum mechanics.* New York: Oxford University Press.

Hoyer, Ulrich
1973        "Über die Rolle der Stabilitätbetrachtungen in der Entwicklung der Bohrschen Atomtheorie." *AHES* 10:177–206.
1974        *Die Geschichte der Bohrschen Atomtheorie.* Weinheim: Physik.
1976        Introductions in *BCW* 2:[3]–[10], [103]–[134].
1980        "Von Boltzmann zu Planck," *AHES* 23:49–86.

Hund, Friedrich
1974        *The history of quantum theory.* London: Harrap.

Jammer, Max
1966        *The conceptual development of quantum mechanics.* New York: McGraw-Hill. 2d ed., New York: Thomash, 1989.

Jensen, Carsten
1985        "Two one-electron anomalies in the old quantum theory." *HSPS* 15:81–106.

Jungnickel, Christa, and Russell McCormmach
1986        *Intellectual mastery of nature: Theoretical physics from Ohm to Einstein.* Vol. 1, *The torch of mathematics, 1800–1870.* Vol. 2, *The now mighty theoretical physics, 1870–1925.* Chicago: University of Chicago Press.

Kangro, Hans
1970        *Vorgeschichte des Planckschen Strahlungsgesetzes.* Wiesbaden: Steiner.

Klein, Martin J.
1962        "Max Planck and the beginnings of the quantum theory." *AHES* 1:459–479.
1963a       "Planck, entropy, and quanta, 1901–1906." *Natural philosopher* 1:83–108.
1963b       "Einstein's first paper on quanta." *Natural philosopher* 2:59–86.
1964a       "Einstein and the wave-particle duality." *Natural philosopher* 3:1–49.
1964b       "The origins of Ehrenfest's adiabatic principle." In: *Tenth International Congress of History of Science* (1962), *Actes* 2:801–804.
1965        "Einstein, specific heats, and the early quantum theory." *Science* 148:173–180.
1967        "Thermodynamics in Einstein's thought." *Science* 157:509–516.

| 1970a | "The first phase of the Bohr-Einstein dialogue." *HSPS* 2:1–39. |

1970a   "The first phase of the Bohr-Einstein dialogue." *HSPS* 2:1–39.
1970b   "Maxwell, his demon, and the second law of thermodynamics." *American scientist* 58:84–97.
1970c   *Paul Ehrenfest. Vol. 1, The making of a theoretical physicist.* Amsterdam: North-Holland; New York: Elsevier.
1972   "Mechanical explanation at the end of the nineteenth century." *Centaurus* 17:58–82.
1973   "The development of Boltzmann's statistical ideas." In *The Boltzmann equation: Theory and applications,* ed. E. G. D. Cohen and W. Thirring, 53–106. Vienna: Springer.
1977   "The beginnings of quantum theory." In *History of twentieth century physics,* ed. C. Weiner. New York: Academic Press.
1979   "Einstein and the development of quantum physics." In *Einstein: A centenary volume,* ed. A. P. French. (London: Heinemann, 1979).
1986   "Great connections come alive: Bohr, Ehrenfest, and Einstein." In *The lesson of quantum theory,* ed. J. de Boer et al., 325–342. Amsterdam: Elsevier.

Klein, Martin J., Abner Shimony, and Trevor J. Pinch
1979   "Paradigm lost? A review symposium." *Isis* 70:429–440.

Konno, Hiroyuki
1978   "The historical roots of Born's probabilistic interpretation." *JSHS* 17:129–145.
1983   "Slater's evidence of the Bohr-Kramers-Slater theory." *Historia scientiarum* 25: 39–52.

Kozhevnikov, Alexei, and Olga Novik
1987   *Analysis of information ties dynamics in early quantum mechanics (1925–1927).* Moscow: Academia Nauk.

Kragh, Helge
1979a   "Niels Bohr's second atomic theory." *HSPS* 10:123–186.
1979b   "Methodology and philosophy of science in Paul Dirac's physics." University of Roskilde (Denmark), text no. 27.
1980   "Anatomy of a priority conflict: The case of element 72." *Centaurus* 23:275–301.
1981   "The concept of the monopole." *HSPS* 12:141–172.
1982   "Cosmophysics in the thirties: Towards a history of Dirac's cosmology." *HSPS* 13(1):69–108.
1985   "The fine structure of hydrogen and the gross structure of the physics community, 1916–1926," *HSPS* 15:67–125.
1990   *Dirac: A scientific biography.* Cambridge: Cambridge University Press.

Krajewski, Wladislaw
1977   *Correspondence principle and growth of science.* Dordrecht: Reidel.

Kramers, Hans A.
1923   "Das Korrespondenzprinzip und der Schalenbau des Atoms." *NW* 11:550–559.

1956        *Collected scientific papers.* Amsterdam.
Kuhn, Thomas S.
1978        *Black-body theory and the quantum discontinuity, 1894–1912.*
            New York: Oxford University Press.
1979        "Metaphor in science." In *Metaphor and thought,* ed. A. Ortony.
            Cambridge: Cambridge University Press.
1983        "Revisiting Planck." *HSPS* 14:231–252.
Kuhn, Thomas S., J. L. Heilbron, Paul Forman, and Lini Allen
1967        *Sources for history of quantum physics: An inventory and report.*
            Philadelphia: American Philosophical Society.
MacKinnon, Edward
1977        "Heisenberg, models and the rise of matrix mechanics." *HSPS*
            8:137–188.
1982        *Scientific explanation and atomic physics.* Chicago: University of
            Chicago Press.
Maxwell, James Clerk
1890        *Scientific papers.* Cambridge: Niven.
McCormmach, Russell. *See* Jungnickel, Christa.
Mehra, Jagdish, and Helmut Rechenberg
1982–       *The historical development of quantum theory.* New York:
            Springer.
            Vol. 1, *The quantum theory of Planck, Einstein, Bohr, and Som-
            merfeld.* 1982.
            Vol. 2, *The discovery of quantum mechanics.* 1982.
            Vol. 3, *The formulation of matrix mechanics and its modifications,
            1925–1926.* 1982.
            Vol. 4, *The fundamental equations of quantum mechanics: The
            reception of the new quantum mechanics.* 1982.
            Vol. 5, *Erwin Schrödinger and the rise of wave mechanics.* 1987.
Merleau-Ponty, Jacques
1965        *Philosophie et théorie physique chez Eddington.* Annales littéraires
            de l'Université de Besançon, vol. 75. Paris.
Meyenn, Karl von
1980        "Pauli's Weg zum Ausschliessungspinzip." *Physikalische Blätter*
            36:293–298; 37:13–20 (vol. 37 is 1981).
Meyer-Abich, Klaus M.
1965        *Korrespondenz, Individualität und Komplementarität: Eine Studie
            zur Geistgeschichte der Quantentheorie in den Beiträgen Niels
            Bohrs.* Wiesbaden: Franz Steiner.
1967        "Die Sprache in der Philosophie Niels Bohrs." In *Das Problem der
            Sprache,* ed. H. G. Gadamer, 97–105.
Miller, Arthur I.
1978        "Visualization lost and regained: The genesis of the quantum
            theory in the period 1913–1927." In *On aesthetics in science,* ed.
            J. Wechsler. Cambridge: MIT Press.
1984        *Imagery in scientific thought: Creating twentieth century physics.*
            Boston: Birkhäuser.

Moore, Ruth
  1966      *Niels Bohr: The man, his science and the world they changed.*
            New York: Alfred Knopf.
Murdoch, Dugald
  1987      *Niels Bohr's philosophy of physics.* Cambridge: Cambridge University Press.
Needell, Allan A.
  1980      "Irreversibility and the failure of classical dynamics: Max Planck's work on the quantum theory, 1900–1915." Diss., University of Michigan, Ann Arbor.
  1988      Introduction to Max Planck, *The theory of heat radiation,* xi–xliii. Los Angeles: Tomash. *See* Planck 1906.
Nisio, Sigeko
  1969      "X-rays and atomic structure in the early age of the old atomic theory." *JSHS* 8:55–75.
  1973      "The formation of the Sommerfeld quantum theory of 1916." *JSHS* 12:39–78.
Pais, Abraham
  1982      *"Subtle is the Lord—": The science and the life of Albert Einstein.* Oxford: Oxford University Press.
Paty, Michel
  1988      *La matière dérobée.* Paris: Archives contemporaines.
Pauli, Wolgang
  1964      *Collected scientific papers.* Ed. R. Kronig and V. Weisskopf. 2 vols. New York.
  1979      *Wissenschfatlicher Briefwechsel.* Vol. 1. Ed. A. Hermann, K. von Meyenn, and V. Weisskopf. New York: Springer.
  1985      *Wissenschaftlicher Briefwechsel.* Vol. 2. Ed. K. von Meyenn. New York: Springer.
Petersen, Aage
  1963      "The philosophy of Niels Bohr." *Bulletin of the atomic scientist,* 8–14.
  1968      *Quantum physics and the philosophical tradition.* Cambridge: MIT Press.
  1969      "On the philosophical significance of the correspondence argument." In: Boston Colloquium for the Philosophy of Science, 1961/62–1966/68, *Proceedings* 5:242–252. Boston Studies in the Philosophy of Science, vol. 5. Dordrecht: Reidel.
Petruccioli, Sandro
  1988      *Atomi metafore paradossi: Niels Bohr e la costruzione di una nuova fisica.* Rome: Theoria.
Planck, Max
  1920      *Die Entstehung und bisherige Entwicklung der Quantentheorie.* Leipzig: Barth. Also in *PAV* 3:121–134.
  1958      *Physikalische Abhandlungen und Vorträge.* 3 vols. Braunschweig.
Poincaré, Henri
  1902      *La science et l'hypothèse.* Paris. Reprint. Paris: Flammarion, 1968.

Radder, Hans
1982       "Between Bohr's atomic theory and Heisenberg's matrix mechanics: A study of the role of the Dutch physicist H. A. Kramers." *Janus* 69:223–252.

Radzabov, U.A.
1980       "The correspondence principle—history and present state." *Danish Yearbook of Philosophy* 17:59–82.

Ramunni, Jérôme
1981       *Les conceptions quantiques de 1911 à 1927.* Paris: Vrin.

Robertson, Peter
1979       *The early years: The Niels Bohr Institute, 1921–1930.* Copenhagen: Akademisk Forlag Universitetsforlaget.

Robotti, Nadia
1986       "The hydrogen spectroscopy and the old quantum theory." *Rivista di storia della scienza* 3:45–102.

Röseberg, Ulrich
1985       *Niels Bohr: Leben und Werk eines Atomphysikers.* Stuttgart: Wissenschaftliche Verlagsgesellschaft.

Rosenfeld, Léon
1936       "La première phase de l'évolution de la théorie des quanta." *Osiris* 2:149–196.
1971       "Men and ideas in the history of atomic theory." *AHES* 7:69–90.

Rosenfeld, Léon, and Erik Rüdinger
1967       "The decisive years 1911–1918." In Rozental 1967.

Rozental, Stefan, ed.
1967       *Niels Bohr: His life and work as seen by his friends and colleagues.* Amsterdam: North-Holland; New York: John Wiley & Sons.

Rud Nielsen, J.
1963       "Memories of Niels Bohr." *Physics Today* 16:22–30.
1972       Introduction to part II, *BCW* 1:93–123.
1977a     Introduction to part I, *BCW* 3:3–46.
1977b     Introduction to part I, *BCW* 4:3–42.

Rüdinger, Erik
1985       "The correspondence principle as a guiding principle." *Rivista di storia della scienza* 2:357–367.

Rüdinger, Erik, and Klaus Stolzenburg
1984       Introduction to part II, *BCW* 5:219–240.

Salam, Abdus, and Eugen Wigner
1972       *Aspects of quantum theory.* Cambridge: Cambridge University Press. Essays in honor of P. A. M. Dirac.

Serwer, Daniel
1977       "*Unmechanischer Zwang*: Pauli, Heisenberg, and the rejection of the mechanical atom, 1923–1925." *HSPS* 8:189–256.

Shimony, Abner
1983       "Reflections on the philosophy of Bohr, Heisenberg and Schrödinger." In *Philosophy, physics, psychoanalysis*, ed. R. S. Cohen et al., 209–221. Dordrecht: Reidel.

Sommerfeld, Arnold
  1968a    *Gesammelte Schriften.* Ed. F. Sauter. 4 vols. Braunschweig.
  1968b    *Briefwechsel.* Ed. A. Hermann. Basel.
Stachel, John
  1983    "Einstein and the quantum: Fifty years of struggle." In *From quarks to quasars,* ed. R. Colodny. Pittsburgh: University of Pittsburgh Press.
  1986    "Eddington and Einstein." In *The prism of science,* ed. E. Ulmann-Margalit, 225–250. Dordrecht.
Stolzenburg, Klaus
  1977    "Die Entwicklung des Bohrschen Komplemetaritätsgedankens in den Jahren 1924 bis 1929." Diss., Stuttgart University.
  1985    Introduction to *BCW* 5:1–96.
Stuewer, Roger H.
  1975    *The Compton effect: Turning point in physics.* New York: Science History Publications.
ter Haar, Dirk
  1967    *The old quantum theory.* London: Pergamon Press.
van der Waerden, Bartel L.
  1960    "Exclusion principle and spin." In *Theoretical physics in the twentieth century,* ed. M. Fierz and V. Weisskopf. New York: Interscience.
  1967    *Sources of quantum mechanics.* Amsterdam: North-Holland.
  1973    "From matrix mechanics to unified quantum mechanics." In *The physicist's conception of nature,* ed. J. Mehra. Dordrecht: Reidel.
Wassermann, Neil Henry
  1981    "The BKS paper and the development of the quantum theory of radiation in the work of Niels Bohr." Diss., Harvard University.
Wessels, Linda
  1979    "Schrödinger's route to wave-mechanics." *Studies in the History and Philosophy of Science* 10:311–340.
Wheaton, Bruce R.
  1983    *The tiger and the shark: Empirical roots of wave-particle dualism.* Cambridge: Cambridge University Press.
Whittaker, Edmund Taylor
  1960    *A history of the theories of aether and electricity.* Vol. 1, *The classical theories.* Vol. 2, *The modern theories, 1900–1926.* Reprint. New York: Dover, 1960.
Wise, Norton
  1987    "How do sums count? On the cultural origins of statistical causality." In *The probabilistic revolution,* ed. L. Krüger et al., vol. 1, *Ideas in history,* 395–425. Cambridge: MIT Press.

# Bibliography of
# Primary Literature

Baker, Henry Frederick
 1922 *The principles of geometry.* Cambridge.
Becker, Richard
 1924 "Über Absorption und Dispersion in Bohrs Quantentheorie." *ZP* 27:173–188.
Birtwistle, George
 1928 *The new quantum mechanics.* Cambridge.
Bohr, Niels
 1913 "On the constitution of atoms and molecules." *PM* 26:1–25, 476–502, 857–875.
 1914 "On the effect of electric and magnetic fields on spectral lines." *PM* 27:506–524.
 1915a "On the series spectrum of the hydrogen and the structure of the atom." *PM* 29:332–335.
 1915b "On the quantum theory of radiation and the structure of the atom." *PM* 30:394–415.
 [1916] "On the application of the quantum theory to periodic systems." Unpublished paper, intended for publication in *PM*, April 1916. In *BCW* 2:[431]–[461].
 1918a "On the quantum theory of line spectra, part I: On the general theory." *KDM* 4(1):1–36. Ready for printing on 27 Apr. 1918.
 1918b "On the quantum theory of line spectra, part II: On the hydrogen spectrum." *KDM* 4(1):36–100. Ready for printing on 30 Dec. 1918.
 [1920a] "On the interaction between light and matter." Translation of a lecture given on 13 Feb. 1920 before the Royal Danish Academy. In *BCW* 3:[227]–[240].

1920b    "On the series spectra of the elements." Lecture before the German Physical Society in Berlin, 27 Apr. 1920. Translated by A. D. Udden. In Bohr 1922d, 20–60.

[1920c]    "Some considerations of atomic structure." Translation of a lecture given before the Physical Society of Copenhagen, 15 Dec. 1920. In BCW 4:[43]–[69].

1921a    "Atomic structure." Nature 107 (24 Mar.):104–107.

1921b    "Zur Frage der Polarisation der Strahlung in der Quantentheorie." ZP 6:1–9.

1921c    "Atomic structure." Nature 108 (13 Oct.):208–209.

[1921d]    "Constitution of atoms." Unpublished manuscript for the Solvay congress (1921). In BCW 4:[99]–[174].

1922a    Atomernes Bygning og Stoffernes fysiske og kemiske Egenskaber. Copenhagen: Gjellerup. Lecture before the Physical Society, Copenhagen, 18 Oct. 1921. German translation: "Der Bau der Atome und die physikalischen und chemischen Eigenschaften der Elemente," ZP 9:1–67. English translation (with slight modifications) in The theory of spectra and atomic constitution (Cambridge, 1922).

1922b    "On the quantum theory of line spectra, part III: On the spectra of elements of higher atomic number." KDM 4(1):101–118. Written in spring 1918, printed in November 1922, with an updating appendix of September 1922.

[1922c]    "Seven lectures on the theory of atomic structure." Göttingen, 1922. Unpublished. In BCW 4:[341]–[419].

1922d    The theory of spectra and atomic constitution. Cambridge.

1923a    "The effect of electric and magnetic fields on spectral lines" (seventh Guthrie lecture). Physical Society (London), Proceedings 35:275–302.

1923b    "Über die Anwendung der Quantentheorie auf dem Atombau. I. Die Grundpostulate der Quantentheorie." ZP 13:117–165. Translated as On the application of the quantum theory to atomic structure, part I: The fundamental postulates, supplement to PCPS (1924).

1923c    "On Atomernes Bygning." FT 21:6–44. Translated as "The structure of the atom," Nature 112:29–44.

1923d    "L'application de la théorie des quanta aux problèmes atomiques." In Atomes et électrons: Rapports et discussions du conseil de physique tenu à Bruxelles du 1er au 6 avril 1921, 364–380. Paris.

1923e    "Linienspektren und Atombau." AP 71:228–288.

[1923f]    "Über die Anwendung der Quantentheorie auf dem Atombau. II. Theorie der Serienspektren." Unpublished manuscript. In BCW 3:[502]–[531].

[1924a]    "Atomteoretiske Problemer." Unpublished manuscript, BMSS. English in BCW 3:[569]–[574].

1924b    "Zur Polarisation des Fluorescenzlichtes." NW 12:1115–1117.

1925a    "Über die Wirkung von Atomen bei Stössen." ZP 34:142–157.

1925b    "Grundlaget for den moderne Atomforskning" (lecture on the award of the Ørsted medal, 22 Oct. 1924), *FT* 23:10–17. Translated in *BCW* 5.

1925c      "Atomic theory and mechanics." Supplement to *Nature* 116: 845–852. Also in *BCW* 5:269–280.

[1962]      Interview with T. S. Kuhn, A. Petersen, and E. Rüdinger. AHQP.

Bohr, Niels, and Dirk Coster
1923      "Röntgenspektrum und periodisches System der Elemente." *ZP* 12:342–374.

Bohr, Niels, Hendrik Kramers, and John Clarke Slater
1924      "The quantum theory of radiation." *PM* 47:785–822.

Boltzmann, Ludwig
1872      "Weitere Studien über das Wärmegleichgewicht unter Gasmolekülen." *WB* 66:275. Also in *BWA* 1:316–402.

1877a      "Bemerkungen über einige Probleme der mechanischen Wärmetheorie." *WB* 75:62–100. Also in *BWA* 2:112–148.

1877b      "Über die Beziehung zwischen den zweiten Hauptsatz der mechanischen Wärmetheorie und der Wahrscheinlichkeitsrechnung, respective den Sätzen über das Wärmegleichgewicht." *WB* 76: 373–435. Also *BWA* 2:164–223.

1884a      "Über eine von Hrn. Bartoli entdeckte Beziehung der Wärmestrahlung zum zweiten Hauptsatze." *AP* 22:31–39.

1884b      "Ableitung des Stefan'schen Gesetzes, betreffend die Abhängigkeit der Wärmestrahlung von der Temperatur aus der elektromagnetischen Lichttheorie." *AP* 22:291–294.

1894      "Über den Beweis des Maxwellschen Geschwindgkeitsverteilungsgesetzes unter Gasmolekülen." *AP* 53:955–958.

1895      "Nochmals das Maxwellsche Verteilungsegesetz der Geschwindigkeiten." *AP* 55:223–224.

1896a      "Entgegnung über die Wärmetheoretischen Betrachtungen des Hrn. E. Zermelo." *AP* 57:773–784.

1896b      *Vorlesungen über die Gastheorie.* Vol. 1. Leipzig. Translated by S. G. Brush with vol. 2 (1898), reference below.

1897a      "Zur Hrn. Zermelo's Abhandlung 'Über die mechanische Erklärung irreversibler Vorgänge.'" *AP* 60:392–398. Also in *BWA* 3:579–586.

1897b      "Über irreversible Strahlungsvorgänge, I." *BB*, 660–662. Also in *BWA* 3:615–617.

1897c      "Über irreversible Strahlungsvorgänge, II." *BB*, 1016–1018. Also in *BWA* 3:618–621.

1898a      *Vorlesungen über die Gastheorie.* Vol. 2. Leipzig. Translated by S. G. Brush as *Lectures on gas theory* (Berkeley and Los Angeles, 1964).

1898b      "Über vermeintliche irreversible Strahlungs-vorgänge." *BB*, 182–187. Also in *BWA* 3:622–628.

1904      *Vorlesungen über die Prinzipien der Mechanik.* Leipzig.

Born, Max
1922      "Quantentheorie und Störungsrechnung." *NW* 11:537–542.

1924      "Über Quantenmechanik." *ZP* 26:379–395.

1925 *Vorlesungen über Atommechanik.* Berlin.
1926a "Zur Quantentheorie der Stossvorgänge." *ZP* 37:863–867.
1926b "Quantenmechanik der Stossvorgänge." *ZP* 38:803–827.

Born, Max, and Werner Heisenberg
1923a "Über Phasenbeziehungen bei den Bohrschen Modellen von Atomen und Molekeln." *ZP* 14:44–55.
1923b "Die Elektronenbahnen im angeregten Heliumatom." *ZP* 16:229–243.

Born, Max, Werner Heisenberg, and Pascual Jordan
1926 "Zur Quantenmechanik II." *ZP* 35:557–615.

Born, Max, and Pascual Jordan
[1925a] "Zur Strahlungstheorie." Unpublished manuscript, in the Bohr collection of manuscript by other authors.
1925b "Zur Quantentheorie aperiodischer Vorgänge." *ZP* 33:479–505.
1925c "Zur Quantenmechanik." *ZP* 34:858–888.

Born, Max, and Wolfgang Pauli
1922 "Über die Quantelung gestörter mechanischer Systeme." *ZP* 10:137–158.

Bothe, Walther, and Hans Geiger
1925a "Experimentelles zur Theorie von Bohr, Kramers und Slater." *NW* 13:440–441.
1925b "Über das Wesen des Comptoneffekts: Ein experimenteller Beitrag zur Theorie der Strahlung." *ZP* 32:639–663.

Breit, Gregory
1924 "The quantum theory of dispersion." *Nature* 114:310.
1926 "A correspondence principle in the Compton effect." *PR* 27:362–372.

Brillouin, Léon
1926 "La nouvelle mécanique atomique." *Journal de physique* 7:134–160.

Broad, Charlie Dunbar
1923 *Scientific thought.* London.
1959 "Autobiography." In *The philosophy of C. D. Broad,* ed. P. A. Schilpp. New York.

Broglie, Louis de
1923 "Waves and quanta." *Nature* 112:540.
1924 "A tentative theory of light quanta." *PM* 47:446–458.

Burbury, Samuel Hawksley
1894 "Boltzmann's minimum function." *Nature* 51:78.

Burgers, Johannes Martinus
1917 "Adiabatic invariant of mechanical systems." Koninklijke Akademie van Wetenschappen te Amsterdam, *Proceedings* 20:149–157, 158–162, 163–169.
1918 "Het Atommodel van Rutherford-Bohr." Diss., Leiden, 1918.

Campbell, Norman R.
1920 "Atomic structure." *Nature* 106:408–409.

Charlier, Carl Wilhelm Ludwig
1907 *Die Mechanik des Himmels: Vorlesungen.* Leipzig.

Christiansen, Christian
  1884     "Über die Emission der Wärme von unebenen Oberflächen." *AP* 21:31–39.
Compton, Arthur Holley
  1922     "The spectrum of secondary X-rays." *PR* 19:267–268.
  1923     "A quantum theory of the scattering of X-rays by light elements." *PR* 21:483–502.
Debye, Peter
  1915     "Die Konstitution des Wasserstoff-Moleküls." *BB,* 1–26.
  1916     "Quantenhypothese und Zeeman-Effekt." *PZ* 17:507–512.
Dingle, Herbert
  1937     "Modern Aristotelianism." *Nature* 139:784–790.
Dirac, Paul Adrien Maurice
  [1924?a]  "The correspondence principle for integrable nonperiodic dynamical systems." Unpublished manuscript, AHQP.
  [1924?b]  "The validity of Liouville's theorem in all frames of reference." Unpublished manuscript, AHQP.
  1925a    "The adiabatic invariance of the quantum integrals." *PRS* 107: 725–734.
  1925b    "The adiabatic hypothesis for magnetic fields." *PCPS* 23:69–72.
  [1925c]  "Einstein-Bose statistical mechanics." Most probably the unpublished text of Dirac's talk of 4 Aug. 1925 at the Kapitza club, AHQP.
  [1925d]  "Radiation from a moving Planck oscillator." Unpublished manuscript, AHQP.
  [1925e]  "Heisenberg's quantum mechanics and the principle of relativity." Unpublished manuscript, AHQP.
  [1925f]  "Virtual oscillators." Unpublished manuscript, AHQP.
  1925g    "The fundamental equations of quantum mechanics." *PRS* 109: 642–653.
  1926a    "Quantum mechanics and a preliminary investigation of the hydrogen atom." *PRS* 110:561–579.
  1926b    "The elimination of the nodes in quantum mechanics." *PRS* 111: 281–305.
  1926c    "Relativity quantum mechanics with an application to Compton scattering." *PRS* 111:405–423.
  [1926d]  "Quantum mechanics." Diss., Cambridge University, May 1926, AHQP.
  1926e    "On quantum algebra." *PCPS* 23:412–418.
  1926f    "On the theory of quantum mechanics." *PRS* 112:661–677.
  1926g    "The Compton effect in wave mechanics." *PCPS* 23:500–507.
  1927a    "The physical interpretation of the quantum dynamics." *PRS* 113:621–641.
  1927b    "The quantum theory of the emission and absorption of radiation." *PRS* 114:243–265.
  [1927c]  "Lectures on modern quantum mechanics" (starting in October 1927). Unpublished manuscript, AHQP.

| 1929 | "Quantum mechanics of many electron systems." *PRS* 123: 714–733. |
|---|---|
| 1930 | *The principles of quantum mechanics.* Oxford. |
| 1931 | "Quantized singularities in the electromagnetic field." *PRS* 133: 60–72. |
| 1937 | "Physical science and philosophy." *Nature* 139:1001–1002. |
| 1939a | "The relation between mathematics and physics." Royal Society of Edinburgh, *Proceedings* 59:122–129. |
| 1939b | "A new notation for quantum mechanics." *PCPS* 35:416–418. |
| [1962] | AHQP interview by T. S. Kuhn and E. Wigner. |
| 1977 | "Recollections of an exciting era." In *History of twentieth century physics,* ed. C. Weiner, 109–146. New York. |

Eddington, Arthur

| 1920 | *Space, time and gravitation.* Cambridge. |
|---|---|
| 1923 | *The mathematical theory of relativity.* Cambridge. |
| 1927 | *The nature of the physical world.* Gifford Lectures, Edinburgh, January–March 1927, on 207–208. Cambridge. |

Ehrenfest, Paul

| 1906 | "Zur Planck'schen Strahlungstheorie." *PZ* 7:528–532. |
|---|---|
| 1914a | "A mechanical theorem of Boltzmann and its relation to the theory of energy quanta." Amsterdam Academy, *Proceedings* 16:591–597. |
| 1914b | "Zum Boltzmannschen Entropie-Wahrscheinlichkeits-Theorem." *PZ* 15:657–663. |
| 1916 | "Adiabatische Invarianten und Quantentheorie." *AP* 51:327–352. |
| 1923a | "Le principe de correspondance." In *Atomes et Electrons: Rapports et discussions du conseil de physique tenu à Bruxelles du 1er au 6 avril 1921,* 248–254. Paris. |
| 1923b | "Adiabatische Transformationen in der Quantentheorie und ihre Behandlung durch Niels Bohr." *NW* 11:543–550. |

Ehrenfest, Paul, and Heike Kamerlingh Onnes

| 1915 | "Vereinfachte Ableitung der kombinatorischen Formel, welche der Planckschen Strahlungs-theorie zugrunde liegt." *AP* 46:1021–1024. |
|---|---|

Einstein, Albert

| 1905 | "Über eine die Erzeugung und Verwandlung des Lichtes betreffenden heuristischen Gesichtspunkt." *AP* 17:132–148. |
|---|---|
| 1906 | "Zur Theorie der Lichterzeugung und Lichtabsorption." *AP* 20:199–206. |
| 1916a | "Strahlungs-emission und -absorption nach der Quantentheorie." *PGV* 18:47–62. |
| 1916b | "Zur Quantentheorie der Strahlung." Physikalische Gesellschaft (Zürich), *Mitteilungen* 16:47–62. |
| 1917 | "Zur Quantentheorie der Strahlung." *PZ* 18:121–128. |

Epstein, Paul Sophus

| 1916a | "Zur Theorie des Starkeffektes." *AP* 50:489–520. |
|---|---|
| 1916b | "Zur Quantentheorie." *AP* 51:168–188. |

1922      "Die Störungsrechnung im Dienste der Quantentheorie. III.
          Kritische Bemerkungen zur Dispersionstheorie." *ZP* 9:92–110.
Franck, James, and Paul Knipping
1919      "Die Ionisierungsspannungen des Heliums." *PZ* 20:481–488.
1920      "Über die Anregungsspannungen des Heliums." *ZP* 1:320–332.
Franck, James, and F. Reiche
1920      "Über Helium und Parhelium." *ZP* 1:154–160.
Gibbs, Josiah Willard
1878      "On the equilibrium of heterogeneous substances." Connecticut
          Academy, *Transactions* 3 (1876–1878):108–248, 343–524. Also in
          *Collected works* (New York, 1928), 1:55–353.
Goldstein, Herbert
1950      *Classical mechanics.* Cambridge, Mass.: Addison-Wesley.
Heaviside, Oliver
1893      *Electromagnetic theory.* London. Reprint. Dover, 1950.
Heisenberg, Werner
1921      "Zur Quantentheorie der Linienstruktur und der anomalen
          Zeemaneffekte." *ZP* 8:273–297.
1924      "Über eine Abänderung der formalen Regeln der Quantentheorie
          beim Problem der anomalen Zeemaneffekte." *ZP* 26:291–307.
1925a     "Über eine Anwendung des Korrespondenzprinzips auf die Frage
          nach der Polarisation des Fluoreszenzlichtes." *ZP* 31:617–626.
1925b     "Zur Quantentheorie der Multiplettstruktur und der anomalen
          Zeemaneffekte." *ZP* 32:841–860.
1925c     "Über die Quantentheoretische Umdeutung kinematischer und
          mechanischer Beziehungen." *ZP* 33:879–893.
1926a     "Mehrkörperproblem und Resonanz in der Quantentheorie." *ZP*
          38:411–426.
1926b     "Über die Spektra von Atomsystemen mit zwei Elektronen." *ZP*
          39:499–518.
1926c     "Schwankungserscheinungen und Quantenmechanik." *ZP* 40:
          501–506.
1927      "Über den anschaulichen Inhalt der Quantentheoretischen Kine-
          matik und Mechanik." *ZP* 43:172–198.
1929      "Die Entwicklung der Quantentheorie, 1918–1928." *NW* 17:
          490–496.
Hertz, Heinrich
1889      "Die Kräfte elektrischer Schwingungen, behandelt nach der
          Maxwellschen Theorie." *AP* 36:1–22.
Hilbert, David
1899      *Grundlagen der Geometrie.* Leipzig.
Horton, Frank, and Ann Catherine Davies
1919      "An experimental determination of the ionisation potential for
          electrons in helium." *PRS* 95:408–429.
Hund, Friedrich
1923      "Theoretische Betrachtungen über die Ablenkung von freien
          langsamen Elektronen in Atomen." *ZP* 13:241–263.

Jordan, Pascual
1927a     "Über eine neue Begründung der Quantenmechanik." *ZP* 40: 809–838.
1927b     "Über eine neue Begründung der Quantenmechanik. Teil II." *ZP* 44:1–25.
Kelland, Paul, and Peter Guthry Tait
1882      *Introduction to quaternions.* 2d ed. London.
Kemble, Edwin C.
1921      "The probable normal state of the helium atom." *PM* 42:123–133.
Kirchhoff, Gustav
1859      "Über den Zusammenhang zwischen Emission und Absorption von Licht und Wärme." Akademie der Wissenschaften zu Berlin, *Monatsberichte,* 783–787.
1860      "Über das Verhältnis zwischen dem Emissionsvermögen der Körper für Wärme und Licht." *AP* 109:275–301.
1894      *Vorlesungen über die Theorie der Wärme.* Ed. M. Planck. Leipzig.
Kramers, Hendrik Anthony
1919      "Intensities of spectral lines." *KDM* 8.III.3:284–388.
1920      "Über den Einfluss eines elektrischen Feldes auf die Feinstruktur der Wasserstofflinien." *ZP* 3:199–223.
1923a     "Über das Modell des Heliumatoms." *ZP* 13:312–341.
1923b     "Das Korrespondenzprinzip und der Schalenbau des Atoms." *NW* 11:550–559.
1923c     "Theory of X-ray absorption and of the continuous X-ray spectrum." *PM* 46:836–871.
1924a     "The law of dispersion and Bohr's theory of spectra." *Nature* 113 (25 March):673–674.
1924b     "The quantum theory of dispersion." *Nature* 114 (22 July): 310–311.
1925a     "On the behaviour of atoms in an electromagnetic wave field." In *6e Skand. Math. Kongress,* 143–153. Also in H. A. Kramers, *Collected scientific papers* (Amsterdam, 1956).
[1925b]   "Über die Eigenschaften von Atomen in einem Strahlungsfelde." Unpublished manuscript (February), AHQP.
1925c     "Eenige Opmerkingen over de Quantenmechanica van Heisenberg." *Physica* 5:369–376.
Kramers, Hendrik A., and Werner Heisenberg
1925      "Über die Streuung von Strahlung durch Atome." *ZP* 31:681–708.
Kronig, Ralph
1925a     "Über die Intensität der Mehrfachlinien und ihrer Zeemankomponenten." *ZP* 31:885–897.
1925b     "Über die Intensität der Mehrfachlinien und ihrer Zeemankomponenten. II." *ZP* 33:261–272.
Ladenburg, Rudolf
1921      "Die quantentheoretische Deutung der Zahl der Dispersionselektronen." *ZP* 4:451–468.

Ladenburg, Rudolf, and Fritz Reiche
1923        "Absorption, Zerstreuung und Dispersion in der Bohrschen Atom-
            theorie." NW 11:584–598.
Landé, Alfred
1919a       "Das Serienspektrum des Heliums." PZ 20:228–234.
1919b       "Adiabatenmethode zur Quantelung gestörter Elektronen-
            systeme." PGV 21:578–584.
1920        "Störungstheorie des Heliums." PZ 21:114–122.
1921a       "Über den anomalen Zeemaneffekt (Teil I)." ZP 5:231–241.
1921b       "Über den anomalen Zeemaneffekt (II. Teil)." ZP 7:398–405.
1922        "Zur Theorie der anomalen Zeeman- und magnetooptischen
            Effekte." ZP 11:353–363.
1923a       "Termstruktur und Zeemaneffekt der Multipletts." ZP 15:189–
            205.
1923b       "Termstruktur und Zeemaneffekt der Multipletts: Zweite Mit-
            teilung." ZP 19:112–123.
1923c       "Das Versagen der Mechanik in der Quantentheorie." NW 11:
            725–726.
1924a       "Das Wesen der relativistischen Röntgendubletts." ZP 24:88–
            97.
1924b       "Die absoluten Intervalle der optischen Dubletts und Tripletts."
            ZP 25:46–57.
1926        "Neue Wege der Quantentheorie." NW 14:455–458.
Landé, Alfred, and Werner Heisenberg
1924        "Termstruktur der Multipletts höherer Stufe." ZP 25:279–286.
Lorentz, Hendrik Antoon
1910        "Alte und neue Fragen der Physik." PZ 11:1234–1257.
1927        Problems of modern physics: A course of lectures delivered in the
            California Institute of Technology [in 1922]. Boston: Ginn.
Loschmidt, Joseph
1876        "Über den Zustand des Wärmegleichgewichtes eines Systems von
            Körpern mit Rücksicht auf die Schwerkraft." WB 73:128–142.
Lummer, Otto, and Wilhelm Wien
1895        "Methode zur Prüfung des Strahlungsgesetzes absolut schwarzer
            Körper." AP 56:451–456.
Lyman, Theodore
1922        "The spectrum of helium in the extreme ultraviolet." Science 56:
            167–168.
Maxwell, James Clerk
1867        "On the dynamical theory of gases." Royal Society of London,
            Philosophical transactions 157:49–88. Also in The scientific
            papers of James Clerk Maxwell (Cambridge, 1890).
1871        Theory of heat. London.
Millikan, Robert A., and Ira S. Bowen
1924a       "Extreme ultraviolet spectra." PR 23:1–34.
1924b       "The extension of the X-ray-doublet laws into the field of optics."
            PR 24:209–222.

1924c        "Some conspicuous successes of the Bohr atom and a serious
             difficulty." *PR* 24:223–228.

Minkowski, Rudolph, and Hertha Sponer
1924         "Über den Durchgang von Elektronen durch Atome." *Ergebnisse
             der exacten Naturwissenschaften* 3:67–85.

Ornstein, Leonard, and Herman Burger
1924a        "Strahlungsgesetz und Intensität von Mehrfachlinien." *ZP* 24:
             41–47.
1924b        "Intensitäten der Komponenten im Zeemaneffekt." *ZP* 28:
             135–141.

Oseen, Carl Wilhelm
1915         "Das Bohrsche Atommodell und die Maxwellschen Gleichungen."
             *PZ* 16:395–405.

Pauli, Wolfgang
1919         "Merkurperihelbewegung und Strahlenablenkung in Weyls Gravi-
             tationtheorie." *PGV* 21:742–750.
1922         "Über das Modell des Wasserstoffmolekülions." *AP* 68:177–240.
1923a        "Über die Gesetzmässigkeiten des anomalen Zeemaneffektes." *ZP*
             16:155–164.
1923b        "Über das thermische Gleichgewicht zwischen Strahlung und freie
             Elektronen." *ZP* 18:272–286.
1924         "Zur Frage der Zuordnung der Komplexstrukturterme in starken
             und schwachen äusseren Feldern." *ZP* 20:371–387.
1925a        "Über den Einfluss der Geschwindigkeitsabhängigkeit der Elek-
             tronenmasse auf den Zeemaneffekt." *ZP* 31:373–385.
1925b        "Über den Zusammenhang des Abschlusses der Elektronengrup-
             pen im Atom mit der Komplexstruktur der Spektren." *ZP* 31:765–
             783.
1925c        "Über den Intensitäten der im Elektrischen Feld erscheinenden
             Kombinationslinien." *KDM* 7(3):3–20.
1926a        "Quantentheorie." In *Handbuch der Physik,* ed. H. Geiger and
             K. Scheel, 23:1–278.
1926b        "Über das Wasserstoffspektrum vom Standpunkt der neuen Quan-
             tenmechanik." *ZP* 36:336–363.

Planck, Max
1882         "Verdampfen, Schmeltzen und Sublimieren." *AP* 15:446–475.
1891         "Allgemeines zur neueren Entwicklung der Wärmetheorie." *Zeit-
             schrift für physikalische Chemie* 8:647–656.
1895         "Über den Beweis des Maxwellschen Geschwindigkeitsverteilungs-
             gesetzes unter Gasmolekülen." *AP* 55:220–222.
1896         "Absorption und Emission elektrischer Wellen durch Resonanz."
             *AP.* 57:1–14.
1897a        "Über elektrische Schwingungen, welche durch Resonanz erregt
             und dadurch durch Strahlung gedämpft werden." *AP* 60:577–599.
             Also *PAV* 1:466–488.
1897b        "Über irreversible Strahlungsvorgänge." *BB,* 4 February, 57–68.
             Reprinted, with the sequels, in *PAV* 1:493–600.

1897c    "Über irreversible Strahlungsvorgänge. II." *BB*, 8 July, 715–717.
1897d    "Über irreversible Strahlungsvorgänge. III." *BB*, 16 December, 1122–1145.
1898     "Über irreversible Strahlungsvorgänge. IV." *BB*, 7 July, 449–476.
1899     "Über irreversible Strahlungsvorgänge. V." *BB*, 18 May, 440–480.
1900a    "Über irreversible Strahlungsvorgänge." *AP* 1:69–122. Also *PAV* 1:614–667.
1900b    "Entropie und Temperatur Strahlender Wärme." *AP* 1:719–737. Also *PAV* 1:668–686.
1900c    "Über eine Verbesserung des Wienschen Spektralgleichung." *PGV* 2:202–204. Also *PAV* 1:687–689.
1900d    "Zur Theorie des Gesetzes der Energievertheilung im Normalspektrum." *PGV* 2:237–245. Also *PAV* 1:698–706.
1901     "Über das Gesetz der Energieverteilung im Normalspektrum." *AP* 4:553–563 (received 7 Jan. 1901). Also *PAV* 1:717–727.
1902     "Über die Verteilung der Energie zwischen Aether und Materie." *AP* 9:629–641. Also *PAV* 1:731–743.
1906     *Vorlesungen über die Theorie der Wärmestrahlung.* Leipzig. 2d ed., 1913. Translated from 2d ed. by Morton Masius as *The theory of heat radiation* (Philadelphia: Blakiston, 1913, reprinted 1988 with introduction by Allan A. Needell [*see* Needell 1988]).

Poincaré, Henri
1893     *Les méthodes nouvelles de la mécanique céleste.* Paris.

Ramsauer, Carl
1921a    "Über den Wirkungsquerschnitt der Gasmoleküle gegenüber langsamen Elektronen." *AP* 64:513–540.
1921b    "Über den Wirkungsquerschnitt der Gasmoleküle gegenüber langsamen Elektronen, I. Fortsetzung." *AP* 66:546–558.
1923     "Über den Wirkungsquerschnitt der Gasmoleküle gegenüber langsamen Elektronen, II. Fortsetzung." *AP* 72:345–352.

Richardson, Owen W.
1916     *The electron theory of matter.* Cambridge.

Rubinowicz, Adalbert
1917     "Zur Quantelung der Hohlraumstrahlung." *PZ* 18:96–98.
1918     "Bohrsche Frequenzbedingung und Erhaltung des Impulsmoments." *PZ* 19:441–445, 465–474.
1921     "Zur Polarisation der Bohrschen Strahlung." *ZP* 4:343–346.

Schrödinger, Erwin
1921     "Versuch zur modelmässigen Deutung des Terms der scharfen Nebenserien." *ZP* 4:347–354.
1924     "Bohrs neue Strahlungshypothese und der Energiesatz." *NW* 12:720–724.
1926a    "Quantisierung als Eigenwertproblem, Erste Mitteilung." *AP* 79:361–376.
1926b    "Quantisierung als Eigenwertproblem, Zweite Mitteilung." *AP* 79:489–527.

1926c      "Über das Verhältnis der Heisenberg-Born-Jordanschen Quan-
           tenmechanik zu der meinen." *AP* 79:734–756.
1926d      "Quantisierung als Eigenwertproblem, Dritte Mitteilung:
           Störungstheorie, mit Anwendung auf den Starkeffekt der
           Balmerserien." *AP* 80:437–490.
1926e      "Quantisierung als Eigenwertproblem, Vierte Mitteilung." *AP*
           81:109–139.
Schwarzschild, Karl
1916       "Zur Quantenhypothese." *BB*, 548–568.
Slater, John Clarke
1924       "Radiation and atoms." *Nature* 113:307–308.
1925a      "A quantum theory of optical phenomena." *PR* 25:395–428.
1925b      "The nature of radiation." *Nature* 116:278.
Smekal, Adolf
1923       "Zur Quantentheorie der Dispersion." *NW* 11:873–875.
Sommerfeld, Arnold
1915a      "Die allgemeine Dispersionsformel nach dem Bohrschen Modell."
           In *Festschrift Julius Elster und Hans Geitel*, ed. K. Bergwitz,
           549–584. Braunschweig.
1915b      "Zur Theorie der Balmerschen Serie." *MB*, 425–458.
1915c      "Die Feinstruktur der wasserstoff- und wasserstoffähnlichen
           Linien." *MB*, 459–500.
1916a      "Zur Quantentheorie der Spektrallinien." *AP* 51:1–94, 125–167.
1916b      "Zur Theorie des Zeeman-Effekts der Wasserstofflinien mit einem
           Anhang über den Stark-Effekt." *PZ* 17:491–507.
1919       *Atombau und Spektrallinien*. Braunschweig. 2d ed., 1921; 3d ed.,
           1922; 4th ed., 1924.
1920a      "Ein Zahlenmysterium in der Theorie des Zeemaneffekts." *NW*
           8:511–514.
1920b      "Allgemeine spektroskopische Gesetzte, insbesondere ein magne-
           tooptischer Zerlegungssatz." *AP* 63:221–263.
1922       "Quantentheoretische Umdeutung der Voigt'schen Theorie des
           anomalen Zeemaneffektes vom D-Linientypus." *ZP* 8:257–272.
1923a      "Über die Deutung verwickelter Spektren (Mangan, Chrom usw.)
           nach der Methode der inneren Quantenzahlen." *AP* 70:32–62.
1923b      "Spektroskopische Magnetonzahlen." *PZ* 24:360–364.
1924       "Zur Theorie der Multipletts und ihrer Zeeman effekte." *AP*
           73:209–227.
Sommerfeld, Arnold, and Werner Heisenberg
1922       "Die Intensität der Mehrfachlinien und ihrer Zeemankomponen-
           ten." *ZP* 11:131–154.
Sommerfeld, Arnold, and Gregor Wentzel
1921       "Über reguläre und irreguläre Dublette." *ZP* 7:86–92.
Stoner, Edmund
1924       "The distribution of electrons among atomic levels." *PM* 48:
           719–736.

van Vleck, John H.
1922        "The dilemma of the helium atom." *PR* 19:419–420.
1923        "The normal helium atom and its relation to the quantum theory."
            *PM* 44:842–869.
1925        "Virtual oscillators and scattering in the quantum theory." *PR*
            25:242–243.
Whitehead, Alfred North
1898        *A treatise on universal algebra.* Cambridge.
1906        *The axioms of projective geometry.* Cambridge.
1919        *An enquiry concerning the principles of natural knowledge.*
            Cambridge.
1922        *The principle of relativity with applications to physical sciences.*
            Cambridge.
Whittaker, Edmund T.
1904        *A treatise on the analytical dynamics of particles and rigid bodies.*
            Cambridge. 2d ed. 1917.
1922        "On the quantum mechanism in the atom." Royal Society of
            Edinburgh, *Proceedings* 42:129–142.
Wien, Wilhelm
1894        "Temperatur und Entropie der Strahlung." *AP* 52:132–165.
1896        "Über die Energieverteilung im Emissionsspectrum eines schwar-
            zen Körpers." *AP* 58:662–669.
Wilson, William
1915        "The quantum-theory of line spectra." *PM* 29:795–802.
Wood, Robert W., and Alexander Ellet
1923        "On the influence of magnetic fields on the polarization of reso-
            nance radiation." *PRS* 103:396–403.
Zermelo, Ernst
1895        "Über einen Satz der Dynamik und die mechanische Wärme-
            theorie." *AP* 57:485–494.
1896        "Über mechanische Erklärungen irreversibler Vorgänge: Eine Ant-
            wort auf Hrn. Boltzmann's 'Entgegnung.'" *AP* 59:793–801.

# Index

Lightning Source UK Ltd.
Milton Keynes UK
UKHW010203140921
390518UK00001B/7